現代 商用英文

MODERN BUSINESS ENGLISH

黃瑪莉 著

智勝文化

L/C

ICC(A), ICC(B), ICC(C)

SRCC

OEM
ODM
OBU

Credite Note FOB TPND CY ETD DAP

Statement Payment: by irrevocable and transferabale L/C at sight in our favor.

Full set of clean on board ocean Bill of Lading made out to order
with freight prepaid blank endorse and notifiy applicant.

QC P/I (Proforma Invo
QA
IPQC Debit Note
R&D

P/O (Purchase Order)

Please ship the goods of our order no.001 by courier.

四版序

鑒於台灣和大陸進出口貿易量遽增，同時為因應目前國際趨勢及便於讀者學習並取得各種英文／國貿相關的證照考試，且能將所學實際迅速運用到實務工作上，四版信函範例內容完全取材自實務工作上買賣雙方實際往來聯絡的各種商用英文信函及進出口業務所需用到的相關文件資料。同時，為方便讀者學習參考，商用英文信函皆按照進出口流程順序編排，並於文章中增加許多產品圖片以增加其生動性。

因此，新版中除了更新許多新的文章範例外，也增加了許多和證照考試相關的資料，例如：NEW TOEIC 文法選擇試題範例、國貿大會考歷屆英文考題範例、國貿業務丙級學科基礎貿易英文題庫及範例等。文件方面也增加了最新的:煙燻證明、原廠證明書、校正證明書、兩岸三地產證（ECFA用）、授權書、經銷合約、稽核報告書。此外亦提供了 ECFA 整體經濟效益／ ECFA 項目及進口稅率、INCOTERMS 2010 最新版本內容及應用；並於「產品類別名稱表」中增加許多台灣目前外銷的高科技產品及零組件名稱；於「國際貿易常用專有縮寫名詞」中亦增加了一些考試常出現的縮寫字等資料供參考。

為便於學習參考，每種信函前面亦增加兩部分：常用生字／片語及常用主題／說明／結論句；後面的練習習題改變或增加成國貿相關證照方面的考題題型。此外，亦增加了「要求寬鬆付款函」的範例、信用狀上常見的文件要求及額外條款、及和大陸簽約時訂單上常見到的「一般訂單條款」的內容 。

本人以國貿實務 30 多年工作經驗及 20 多年相關國貿及英文教學經驗，特將本書中的信函範例及所有內容依實際需要全面更新。全書內容完全以中英文對照方式呈現，文辭簡潔易懂，內容豐富完整，排版清晰易讀，考題、範例及練習最多，為現代商用英文最佳學習範本及國貿業界必備參考工具書；完全適用於學校的授課老師、學習商用英文的學生、進出口貿易業務從業人員、以及對英文有興趣的自修的學習者。「Practice makes perfect」熟能生巧——相信只要根據本書文法重點的練習步驟及各種商用英文信函單元中的練習多加練習，定可達到完美商用英文寫作的境界！

感謝智勝文化事業有限公司出版部、行銷部及各相關部門人員的大力協助，也感謝各大專院校師生及國際貿易實務界人士的厚愛，使本書廣受歡迎，需求量日增。深信此四版豐富完整的資料及符合最新趨勢的實用性，將更有助於讀者的學習參考，並能方便輕鬆地實際應用於實務工作及相關證照的考試上。

Mary Huang 謹識

2012 年 8 月

初版序

　　對於許多在校或剛畢業想從事國際貿易、進出口業務、轉口貿易或三角貿易的同學，以及剛開始從事此行業的朋友們，常會有下列的疑惑：(1)國際貿易的流程是什麼？(2)在貿易公司或進出口公司要做些什麼工作？(3)和國外客戶或國外廠商聯絡的各種信函要如何處理？內容怎麼寫才得體並合乎現代商業英文的寫作文體？(4)貿易及商業上有哪些專門術語及特別的縮寫字？

　　然而，即使具備了以上的知識或概念，仍會面對最後一個最大的困惑：無法確定自己寫出來的英文句子，文法上是否完全正確無誤？國外客戶或廠商是否真正明瞭自己的意思？在溝通上是否會產生誤解？

　　英文寫作真的很難嗎？其實不然，和中文的文法架構及散漫的文體比較起來，英文的文法結構和文章的組織是簡單容易太多了──中文詞類有 30 多種，句型不含文言文可歸納的已有 500 多種，甚至有些無法歸納的句型例如：對不起、不客氣、貴姓、不要緊、怎麼樣、一路平安……等句子，只好歸入慣用語(Idiom Expressing)；相對地，英文就只有 9 種詞類、8 種基本簡單句型（複合或複雜句亦是由此 8 種句型擴大延伸而來）和 2 種問句，而文章的組織亦嚴格要求只分三部分：(1)主題；(2)說明；(3)結論。

　　此外，中文很多詞類用法及句型結構和英文不一樣：中文的名詞沒有單複數之分，例如：一個樣品或兩個樣品，名詞相同，英文則需注意單數為 one sample，複數為 two samples；中文的動詞沒有時態之分，主被動不清楚，例如：他昨天來、今天來或明天來，動詞都一樣，英文則要分 He comes，He came，He will come，形容詞及副詞放的位置和英文更是大不相同。因此，在中翻英或英翻中時，就要特別注意英文各個詞類的位置和中文的不同處，並思考如何將中文散漫複雜的句子重新排列組織，以英文八個句型表達出來，千萬不要抓到中文就一成不變照翻，那就會造出所謂的 Chinese English，例如：「你說什麼？」翻成 You say what？「我們收到信用狀前不出貨」翻成 We receive L/C before not ship the goods，而讓國外客戶看得一頭霧水。

　　筆者以從事國際貿易二十多年實務經驗，加上十多年專業教授貿易英文書信寫作以及多年教授外國人華語經驗，特將中英文法架構及不同處比較列出，並將貿易流程中各種信函寫作要點及應對內容加以說明，希望能讓有興趣英文寫作及從事國際貿易者，在輕鬆簡易學習下，寫出正確簡潔優美有力的商用英文。

Mary Hwang

1994 年 4 月

目錄

MODERN BUSINESS ENGLISH

MODERN
BUSINESS
ENGLISH

CONTENTS

其他雜務聯絡接洽信函　451
Miscellaneous Letters

推薦／應徵函及履歷表　465
Recommendation Letters/Application Letters & Resumes

國際貿易實用參考附表（請參閱附贈光碟）　485
Appendices

一、國際貿易常用專有縮寫名詞

CONTENTS MODERN BUSINESS ENGLISH

略語表
ABBREVIATIONS

S	(Subject)	主詞
V	(Vert)	動詞
FV	(Function Verb)	有動作的動詞
LV	(Linking Verb)	聯綴動詞
Aux.	(Auxiliary)	助動詞
Vi	(Intransitive Verb)	不及物動詞
Vt	(Transitive Verb)	及物動詞
V-root	(Root Verb)	原形動詞
O	(Object)	受詞
O.C.	(Object Complement)	受詞補語
S.C.	(Subject Complement)	主詞補語
D.O.	(Direct Object)	直接受詞
I.O.	(Indirect Object)	間接受詞
N	(Noun)	名詞
Pron.	(Pronoun)	代名詞
Adj.	(Adjective)	形容詞
Adv.	(Adverb)	副詞
Prep.	(Preposition)	介系詞
Conj.	(Conjunction)	連接詞
Art.	(Article)	冠詞
Sing.	(Singular)	單數
Pl.	(Plural)	複數
Ph.	(Phrase)	片語
Cl.	(Clause)	子句
p.t.	(Past Tense)	過去式
p.p.	(Past Participle)	過去分詞
V-ing	(Present Participle)	現在分詞
To + V-root	(Infinitive Phrase)	不定詞片語

如何寫出完整
正確的英文句子
How to Write a Correct Sentence

一、英文文法重點複習(Grammar Review)

I. 英文文法內容主要包括：

1. 4 種句子：簡單句（基本句型）、複合句、複雜句、複合複雜句。

2. 8 種基本句型 (Basic Sentences)。

3. 2 種問句：(1)Yes/No 問句（直接問句／附加問句）。
   (2)WH 問句（直接問句／間接問句）。

4. 8 種主要詞類：注意其位置及用法
   (1) 名詞(Noun)：注意單／複數，可數／不可數。
   (2) 代名詞(Pronoun)：注意人稱及主格／受格／所有格變化。
   (3) 形容詞(Adjective)：注意位置和中文不同，形容詞單字放 N 前，片語／子句放 N 後；冠詞(Article)：a, an, the ＝形容詞 Adjective 的一種，a/an 放普通名詞單數前。
   (4) 副詞(Adverb)：注意位置和中文不同（中文副詞放 S 和 V 之間，英文大都放句尾）。
   (5) 介系詞(Preposition)：注意介系詞＋(the)＋ N/V-ing。
   (6) 連接詞(Conjunction)：注意對等。
   (7) 感嘆詞 (Interjection)：Alas! Oh! Wow!, etc.
   (8) 動詞(Verb)：注意時態（主動 12 種、被動 9 種），5 種假設語氣及不規則變化。
   (a)助動詞(Auxiliary Verb)；
   (b)動作動詞(Function/Action Verb)；
   (c)聯綴動詞(Linking Verb)。

5. 4 種組合片語(Phrases)（結構及用法）：
   (1) 介系詞片語(Prepositional Phrase)：當形容詞／副詞用。
   (2) 不定詞片語(Infinitive Phrase)：當名詞／形容詞／副詞用。
   (3) 現在分詞片語(Present Participle Phrase)：當名詞／形容詞用。
   (4) 過去分詞片語(Past Participle Phrase)：當形容詞用。

6. 3 種子句 (Clauses)：

    (1) 對等子句(Coordinate Clause)。

    (2) 主要子句(Main Clause)。

    (3) 附屬子句(從屬子句)(Subordinate Clause)：

      (a)名詞子句(Noun Clause)；

      (b)形容詞子句(Adjective Clause)；

      (c)副詞子句(Adverb Clause)。

    II. 因此，要寫出一個完整正確的英文句子，一定要先瞭解：

1. 英文的句子只有四種

    (1) 簡單句／基本句型(Simple Sentence/Basic Sentence Pattern)

    (2) 複合句(Compound Sentence)

    (3) 複雜句(Complex Sentence)

    (4) 複合複雜句(Compound Complex Sentence)

2. 簡單句：即基本句型，也稱「會話語言」(Spoken Language)，多用於英文會話或商業信函寫作上，因為商業信函為了便於溝通，一般都以較接近口語的簡單句來表達。基本句型只有八種，結構如下：

    (1) S ＋ FV

    (2) S ＋ FV ＋ Adv. or Adv. Ph.

    (3) S ＋ FV ＋ O

    (4) S ＋ FV ＋ I.O.＋ D.O. or S ＋ V ＋ D.O.＋ Prep.＋ I.O.

    (5) S ＋ FV ＋ O ＋ O.C.

    (6) S ＋ LV ＋ S.C.

    (7) There ＋ be ＋ S(N)＋ Adv.

    (8) It ＋ be ＋ N/Adj./Adv./Prep. Ph.

    注意：英文句子嚴格要求第一個字一定要是主詞(S)，第二個字一定要放動詞(FV/LV)，動詞後面再放受詞(O)或補語(S.C./O.C.)。此外，第(2)和(7)的句型不可照中文順序直接翻（詳見句型解說）。

3. 複合句、複雜句和複合複雜句：這三種句子又稱為「寫作語言」(Written Language)，多用於寫作上，在商業信函書寫上也參差使用，但盡量不要寫出像英文文學作品裡，過於複雜、冗長、多修飾語的長句子，或像新聞寫作的句子，其中又插入過多的形容詞修飾語句。商業寫作上，一個複雜句最多帶 1～2 個修飾句即可，不要寫得太長或太複雜。

注意：此 3 種句子即是由簡單句的句型加上連接詞或名詞、形容詞、副詞片語和子句，擴大延伸變化而來的。因此對於片語和子句的種類與放置的位置就要有所了解，才能書寫正確（詳見片語子句解說）。

公式：簡單句 ＋ 連接詞(and/or) ＋ 簡單句 ＝複合句

公式：簡單句 ＋ $\left.\begin{array}{l}名詞 \\ 形容詞 \\ 副詞\end{array}\right\}$ 片語 / 子句 ＝複雜句

---

4. 問句類型：了解句型結構之後，再學習如何將句子變成問句，直接問句只有兩種：Yes/No 問句和 WH 問句（又稱 Information 問句）。

(1) Yes/No 問句的造法

將句中的 Be 動詞或現有的助動詞(Aux.：can, could, may, might, will, would, shall, should, have, has, had, etc.)搬到主詞(S)前面，句中如果沒有 Be 動詞或助動詞，則只能在句首加上 Do, Does, Did 其一。

注意：不可在句首亂加別的字，加上 Do, Does, Did 後面的動詞要用原形動詞(V-root)。

公式：Be ＋ S ＋ S.C.
Aux. ＋ S ＋ V-root
$\left.\begin{array}{l}Do \\ Does \\ Did\end{array}\right\}$ ＋ S ＋ V-root

---

(2) WH 問句的造法

句首先放 WH 或 How 的疑問字，再將句中的 Be 動詞或助動詞搬到主詞前，或在主詞前補上 Do, Does, Did；換言之，即在 Yes/No 問句前加上 WH 或 How 的字。

① WH 的字只有六組：Who, Whom, Whose（問人）；When（問時間）；Where（問地點）；Which（作選擇）；Why（問原因）；What（什麼），如下所示：

```
公式：Who        ＋ Be     ＋ S ＋ S.C.              （問人）
     Whom       ＋ Aux.   ＋ S ＋ V-root
     Whose ＋ N  ＋ do    ⎫
                  does   ⎬ ＋ S ＋ V-root
                  did    ⎭

     When                                          （問時間）
     Where                                         （問地點）
     Which(＋ N)                                    （作選擇）
     Why                                           （問原因）
     What(＋ N)                                     （什麼）
```

② How 的組合，常用的有下列七種：

```
公式：How many ＋ N      ＋ Be    ＋ S. ＋ S.C.      （多少）
     How much (＋ N)    ＋ Aux.  ＋ S ＋ V-root     （多少）
     How often          do     ⎫                   （多常）
     How long         ＋ does   ⎬ ＋ S ＋ V-root    （多久）
     How far            did    ⎭                   （多遠）
     How about ＋ N/V-ing                           （如何）
     How ＋ Be ＋ S                                  （問候）
```

5. 間接問句造法(Indirect Question)

在 WH 問句前放一個句子或 Yes/No 問句，WH 字後面的 Be 動詞和助動詞放回主詞後面，也不需加任何 Do, Does, Did 的助動詞，公式如下：

```
公式：(1)YES/NO 問句＋WH 字＋主詞(S)＋Be 動詞 / 助動詞(Aux.)
     (2)句子(Sentence)＋WH 字＋主詞(S)＋Be 動詞 / 助動詞(Aux.)
```

6. 否定句造法(Negative Sentence)

在 Be 動詞或助動詞後面加 not；句中如沒有 Be 動詞或助動詞，則在動詞前面補上 don't, doesn't 或 didn't，後面動詞變原形；或在動詞後面，名詞前面加no，公式如下：

公式：(1)主詞(S)＋Be 動詞＋ not ＋主補(S.C.)

(2)主詞(S)＋助動詞(will/shall/can...)＋ not ＋原型動詞(V-root)

(3)主詞(S)＋助動詞(have/has/had)＋ not ＋過去分詞(p.p.)

(4)主詞(S)＋ don't/doesn't/didn't ＋原型動詞(V-root)

(5)主詞(S)＋動詞(FV)＋ no ＋名詞(N)

---

7. 學習英文句子寫作的正確方法是：(1)記清楚英文的基本句型結構；(2)背常用的生字片語；(3)記住單字的詞性及 8 種詞類的功用及位置；(4)特別注意在中英文對照翻譯時，其排放的位置和中文不同的地方，將英文單字／片語／子句，正確的放入句型中，即可造出正確的句子。

8. 擴大簡單句型的方法

在基本句型中，加入修飾的形容詞／副詞單字，或名詞／形容詞／副詞的片語和子句，即可擴大句型結構。

例：基本句型：He is our customer.加上形容詞 best 及副詞 in the States

→擴大句型：He is our best customer in the States.

He is our best customer who imports Toys.（加入一個 Adj. Cl.）

基本句型：It is our policy.加入一個 to 的 Adj. Ph.和 when 的 Adv. Cl.

→擴大句型：It is our policy to require payment by L/C when we deal with a new customer.

9. 9 種詞類的功用、位置及用法(Parts of Speech)

(1) 名詞(N)

功用：當主詞(S)，受詞(O)，主補(S.C.)，受補(O.C.)

位置：放在 S, O, S.C.或 O.C.的位置

種類：普通／專有／抽象／物質／集合名詞 5 種

注意：英文名詞分為可數（分單數及複數）與不可數（只有單數）名詞，中文則沒有單複數的問題。

(2) 代名詞(Pron.)

　　功用：代替名詞

　　位置：放在 S, O 的位置

　　種類：人稱代名詞(we, you, they, us, them)／不定代名詞(any, some)／指示代名詞(this, that)／疑問代名詞(who, whom, which)4 種

(3) 動詞(V)

　　功用：當動詞（是句子的靈魂，不可缺少）

　　位置：放在 FV(Function V)或 LV(Linking V) 的位置

　　種類：動作動詞(FV)：buy, send, receive, offer…

　　　　　聯綴動詞(LV)：Be 動詞、感官動詞，become, seem…

　　注意：時態、人稱、語氣、主動／被動要跟著變化，共 26 種變化，但常用的只有 12 種左右，要熟記，不可自創一態（中文沒有時態人稱變化，主被動也不清楚，有時句中也沒有動詞）。

(4) 助動詞(Aux.)

　　功用：幫助動詞

　　位置：放在動詞前面

　　注意：助動詞不是動詞，不可單獨使用，要和原形動詞或過去分詞（當被動式或完成式）一起用，例如：

　　　　　① will, shall, may, can, must, do, does, did ＋ V-root

　　　　　② have, has, had, is, are, am, was, were ＋ p.p.

(5) 形容詞(Adj.)

　　功用：只修飾名詞

　　位置：放在所要修飾的名詞的前面或後面

　　種類：代名形容詞(my, your, his, her, our, their, its)／數量形容詞(one, two, three, first, second…)／性狀形容詞(big, red, good, old, new, iron…)3 種

　　注意：排放位置及比較級數，中文全部前修飾，英文要分前修飾和後修飾。

　　　　　中文：形容詞單字／片語／子句 ＋的＋名詞

英文： 形容詞單字 ＋名詞＋ 形容詞片語／子句
　　　　前修飾　　　　　後修飾

(6) 副詞(Adv.)

　　功用：修飾動詞(V)、形容詞(Adj.)、副詞(Adv.)及子句或句子

　　位置：放在想要修飾的字的前面或後面

　　種類：表示時間／地點／目的／方法／原因／條件／態度／結果／讓步／程度／
　　　　　次數／肯定／否定

　　注意：除了名詞以外，如要補充說明其他詞類，全都用副詞來修飾。
　　　　　　中文位置：大都放在主詞(S)和動詞(V)中間
　　　　　　英文位置：如不強調，大都放句尾

(7) 介系詞(Prep.)

　　功用：介系詞＋(the)＋名詞(N)＝介系詞片語，當做形容詞片語(Adj. Ph.)或副詞片
　　　　　語(Adv. Ph.)用

　　位置：放在想要修飾的字後面或句尾居多

(8) 連接詞(Conj.)

　　功用：連接兩個對等的字、片語或子句

　　位置：在所連接的兩個字、片語或子句之間

　　注意：對等，例如：名詞 and 名詞，動詞 or 動詞，要一致。

(9) 冠詞(Art.)

　　功用：屬形容詞的一種，形容普通名詞，只有 a, an, the 三個字

　　位置：放在普通名詞單數前面

　　a ＋子音開頭的單數名詞，例如：a sample, a catalogue

　　an ＋母音開頭的單數名詞，例如：an order

　　the ＋特定名詞（單數／複數皆可）

　　注意：句中如果有所有格代名詞 our, your, their 或指示代名詞 this, that 就不要再
　　　　　　放冠詞，例如：This is a new sample.或 This is our new sample.不要寫成 This
　　　　　　is our a new sample.（錯誤）。

10.動詞時態變化有主動 12 種、被動 9 種、假設語氣 5 種，共 26 種。但是常用的動詞只有下列 12 種（參下表），其用法及動詞出現的樣式要背記，不可自創一態。

注意：(1)動詞最少 1 個字，最多 5 個字，放 V 位置。

(2)一個句子只能有一個動詞，第二個動詞出現時，要用不定詞 to 隔開，或改為現在分詞：V to V 或 V + Ving。

12 種常用時態（Verb Tense）

Tense 時態	Active 主動	Passive 被動
Simple Pesent 簡單現在式	用現在式動詞（1 個字）like (s)/go (es)	is/are/am + p.p.
Simple Past 簡單過去式	用過去式動詞（1 個字）liked/went	was/were + p.p.
Simple Future 簡單未來式	（動詞 2 個字）shall/will + V-root	shall/will + Be + p.p.
Present Progressive 現代進行式	（動詞 2 個字）is/are/am + V-ing	is/are/am + being + p.p.
Present Perfect 現在完成式	（動詞 2 個字）have/has + p.p.	have/has + been + p.p.
Present Perfect Progressive 現在完成進行式	（動詞 3 個字）have/has + been + V-ing	—
假設語氣：可能的未來	If + S +現在式動詞，S +簡單未來式動詞	

時態用法

(1)簡單現在式：句子所述為事實、真理、習慣（句中有 usually, always, everyday）

(2)簡單過去式：句中有過去時間(yesterday, just, last week, before)

(3)簡單未來式：句中有未來時間(tomorrow, next month, soon)

(4)現在進行式：中文有「現在正在」或英文有 now 時

(5)現在完成式：剛完成或已完成的動作，中文有「已經、曾經、尚未、未曾」

(6)現在完成進行式：從過去到現在，仍在繼續的動作

限定／非限定動詞

(1)限定動詞(finite)：現在式／過去式，可以單獨一個字放動詞位置

　　例：I <u>work</u> for this company now.（現在式）

　　　　I <u>worked</u> for this company before.（過去式）

(2)非限定動詞(non-finite)：原形 V (V-root)／過去分詞(p.p.)／現在分詞(V-ing)不可單獨一個字放 V 位置，要和其他字一起使用

　　①原形動詞三個位置：Please ＋ V-root; To ＋ V-root; Aux. ＋ V-root

　　　例：<u>Please send</u> us your new sample.（please ＋原形 V）

　　　　　We like <u>to do</u> business with you.（不定詞 to ＋原形 V）

　　　　　We will <u>ship</u> the goods to you next week.（助動詞＋原形 V）

　　②過去分詞兩個位置：be ＋ p.p.（被動式），have ＋ p.p.（完成式）

　　　例：The sample <u>was received</u>.（被動式）

　　　　　We <u>have received</u> your sample.（完成式）

　　③現在分詞兩個位置：be ＋ V-ing（進行式）；V ＋ V-ing（當動名詞）

　　　例：We <u>are sending</u> you our new catalogue.（進行式）

　　　　　We like <u>doing</u> business with you.（當動名詞）

II. 注意中翻英時，有下列幾點最大不同句型結構位置，不可照中文順序直接翻譯，要先調整成英文句型結構的位置，再翻成英文

(1) 英文第 2 個句型：S ＋ V ＋ Adv.→中文句型：S ＋ Adv. ＋ V

中文的副詞（時間／地點／目的／方法／原因／和…人／根據…／按照…）都放在主詞(S)和動詞(V)中間，英文則將副詞放在任何一個句型的句尾，排列順序如下：①和…人＋②方法＋③地點＋④時間（由小至大）＋⑤目的。

注意：中文句型：S ＋時間（大→小）＋和…人＋方法＋地點＋ V ＋ O

　　　　英文句型：S ＋ V ＋ O ＋和…人＋方法＋地點＋時間（小→大）

例：中文句型：我　明天下午　2：00　和我老闆　開車　到飯店　接　你

英文排序：　1　　8　　　7　　　4　　　5　　6　　2　3

= I will pick you up with my boss by car at the hotel

at 2：00 tomorrow afternoon.

(2) 英文第 6 個句型：S ＋ LV ＋ S.C. (N/Adj.)

主詞補語(S.C.)如果是 Adj.時，要注意中文句中的動詞「是」會省略，英文則要記得補上一個 Be 動詞

例：①你們的價錢（是）太貴

Your price is too high.

②我們的品質（是）很好

Our quality is very good.

(3) 英文第 7 個句型：There ＋ Be ＋ N(＝ S)＋ Adv.（地點／時間）

中文的「有」，英文有三個表達方式：

①人＋有，用 have

例：我們有 30 個業務人員

We have 30 sales.

②具有，用 with

例：具有 30 個業務人員，我們可將貴產品推銷得很好。

With 30 sales, we can promote your products well.

③地點／時間＋有，用 There ＋ Be 的句型

例：①樣品室裡有很多樣品

There are many samples in the showroom.

②這週沒有船

There is no vessel available this week.

(4) 英文形容詞的位置分為前後修飾：Adj.單字 ＋ N ＋ Adj.片語 / 子句

中文全部皆為前修飾：Adj.單字 / 片語 / 子句 ＋的＋ N

例：①這　是　一本　好的　英文　書

　　1　2　　3　　4　　5　6

This is a good English Book.（Adj.單字放 N(book)前）

②樣品室裡的　樣品　是　新的

　　　 2　　　 1　 3　　4

The sample in the showroom is new.（Adj.片語放 N(sample)後）

③我們　收到　你們寄來的　信

　 1　　 2　　　 4　　　 3

We received the letter which you sent.（Adj.子句放 N(letter)後）

(5) 注意中文常用「N＋的」形容第二個 N：N1 的 N2，英文則要倒過來寫：

N2 of/for/in N1

中文：N1 的 N2

英文：N2 of/for/in N1

例：①樣品的品質→品質 of 樣品(the quality of the sample)

　　②台北的交通→交通 in 台北(the traffic in Taipei)

　　③ 100 元的支票→支票 for 100 元(the check for 100 dollars)

　　④ 1000 個的價錢→價錢 for 1000 個(the price for 1000pcs)

　　⑤鞋子的工廠→工廠 of 鞋子(the manufacturer of shoes)

注意：N1 和 N2 是同樣的東西或有「屬於」之意，的＝ of。

　　　N1 和 N2 是不同的東西或有「為了」之意，的＝ for。

(6) 疑問句不可照中文順序翻譯，要改為英文疑問句型

例：①你住在哪裡？

　　WH＋do＋S＋V＝Where do you live？

　　②你是王先生嗎？

　　Be＋S＋S.C.＝Are you Mr. Wang？

　　③你可以給我們兩個樣品嗎？

　　Aux.＋S＋V-root＋I.O.＋D.O.＝Can you give us two samples？

(7) 中文名詞沒有單複數的問題，英文要注意單數名詞前加冠詞或所有格代名詞，複
數名詞後面加 s 或 es

例：①我們是出口商

　　　We are an exporter.

　　②請寄兩個樣品給我們

　　　Please send us two samples.

(8) 中文動詞沒有時態，人稱變化、主被動不清楚，英文要注意動詞變化

　　例：①客人昨天去我們工廠（過去式）

　　　　The customer went to our factory yesterday.

　　　②客人明天去我們工廠（未來式）

　　　　The customer will go to our factory tomorrow.

　　　③樣品下週寄給你們（未來式被動）

　　　　The sample will be sent to you next week.

　　注意：人（當主詞）＋主動；物（當主詞）＋被動動詞，但使役動詞相反

(9) 英文一個句子只能有一個動詞，如有第二個動詞出現，要用不定詞 to 隔開，或將
　　第二個動詞改為動名詞：V to V 或 V ＋ V-ing

　　例：我們　　決定　　停止　　進口　　你們的產品（中文動詞可無限連放）

　　　　　　　　V1　　V2　　V3

　　　We decided to stop importing your products.

　　　　　　V　　　to V　　V-ing

(10)中文敘述：由大→小，由遠→近

　　英文敘述：由小→大，由近→遠

　　例：①今晚 7：00

　　　　　at 7：00 tonight

　　　②台北的希爾頓飯店

　　　　　at the Hilton Hotel in Taipei

12.英文寫作時，如能注意以上文法重點，即能寫出完整正確的句子。然後將句子按照
　英文寫作方法布局，即可營造出一篇簡潔正確的商用英文書信。

MODERN BUSINESS ENGLISH
現代 商用英文

13. 英文寫作方法(English Writing Method)/ 文章布局方法

  (1) 開頭第一句或第一段：主題

   (Beginning of the sentence/paragraph: main idea)

  (2) 中間句子或段落：說明（支持主題）

   (Middle of the sentence/paragraph: example/support)

  (3) 結尾句子或末段：結論

   (End of the sentence/paragraph: conclusion)

14. 文體種類（Types of Composition)有 4 種

  (1) 描寫文(Description)：描寫某事或某人(describe something or someone)，讓人看起來、聽起來、聞起來或覺得如何（訴諸於感官上的描述）。

  (2) 敘述文(Narration)：說故事方式(tell a story)，敘述某人或某事在特定期間所做或所發生的事。

  (3) 說明文(Exposition)：事實陳述(deal with facts)，提供讀者資訊、陳述實際的報導或說明 一個過程。

  (4) 辯論文(Argumentation)：為某事或某意見辯白(argue with something or some ideas)。

二、八種基本句型結構（簡單句）及例句
(Basic Sentence Patterns/Simple Sentences)

---

1. S ＋ FV　主詞（名詞或代名詞）＋ 動詞
   Subject (Noun/Pronoun) ＋ Function Verb

注意：照中文順序直接翻譯，主詞放名詞或代名詞，動詞時態要變化。

例：(1)The customer is coming.（客人正進來）──現在進行式

(2)The claim happened.（抱怨發生了）──過去式

(3)The sample has been received.（樣品已經收到了）──物（主）＋被動

---

2. S ＋ FV ＋ Adv. or Adv. Ph.　主詞＋動詞＋副詞或副詞片語
   Subject ＋ Function Verb ＋ Adverb or Adverb Phrase

注意：中文副詞位置在 S 和 V 中間，英文要放到句尾。

例：(1)I work for a trading company.（我　在貿易公司　上班）

(2)The customer will come next Monday.（客人　下週一　來）

(3)My boss will come back at 5：00 p.m.（我老闆　下午五點　回來）

(4)Please reply as soon as possible.（請　儘快　回覆）

　　　　　　　　　　　　　　1　　　3　　　2

---

3. S ＋ FV ＋ O　主詞＋動詞＋受詞（名詞或代名詞）
   Subject ＋ Function Verb ＋ Object (Noun/Pronoun)

注意：(1) 照中文順序直接翻譯，但是句中如有副詞（時間／地點），要搬到句尾。

(2) Please 後面的 you 省略，直接跟原形動詞(V-root)。

例：(1)We have two factories.（我們　有　兩個工廠）

(2)We produce computers.（我們　生產　電腦）

(3)We have received your letter.（我們　已經收到　你們的信了）

(4)We will mail two catalogs.（我們　將寄　兩份目錄）

(5)I need two samples. （我　需要　兩個樣品）

(6)Please pay the sample charge. （請　付　樣品費）

(7)Please advise your flight details. （請　告知　你的班機明細）

(8)Please confirm your order. （請　確認　你們的訂單）

(9)I will pick you up at the hotel tomorrow.

　　（ 我　明天將到飯店　接　你 ）
　　　1　　5　　4　　2 3

4.(A)S＋FV＋I.O.＋D.O.　主詞＋動詞＋間接受詞（人）＋直接受詞（物）
Subject ＋ Function Verb ＋ Indirect Object ＋ Direct Object

注意：照中文順序直接翻譯，一樣注意句中如有副詞，要搬到句尾。

例：(1)We can give you two free samples.

　　　（我們可以給你兩個免費樣品）

(2)We can give you 5% commission.

　　　（我們可以給你百分之五的佣金）

(3)We will place you a new order.

　　　（我們將下給你們一張新訂單）

(4)Our cooperation will bring you big profits.

　　　（我們的合作將帶給你們豐厚的利潤）

(5)We will open you the L/C.

　　　（我們將開給你們信用狀）

(6)Please send us your quotation.

　　　（請寄給我們你們的報價單）

(7)We sent you an e-mail yesterday.

　　　（我們　昨天　發給你們一封電子郵件）
　　　　1　　6　　2　3　4　　5

4.(B)S＋FV＋D.O.＋Prep.＋I.O.

主詞＋動詞＋直接受詞（物）＋介詞＋間接受詞（人）

Subject ＋ Function Verb ＋ Direct Object ＋ Prep.＋ Indirect Object

注意：間接受詞（人）放在直接受詞（物）後面，要多放一個介系詞。動詞give, send, open, issue…的介系詞（給）都用 to，但是有兩個動詞例外：place 的給用 with，bring 的給用 for──參例句(2)(3)。

例：(1)We will give two free samples to you.

(2)We will place a new order with you.*

(3)Our cooperation will bring a big profit for you.*

(4)We can give 5% commission to you.

(5)We will open the L/C to you.

(6)Please send your quotation to us.

(7)We sent a fax to you yesterday.

5.S＋FV＋O＋O.C.　主詞＋動詞＋受詞＋受詞補語（名詞或形容詞）

Subject ＋ Function Verb ＋ Object ＋ Object Complement（N/Adj.）

注意：動詞後面需加受詞補語的動詞，稱作「不完全及物動詞」，常用的有call, keep/make（使得）, find, let, ask, think, consider, wish, hear, feel, have, get, elect, choose…

（照中文順序直接翻譯）。

例：(1)We think your price too high.（我們認為你們的價錢太貴）

(2)We find many cartons broken.（我們發現很多箱子破損）

(3)We call him Peter.（我們叫他彼德）

(4)Please make the color dark.（請把顏色加深）

(5)Your visit made us happy.（你的拜訪使得我們很高興）

6.S ＋ LV ＋ S.C.（N/Adj.）　主詞＋聯綴動詞＋主詞補語（名詞／形容詞）
Subject ＋ Linking Verb ＋ Subject Complement (Noun/Adjective)

注意：(1)聯綴動詞(LV)沒有動作，只聯結主詞(S)和主補(S.C.)之間的關係，所以主詞(S)和主補(S.C.)必須是同樣的人或事，可以互換位置，倒過來寫。

　　　例：Our best price is US$10.＝ US$10 is our best price.

　　　（我們的最好價錢是 10 美元＝ 10 美元是我們的最好價錢）

(2)因此要特別注意不可寫出下列的句子：

Our product is good quality.

（我們產品是好品質）──錯誤

因為產品(product)不等於品質(quality)，正確寫法如下：

Our product is (the product) of good quality.

（我們產品是屬於好品質的產品）

要多一個 of，而 the product 可省略。同樣的，要注意下列這句中文的英譯：

中文：我們（是）隨時為你們服務（的人）

英文：We are of the service to you at anytime.（of 不可省去）

(3)聯綴動詞主要是 Be 動詞（Be 動詞中文只有三個意思：是、在、到）和感官動詞（feel, look, sound, taste, smell 及 become, seem, turn, get, grow, appear, keep）。

(4)主補(S.C.)如果是名詞時，照中文順序直接翻譯，但是如果是形容詞時，要注意中文的 Be 動詞（是）會省略，英文要補上一個 Be 動詞。

　　　例：我們的生意（是）很好＝ Our business is very good.

　　　例：S ＋ LV ＋ S.C. (N)

　　　　(1)I am the sales manager.（我是業務經理）

　　　　(2)This is our new sample.（這是我們的新樣品）

　　　　(3)Our best delivery time is 30 days.（我們最快交期是 30 天）

　　　　(4)We are a manufacturer.（我們是工廠）

　　　　(5)Taipei is a big city.（台北是個大都市）

　　　　(6)Our phone number is 87654321.（我們的電話號碼是 87654321）

例：S + LV + S.C. (Adj.)中文動詞「是」省去，英文要補上一個 Be 動詞

(1)Your price is too high.（你們的價錢太貴）

(2)Our quality is excellent.（我們的品質很好）

(3)Our quality control is strict.（我們的品管很嚴格）

(4)I am very busy.（我很忙）

(5)I am tired.（我很累）

(6)The color is too dark.（顏色太深）

(7)The business is turning better.（生意好轉）

7. There + Be + S(Noun) + Adv.（地點／時間）

注意：(1)中文：地點／時間的副詞開頭＋有，就用此句型（There + Be ＝有），There
　　　+ Be 動詞後面，要馬上跟一個名詞（＝主詞），再把時間／地點放句尾，
　　　不可再加 have，寫成 There are have 就錯誤。

(2)此句型不可照中文順序直接翻譯。

例：(1)樣品室裡　有　很多　樣品（地點＋有，用 There + Be 開頭）
　　　　4　　　1　　2　　3

There are many samples in the showroom.

(2)我們公司附近　有　一家　飯店（地點＋有）
　　　　4　　　　1　　2　　3

There is a hotel near our office.

(3)你們的產品上　有　些　不良（地點＋有）
　　　　4　　　1　2　3

There are some defects in your products.

(4)昨天　有　颱風（時間＋有）
　　3　　1　　2

There was a typhoon yesterday.

(5)在開始推銷時　將會有　一些　困難（時間＋有）
　　　　4　　　　1　　　2　　　3

There will be some difficulties at the beginning of promotion.

8. It ＋ Be ＋ N/Adj./Adv.（名詞／形容詞／副詞）

注意：It 可以代替天氣、時間、地點、距離、東西和事情

例：(1)It is cold.（天氣很冷）

(2)It is Dec. 25, 2011.（時間是 2011 年 12 月 25 日）

(3)It is noisy.（地點很吵鬧）

(4)It is far from our office to the factory.（從我們公司到工廠距離很遠）

(5)It is on the table.（東西在桌上）

(6)It is important.（事情很重要）

練習 1　句型練習

注意：翻譯時，名詞注意單複數，動詞注意時態／人稱／主被動變化，副詞時間地
點放到句尾，第 2、6、7 句型注意中文位置要調整，照下列阿拉伯數字順序
翻成英文。

1. S ＋ FV

(1)我們的　客人　離開了　（過去式）　　　客人：customer/client
　　　1　　　2　　　3

(2)他　正在工作　（現在進行式）　　工作：work
　　1　　2

(3)我們的　目錄　已經寄了　（現在完成被動式）　　目錄：catalogue
　　　1　　2　　　3　　　　　　　　　　　　　　　寄：send/sent (p.p.)

2. S ＋ FV ＋ Adv.（時間／地點／目的／方法）

(1)請　儘快　回覆　（原形）　　回覆：reply；儘快：soon
　　1　　3　　2

(2)我們　明天　去　你們的　工廠　（未來式）　　工廠：factory
　　1　　5　　2　　3　　　4

(3)他　上週　去　日本　作生意　（過去式）　　作生意：for business
　　1　　4　　2　　3　　5

(4)請　用空運　裝運　（原形）　　裝運：ship；用空運：by air
　　1　　3　　2

3. S ＋ FV ＋ O

(1)我們　有　100 個　員工　（現在式）　　員工：employee
　　1　　2　　3　　　4

(2)請　報　最好的　價錢　（原形）　　報（價）：quote/offer
　　1　2　　3　　　4

(3)我們　接受　你們的　交期　（現在式）　　接受：accept
　　1　　2　　3　　4　　　　　　　　　交期：delivery time

(4)請　儘快　告知　你們的　意見　（原形）　告知：advise
　　1　5　　2　　3　　4　　　　　　　意見：comment

(5)我們　收到了　你們的　信用狀　（過去式）　收到：receive
　　1　　2　　3　　　4　　　　　　　信用狀：L/C

4. S＋FV＋I.O.（人）＋D.O.（物）或
S＋FV＋D.O.（物）＋介系詞＋I.O.（人）

(1)我們　這週內　會開　信用狀　給貴公司（未來式）　　開：open/issue
　　1　　5　　2　　3　　4

(2)請　儘快　寄　新樣品　給我們　（原形）　　寄：send
　　1　5　　2　3　　4

(3)請　準時　出貨　給我們　（原形）　　出貨：ship the goods
　　1　4　　2　　3　　　　　　　　準時：on time

(4)請　告訴　我們　出貨明細　（原形）　出貨明細：shipping details
　　1　2　　3　　4

5. S＋V＋O＋O.C.（N or Adj.）

(1)我們　認為　你們的品質　比較好　（現在式）　品質：quality
　　1　　2　　3　　　　4　　　　　　　　比較好：better

(2)讓　我　幫你　拿行李　（原形）　　拿行李：with the luggage/
　　1　2　3　　4　　　　　　　　　　　　　　　with the baggage

(3)你們的抱怨　使得　我們　很驚訝　（過去式）　抱怨：complaint
　　1　　　2　　3　　4　　　　　　　驚訝：surprised

6. S ＋ LV ＋ S.C.（N/Adj.）

(1)我們的生意（是）很好　（現在式）　　生意：business
　　　　　1　　　　2　　3

(2)你們的價錢　似乎　很合理　（現在式）　　合理：reasonable
　　　1　　　2　　　3

(3)這　是　我們的　最新的　設計　（現在式）　　設計：design
　1　2　　3　　　4　　　5

(4)這　是　我們的　主要產品　（現在式）　　主要產品：main product
　1　2　　3　　　4

(5)我們的　主要市場　是　日本（現在式）　　主要市場：main market
　　1　　　2　　　3　4

(6)我們的　最低量　是　1000 個　（現在式）　　最低量：min. quantity
　　1　　　2　　3　　4

7. There ＋ Be ＋ N(S)＋ Adv.（時間／地點）
There ＋ Be ＝有

(1)我們的倉庫裡　有　現貨　（現在式）　　現貨：stock
　　　　3　　　1　2　　　　　　倉庫：warehouse

(2)我們公司對面　有　一家飯店　（現在式）　　對面：opposite
　　　　3　　　1　2

(3)上週　沒有　船　（過去式否定）　　船：vessel
　3　　1　　2

(4)展示架上　有　很多樣品　（現在式）　　展示架上：on the display
　　3　　　1　　2

8. It ＋ Be ＋ N/Adj./Adv.
It ＝天氣／時間／地點／距離／東西／事情

(1)外面　　（天氣）　正在下雨　（現在進行式）　　外面：outside
　　3　　　　1　　　　2

(2)現在　　（時間）　是 3：30　（現在式）
　　4　　　　1　　　2　3

(3)這裡　　（地點）　（是）　很安靜　　（現在式）　　安靜：quiet
　　4　　　　1　　　2　　　3

(4)從你的飯店　到我們的辦公室　（距離）　（是）　很遠　（現在式）
　　　4　　　　　5　　　　　　1　　　2　　　3

(5)（東西）　是　金色的　　（現在式）
　　1　　　2　　3

(6)（事情）　（是）　很緊急　（現在式）　　緊急：urgent
　　1　　　2　　　3

三、兩種直接問句造法(Yes/No Question & WH Question)

1. Yes/No 問句

造法：將句中的 Be 動詞或現有助動詞(will/can/have⋯)搬到主詞(S)前面，句中如沒有
Be 動詞或助動詞，則必須在句首加上 Do/Does/Did，後面動詞要變原形。

公式：(1)Be（is/are/am/was/were）＋ Subject

　　　(2)Aux.（will/shall/can/may/must）＋ Subject ＋ V-root

　　　(3)Have/Has/Had ＋ Subject ＋ p.p.

　　　(4)Do/Does/Did ＋ Subject ＋ V-root

注意：進行式(Be ＋ V-ing)和被動式(Be ＋ p.p.)中也有 Be 動詞，直接搬到主詞前面造
成問句。句子中的 we, our, us 問句要改為 you, your。

句　子	Yes/No 問句
例：(1)We are a manufacturer.	→ Are you a manufacturer？
(2)This is our new sample.	→ Is this your new sample？
(3)We are producing your order now.	→ Are you producing our order now？
(4)The shipment was made by air.	→ Was the shipment made by air？
(5)The samples were sent yesterday.	→ Were the samples sent yesterday？
(6)We will place this order with you.	→ Will you place this order with us？
(7)I would like something to drink.	→ Would you like smomething to drink？
(8)We can give you 5% commission.	→ Can you give us 5% commission？
(9)I could do you a favor.	→ Could you do me a favor？
(10)You may help me.	→ May I help you？
(11)We have completed your order.	→ Have you completed our order？
(12)The sample has been received.	→ Has the sample been received？
(13)We accept your price.	→ Do you accept our price？
(14)The customer likes this model.	→ Does the customer like this model？
(15)We shipped the goods on time.	→ Did you ship the goods on time？

2. 附加問句（Yes/No 問句的另一種問法）

造法：在句子後面加一個問句（主詞改代名詞）

注意：(1)前面句子是肯定句時，後面問句就用否定

前面句子是否定時，後面問句就用肯定

(2)前面句子是用 Be 動詞，後面問句就用同樣的 Be 動詞

前面是助動詞，後面就用同樣的助動詞

前面是動詞，後面就依一樣的時態補上 do/does/did

公式：S ＋ Be ＋ S/C　　　　　　Be ＋ not ＋ S？

　　　S ＋ Aux. ＋ V-root ＋ O　　Aux. ＋ not ＋ S？

　　　S ＋ V ＋ O　　　　　　　do/does/did ＋ not ＋ S？

公式：S ＋ Be ＋ not ＋ S/C　　　　　　　　Be ＋ S？

　　　S ＋ Aux. ＋ not ＋ V-root ＋ O　　　Aux. ＋ S？

　　　S ＋ don't/doesn't/didn't ＋ V-root　　do/does/did ＋ S？

例：(1)You are the manufacturer, aren't you？

You are not the manfacturer, are you？

(2)He is your regular customer, isn't he？

He is not your regular client, is he？

(3)You will ship the goods on time, won't you？

You will not ship the goods on time, will you？

(4)You can accept our price, can't you？

You can not accept our price, can you？

(5)You have completed our order, haven't you？

You haven't completed our order, have you？

(6)He has received the sample, hasn't he？

He hasn't received the sample, has he？

(7)You like our products, don't you？

You don't like our products, do you？

(8)The price includes our commission, doesn't it？

The price doesn't include our commission, does it？

(9)You received our sample, didn't you？

You didn't receive our sample, did you？

3. WH(Information)問句

　　造法：句首先放 WH 或 How 的疑問字，再將句中的 Be 動詞／助動詞搬到主詞前，
　　　　　或在主詞前面補上 Do/Does/Did；換言之，即在 Yes/No 問句前，加上 WH 或
　　　　　How 的疑問字。

```
公式：(1)WH/How ＋ Be ＋ Subject
      (2)WH/How ＋ Aux.＋ Subject ＋ V-root
      (3)WH/How ＋ have ＋ Subject ＋ p.p.
      (4)WH/How ＋ do/does/did ＋ Subject ＋ V-root
```

　注意：此句型結構位置和中文不同（中文的疑問字和句子同一位置，例如：問句：
　　　　　你住哪裡？句子：我住台北）

　例：(1)Who is your boss？（你老闆是誰？）

　　　　(2)Whom shall I give this sample to？（我應該將此樣品給誰？）

　　　　(3)Whose sample is this？（這是誰的樣品？）

　　　　(4)Where is your hotel？（你的飯店在哪裡？）

　　　　(5)When can I get your quotation？（我何時能拿到你的報價單？）

　　　　(6)Which model do you want to buy？（你要買哪個型號？）

　　　　(7)Why did you delay the shipment？（你為什麼延遲出貨？）

　　　　(8)What is your packing method？（你們的包裝方式是什麼？）

　　　　(9)How many samples can you give us？（你可以給我們多少樣品？）

　　　　(10)How much commission do you request？（你要求多少佣金？）

　　　　(11)How long is your delivery time？（你們的交期是多久？）

　　　　(12)How far is it from your office to the factory？（從你們公司到工廠是多遠？）

　　　　(13)How often do you come here？（你多常來這兒？）

　　　　(14)How large is your factory？（你們工廠是多大？）

　　　　(15)How will you pay us, by L/C or T/T？

　　　　　　（你們將如何付款給我們，用信用狀或電匯？）

練習 2 直接問句練習

1. 將下列句子改成 Yes/No 問句

(1)The samples are made of the costly material.

(2)The price is based on CIF New York by sea.

(3)I am looking for hand tools.

(4)The customer will come here tomorrow.

(5)This mistake will never happen again.

(6)You may know my name.

(7)I can meet you at the hotel.

(8)We can give you 5% discount.

(9)We have 10 sales persons.

(10)We like your new model.

(11)The price includes your commission.

(12)We sent you the sample last week.

(13)I have ever been to Hong Kong.

(14)We have received your L/C.

(15)The customer has confirmed the order.

2. 造下列 WH 問句（中翻英）

注意：句中如沒有 Be 動詞／助動詞，則在主詞前要補加 do/does/did。

(1)誰　將付款　給我們？(Who)　　付款：pay
　　1　　2　　　3

(2)你　昨天　遇到　誰？(Whom)
　　2　　4　　3　　1

(3)這　是　誰的　報價單？(Whose)　　報價單：quotation
　　4　　3　　1　　2

(4)你們的　工廠　在　哪裡？(Where)　　工廠：factory
　　　3　　　4　　2　　1

(5)您　何時　將　來　這裡？(When)
　　3　　1　　2　　4　　5

(6)你　喜歡　哪種顏色？(Which)
　　2　　3　　　　1

(7)你們的樣品　為什麼　看起來　很髒？(Why)
　　　　2　　　　1　　　3　　　4

(8)你們　提供　什麼服務？(What)　　提供：offer　服務：service
　　2　　3　　　1

(9)你們　有　多少　客戶？(How many)　　客戶：customer/client
　　3　　4　　1　　2

(10)你們　可以　給　我們　多少　折扣？(How much)　　折扣：discount
　　4　　3　　5　　6　　1　　2

(11)你　將　在台北　停留　多久？(How long)　　停留：stay
　　3　　2　　5　　　4　　　1

(12)從這兒　到飯店（距離）是　多遠？(How far)
　　4　　5　　3　　2　　1

(13)你們　一個月　出　幾次　貨？(How often)　　出貨：ship the goods/
　　2　　5　　3　　1　　4　　　　　　　　　　　　deliver the goods

(14)台灣　是　多大？(How large)
　　3　　2　　1

(15)你們　將　如何　出貨？(How)　　出貨：make the shipment/
　　3　　2　　1　　4　　　　　　　　　　effect the shipment

(16)你們的　生意　（是）如何？(How)　　生意：business
　　3　　　4　　　2　　1

3. 將下列句子改為附加問句

(1)You are an importer, _____ _____ ?

(2)Mr. Johnson is your boss, _____ _____ ?

(3)Amy will go to the factory, _____ _____ ?

(4)Your factory produces this item, _____ _____ ?

(5)Jim hasn't got this sample, _____ _____ ?

四、間接問句及否定句造法(Indirect Question & Negative Sentence)

1. 間接問句(Indirect Question)

   造法：在 WH 問句前面加一個句子(Sentence)或一個 Yes/No 問句

   注意：WH 問句中的 Be 動詞或助動詞，放回主詞後面，也不需補加任何 do/does/did 的助動詞。如果是句子開頭的間接問句，句尾打句點，如果是 Yes/No 問句開頭，句尾打問號(？)

   > 公式：(1)Sentence ＋ WH ＋ S ＋ Be/Aux.
   >
   > (2)Yes/No Question ＋ WH ＋ S ＋ Be/Aux.

   例：(1)Please advise us when we can receive your L/C.

   (2)Please advise how you will pay us.

   (3)We would like to know where your factory is.

   (4)Can you tell us how much commission you want？

2. 否定句(Negative Sentence)

   造法：在句中的 Be 動詞或助動詞後面，加一個 not；句中如果沒有 Be 動詞或助動詞，則在動詞前面補上 don't/doesn't/didn't，後面動詞要變原形；或在動詞後面，名詞前面加一個 no

   > 公式：(1)S ＋ Be ＋ not
   >
   > (2)S ＋ Aux.＋ not ＋ V-root
   >
   > (3)S ＋ have ＋ not ＋ p.p.
   >
   > (4)S ＋ don't/doesn't/didn't ＋ V-root
   >
   > (5)S ＋ V ＋ no ＋ N

   例：(1)We are not the manufacturer.

   (2)We will not ship the goods before receiving your L/C.

   (3)We can not ship the goods this week due to no vessel available.

   (4)We have not received your sample.

   (5)We don't produce this item.

   (6)The price doesn't include any commission for you.

   (7)We have no stock for this item.

練習 3　間接問句及否定句練習

1. 將下列句子合併成間接問句

(1)We hope to know

How much discount can you give us？

(2)Please don't hesitate to advise us

When will you come to Taiwan？

(3)Can you tell us

How soon can you ship our goods to us？

(4)Please let us know

How long are you going to stay in Taipei？

(5)Would you please tell us

Which model do you like？

2. 將下列句子改成否定句

(1)Your price is higher.

(2)This is our new sample.

(3)We will pick you up at the airport.

(4)We can send you the sample free of charge.

(5)We shipped the goods of your order last week.

(6)The customer has placed a big order with us.

(7)We accept your offer.

(8)Your quality looks good.

(9)The customer requests high quality product.

(10)They opened the L/C yesterday.

(11)The customer has time to meet us.

(12)We have good quality control.

五、片語／子句說明及應用(Phrases & Clauses)

1. 片語(Phrase)
   - 定義：兩個字以上，形成一個意思的一組字
   - 樣式：可分為：(1)固定片語和(2)組合片語兩類

(1)固定片語三種：等於生字，不可改變，要背記

①動詞片語(Verb Phrase)：動詞開頭的一組字，有特定意思，不可改變

位置：放 V 位置

例：

look forward to	盼望
pick up	接……（人）
pay attention to	注意
turn on	打開（開關）
turn off	關上（開關）

②介系詞片語(Prepositional Phrase)：介系詞本身由一組字形成

位置：當介系詞用

例：

according to	按照、依照
due to	由於
in accordance with	按照、依照
in front of	在……前面
in addition to	此外
in order to	為了……
next to	隔壁
owing to	由於
with regard to	有關
with a view to	為了……

③副詞片語(Adverb Phrase)：由一組字形成的副詞，有特定意思，特別表示時間

位置：放副詞位置

例：at once　　　　　馬上

long time ago	很久以前
right away	馬上
right now	現在
the day after tomorrow	後天
the day before yesterday	前天

(2)組合片語四種樣式（長相／結構）：可以自己組合；組合後，可以放三個位置：

　　①放在名詞位置(S, O, S.C., O.C.)，稱名詞片語(N. Ph.)

　　②放在形容詞位置（在 N 後面），稱形容詞片語(Adj. Ph.)

　　③放在副詞位置（在 V，Adj.，Adv.，子句或句子後面），稱副詞片語(Adv. Ph.)

　　四種組合片語樣式，如下所述：

①介系詞片語(Prepositional Phrase)

　　組合方式：介系詞＋ the ＋普通名詞(N)

　　　　　　　　介系詞＋專有名詞或動名詞(V-ing)（不加冠詞）

　　位置：放在形容詞位置（當 Adj. Ph.）或副詞位置（當 Adv. Ph.）

　　例：在桌上　　　on the table

　　　　在台北　　　in Taipei

　　　　在飯店　　　at the hotel

　　應用：①桌上的書是我的

　　　　　　The book on the table is mine.

　　　　　　（on the table 放在名詞 book 後面當 Adj. Ph.）

　　　　　②我們會到飯店接你

　　　　　　We will pick you up at the hotel.

　　　　　　（at the hotel 放句尾，表示地點，當 Adv. Ph.）

②不定詞片語(Infinitive Phrase)

　　組合方式：不定詞＋原形動詞＝ To ＋ V-root

　　位置：放在名詞位置（當 N. Ph.）、形容詞位置（當 Adj. Ph.）或副詞位置（當 Adv. Ph.）

例：賺錢   to make money

和你們作生意 to do business with you

應用：①賺錢很重要

  <u>To make money</u> is important.

  （to make money 放在主詞位置，當 N. Ph.）

②我很高興<u>認識你</u>

  I am glad <u>to meet you</u>.

  （to meet you 放在形容詞 glad 後面，當 Adv. Ph.）

③謝謝<u>你的報價要求</u>

  Thanks for your request <u>to quote you a price</u>.

  （to quote you a price 放在名詞 request 後面，當 Adj. Ph.）

③現在分詞片語(V-ing Phrase)

 組合方式：V-ing ＋ O(N)

 位置：放在名詞位置（當 N. Ph.）或形容詞位置（當 Adj. Ph.）

例：賺錢   making money

和你們作生意 doing business with you

應用：①我們喜歡<u>和你們作生意</u>

  We like <u>doing business with you</u>.

  （doing business with you 放在受詞位置，當 N. Ph.）

②我們收到你們<u>表示對我們產品感興趣的</u>信

  We received your letter <u>showing interest in our product</u>.

  （showing interest in our product 放在名詞 letter 後面，當 Adj. Ph.）

④過去分詞片語(p.p. Phrase)

 組合方式：p.p.＋ Adv.

 位置：放在形容詞位置（當 Adj. Ph.）

例：①我們收到你們<u>上週寄來的</u>信

  We received your letter (which was) <u>sent last week</u>.

  （sent last week 放在名詞 letter 後面，當 Adj. Ph.）

②我們接受<u>你們列在價目表上的</u>價錢

We accept the prices (which were) <u>listed in your pricelist</u>.

（listed in your pricelist 放在名詞 prices 後面，當 Adj. Ph.）

注意：(1)不定詞片語(To ＋ V-root)和現在分詞片語(V-ing)不同處：

To ＋ V-root ＝未來動作或即將要做的動作

V-ing ＝過去動作或即將結束的動作或一件事

例：We like <u>to do</u> business with you.

（表示尚未作生意）

We like <u>doing</u> business with you.

（表示已在作生意）

(2)現在分詞片語(V-ing)和過去分詞片語(p.p.)放名詞後面當形容詞片語，不同用法：

N ＋ V-ing（＝主動或進行式）

N ＋ p.p.（＝被動，省略 which ＋ Be）

2. 子句(Clause)

子句有三種：(1)對等子句；(2)主要子句；(3)附屬子句（或稱從屬子句）

子句結構：子句即句子，所以結構和基本句型相同

(1) 對等子句(Coordinate Clause)

用法：用連接詞 and,or,but 連接兩個基本句型的句子，拿掉連接詞，可以各別獨立使用，亦稱獨立子句

位置：子句（基本句型）＋ and/or/but ＋子句（基本句型）

應用①：用 and 連接兩個意思相同的對等子句

例句：We are an exporter <u>and</u> you are an exporter, too.

句型：S ＋ LV ＋ S.C. and S ＋ LV ＋ S.C.

應用②：用 or 連接兩個選擇性的對等子句

例句：You import computers <u>or</u> (you) export computers.

句型：S ＋ V ＋ O   or   S ＋ V ＋ O

應用③：用 but 連接兩個相反意思的對等子句

例句：We like your product <u>but</u> your price is too high.

句型：S ＋ V ＋ O   but   S ＋ LV ＋ S.C.

(2) 主要子句(Main Clause)

結構：基本句型

用法：和附屬子句一起用，但也可以單獨使用，亦稱獨立子句

(3) 附屬子句(Subordinate Clause)

用法：不可單獨使用，一定要和主要子句一起用，亦稱非獨立子句

位置：有三個：

　　　①放在名詞位置(S/O/S.C./O.C.)，稱名詞子句(N. Cl.)

　　　②放在形容詞位置（N 後面），稱形容詞子句(Adj. Cl.)

　　　③放在副詞位置（V., Adj., Adv.，子句，句子後面），稱副詞子句(Adv. Cl.)

①名詞子句(N. Cl.)

　用法：大多放在受詞位置，用 that 帶出的子句

　位置：S ＋ V ＋ that ＋子句（＝主要子句＋ that 名詞子句）

　例句：(1)We know that you import sport shoes.

　　　　(2)We believe that you will like our products.

　句型：S ＋ V ＋ O(＝ that ＋ S ＋ V ＋ O)

　分析：主要子句＝ S ＋ V ＋ O，但受詞不是一個字，而是一個 that ＋ S ＋ V ＋ O 的附屬名詞子句。

②形容詞子句(Adj. Cl.)

　用法：在 N 後面，放一個 who/whom/whose 或 which 的子句，大多和主要子句糾纏在一起，放在S,O,S.C.,O.C.或介系詞片語(Prep. ＋ the ＋ N)的名詞(N)後面，借用 N 當後面子句的主詞，受詞或所有格代名詞

　位置：N（人）＋ who ＋ V（主格）

　　　　　　　　whom ＋ S ＋ V（受格）

　　　　　　　　whose ＋ N ＋ V（所有格）

　　　　N（物）＋ which ＋子句（主格／受格／所有格）

　例句(1)：The man who just talked to you is my boss.

　句　型：S (N) ＋ who ＋ V ＋ O ＋ LV ＋ S.C.

　分　析：S ＋ LV ＋ S.C.＝主要子句， who ＋ V ＋ O ＝附屬 Adj. Cl.掛在主詞

man (N) 的後面，補充說明 man

例句(2)：We sell our product to the buyer <u>whom you met yesterday</u>.

句　型：S ＋ V ＋ D.O.＋ to ＋ I.O. (N) ＋ whom ＋ S ＋ V ＋ Adv.

分　析：S ＋ V ＋ D.O.＋ to ＋ I.O.＝主要子句，whom ＋ S ＋ V ＋ Adv.＝附
屬 Adj. Cl. 掛在間接受詞 buyer (N)的後面，補充說明 buyer

例句(3)：Our manufacturer is ABC Co. <u>whose quality is excellent</u>.

句　型：S ＋ LV ＋ S.C.(N) ＋ whose ＋ N ＋ LV ＋ S.C.

分　析：S ＋ LV ＋ S.C.＝主要子句，whose ＋ N ＋ LV ＋ S.C.＝附屬 Adj.Cl.掛
在主詞補語 ABC Co. (N) 後面，補充說明 ABC Co.

例句(4)：The sample <u>which you sent us</u> was broken.

句　型：S(N)＋ which ＋ S ＋ V ＋ O ＋ V

分　析：S ＋ V＝主要子句，which ＋ S ＋ V ＋ O ＝附屬 Adj.Cl. 掛在主詞 sample
(N)後面，補充說明 sample

形容詞子句合併方式

　　將兩個簡單句用 who/whom/whose/which 合併成一個複雜句；換言之，兩個句子中
有同樣的名詞時，第二個句子的名詞不再重複，借用前面的名詞當第二個句子中的主
詞，受詞或所有格代名詞。

例句(1)：This man is Mr. Wang. <u>Mr. Wang</u> is my boss.

　　＝ This man is Mr. Wang, <u>who</u> is my boss.

　　（借用前面的 N 當第二句的主詞）

例句(2)：This is the customer.　I talked to <u>the customer</u> yesterday.

　　＝ This is the customer <u>whom</u> I talked to yesterday.

　　＝ This is the customer <u>to whom</u> I talked yesterday.

　　（借用前面的 N 當第二句的受詞）

例句(3)：This is the biggest <u>design company</u>. <u>Their</u> products are elegant.

　　＝ This is the biggest design company <u>whose</u> products are elegant.

　　（借用前面的 N 當第二句的所有格代名詞）

例句(4)：We sent you a sample. The sample was specially made for you.

= We sent you a sample (which was) specially made for you.

（借用前面的 N 當第二句的主詞；which was 可省略改成形容詞片語）

例句(5)：We received your sample. You sent the sample on June 20.

= We received your sample (which) you sent on June 20.

（借用前面的 N 當第二句的受詞；which 後為 S＋V 時，which 可省略）

例句(6)：We sent you a sample. Its material is gold.

= We sent you a sample which material is gold.

（借用前面的 N 當第二句的所有格代名詞）

In which, on which, for which, from which 的用法

借用前面的 N 當第二句的副詞，表示地點或事情

例句(1)：We sent you a quotation.

You will find our best terms on the quotation.

= We sent you a quotation on which you will find our best terms.

（on which 當第二句的地點副詞）

例句(2)：We mailed you a parcel.

In the parcel you will find our new sample.

= We mailed you a parcel in which you will find our new sample.

（in which 當第二個句子的地點副詞）

例句(3)：We sent you a catalog.

From the catalog you will have a clear idea about our products.

= We sent you a catalog from which you will have a clear idea about our products.

（from which 當第二句的地點副詞）

例句(4)：We have very good relationship with many local suppliers.

For this reason it makes us can offer you the best prices.

= We have very good relationship with many local suppliers for which makes us

can offer you the best prices.

（for which 在第二句表示事情：為了此事）

③副詞子句(Adv. Cl.)

用法：用 when/before/after/since（表示時間），if（條件），because/since/as（原因），although/ but（讓步），so/therefore（結果），so that（目的）帶出的子句

位置：和主要子句分開，放在主要子句之前或之後，補充說明主要子句

例句(1)：When we receive your L/C, we will ship the goods.

句　型：When ＋ S ＋ V ＋ O，S ＋ V ＋ O

分　析：When ＋ S ＋ V ＋ O ＝時間副詞子句，S ＋ V ＋ O ＝主要子句

例句(2)：We will send you the sample after we receive your sample charge.

句　型：S ＋ V ＋ I.O.＋ D.O. after ＋ S ＋ V ＋ O

分　析：S ＋ V ＋ I.O.＋ D.O.＝主要子句，after ＋ S ＋ V ＋ O ＝時間副詞子句

例句(3)：Because this is a big order, we hope to get a discount.

句　型：Because ＋ S ＋ LV ＋ S.C.，S ＋ V ＋ O（＝名詞片語）

分　析：Because ＋ S ＋ LV ＋ S.C.＝原因副詞子句，S ＋ V ＋ O ＝主要子句

例句(4)：Although your quality is good, your price is too high.

句　型：Although ＋ S ＋ LV ＋ S.C.，S ＋ LV ＋ S.C.

分　析：Although ＋ S ＋ LV ＋ S.C.＝讓步副詞子句，S ＋ LV ＋ S.C.＝主要子句

例句(5)：Your price is too high, so that we can not place you the order.

句　型：S ＋ LV ＋ S.C.，so that ＋ S ＋ V ＋ I.O.＋ D.O.

分　析：S ＋ LV ＋ S.C.＝主要子句，so that ＋ S ＋ V ＋ I.O.＋ D.O.＝目的副詞子句

例句(6)：If you come here next week, we will give you a sample.

句　型：If ＋ S ＋ V ＋ Adv.，S ＋ V ＋ I.O.＋ D.O.

分　析：If ＋ S ＋ V ＋ Adv.＝條件副詞子句，S ＋ V ＋ I.O.＋ D.O.＝主要子句

練習 4　片語／子句練習

1.～15.寫出下列劃線部分是屬於什麼片語(N. Ph., Adj. Ph. or Adv. Ph.)或屬於什麼子句(N. Cl., Adj. Cl., or Adv. Cl.)；16.～20.選擇

1. To ship the goods on time is important.

2. Our principle is to accept the payment by L/C for the first deal.

3. We have 30 sales persons with very good experience.

4. This is the best way to improve your quality.

5. We just received the sample sent by you last week.

6. We are sending you our samples by air parcel post.

7. We have sent you our letter on April 20.

8. We sent you our latest catalog for your reference.

9. If you place us the order this week, we can ship the goods next week.

10. Because our quality is excellent, our products are welcomed by customers.

11. When we receive your L/C, we will arrange the shipment soon.

12. We trust that we will receive your order very soon.

13. We are interested in the products which you displayed at your booth.

14. The man whom you talked to yesterday is our best customer.

15. We have many sales persons who know the market situation very well.

16. (　) Our factory is located in Kunshan _____ is near Shanghai.

   (a)which　(b)that　(c)where　(d)when

17. (　) We sent you two samples _____ were specially made according to your specification.

   (a)who　(b)which　(c)what　(d)when

18. (　) We met you two years ago _____ you were in charge of the export department.

   (a)where　(b)which　(c)when　(d)that

19. (　) Our sales manager is Mr. Johnson _____ is on a business trip in Europe now.

   (a) which　(b)who　(c)whom　(d) what

20. (　) The man _____ you met at the airport yesterday is our export manager.

   (a)which　(b)who　(c)whom　(d)what

六、複合句／複雜句造法(Compound/Complex Sentences)

1. 複合句(Compound Sentence)

　　造法：用連接詞 and/or/but 連接兩個簡單句（基本句型）的對等子句

　　結構：簡單句＋ and/or/but ＋簡單句

　　例句(1)：We will consider your offer and (we will) reply you soon.

　　句　型：簡單句(S ＋ V ＋ O) and　簡單句(S ＋ V ＋ O ＋ Adv.)

　　例句(2)：You will come to our office or we shall go to your hotel.

　　句　型：簡單句(S ＋ V ＋ Adv.) or　簡單句(S ＋ V ＋ Adv.)

　　例句(3)：We received your shipment but we found 5 cartons broken.

　　句　型：簡單句(S ＋ V ＋ O) but　簡單句(S ＋ V ＋ O ＋ O.C.)

2. 複雜句(Complex Sentence)

　　造法：簡單句（主要子句）加上名詞／形容詞／副詞片語或子句（附屬子句）

　　　　　　　　　　　名詞

　　結構：簡單句＋形容詞　　　片語／子句＝複雜句

　　　　　　　　　　　副詞

　　例句(1)：簡單句：It is important.

　　　　　　　複雜句：To ship the goods on time is important.

　　　　　　　句　型：S（＝放不定詞的名詞片語）＋ LV ＋ S.C.

　　例句(2)：簡單句：We believe it.

　　　　　　　複雜句：We believe (that) you will be satisfied with our offer.

　　　　　　　句　型：S ＋ V ＋ O （＝放一個 that 帶出的名詞子句）

　　例句(3)：簡單句：The sample was broken.

　　　　　　　複雜句：The sample sent by you on July 1 was broken.

　　　　　　　句　型：S ＋ 一個過去分詞的形容詞片語 ＋ V

　　例句(4)：簡單句：We sent you a sample.

　　　　　　　複雜句：We sent you a sample which was specially made for you.

　　　　　　　句　型：S ＋ V ＋ I.O.＋ D.O.＋一個 which 帶出的形容詞子句

例句(5)：簡單句：We sent you a sample.

　　　　複雜句：We sent you a sample by air parcel post on July 1.

　　　　句　型：S＋V＋I.O.＋D.O.＋兩個副詞片語（方法＋時間）

例句(6)：簡單句：We will ship the goods.

　　　　複雜句：We will ship the goods after we receive your L/C.

　　　　句　型：S＋V＋O＋一個表時間的副詞子句

例句(7)：簡單句：The shipment will be delayed.

　　　　複雜句：The shipment will be delayed because there is no vessel available this week.

　　　　句　型：S＋V＋一個表原因的副詞子句

例句(8)：簡單句：We will accept your price.

　　　　複雜句：We will accept your price if your payment is by sight L/C.

　　　　句　型：S＋V＋O＋一個表條件的副詞子句

例句(9)：簡單句：Please advise your shipping method.

　　　　複雜句：Please advise your shipping method so that we can arrange the shipment.

　　　　句　型：Please＋V-root＋O＋一個表目的的副詞子句

例句(10)：簡單句：Our quality is excellent.

　　　　　複雜句：Our quality is excellent, so our products are popular.

　　　　　句　型：S＋LV＋S.C.＋一個表結果的副詞子句

3. 複合複雜句(Compound & Complex Sentence)

　造法：用連接詞 and/or/but 連接一個簡單句和一個複雜句，或連接兩個複雜句

　結構：簡單句＋and/or/but＋複雜句

　　　　複雜句＋and/or/but＋簡單句

　　　　複雜句＋and/or/but＋複雜句

　　　　複合複雜句＝一個複雜句當中，本身又用連接詞 and/or/but 連接二個複合句

例句(1)：We received your order and (we) confirm that we will ship the goods when your L/C is received.

句　型：簡單句＋ and ＋複雜句

分　析：S ＋ V ＋ O ＋ and ＋ S ＋ V ＋ O（＝ that 的名詞子句）＋ when 的副詞
子句

例句(2)：As requested, we sent you a sample for quality test yesterday and please confirm
your approval soon.

句　型：複雜句＋ and ＋簡單句

分　析：副詞子句＋ S ＋ V ＋ I.O.＋ D.O.＋副詞片語（目的＋時間）＋ and ＋ S
＋ V ＋ O ＋副詞單字

例句(3)：We received with thanks your letter and catalog on June 25 and (we) hope
that you can send us three samples free of charge.

句　型：複雜句＋ and ＋複雜句

分　析：S ＋ V ＋副片＋ O ＋副片＋ and ＋ S ＋ V ＋ O（＝ that 的名詞子句＋ Adj.
Ph.）

例句(4)：We received the shipment which you made on May 30, but (we) regret to inform
you that there are 5 cartons broken.

句　型：複雜句＋ but ＋複雜句

分　析：S ＋ V ＋ O ＋ which 的 Adj.Cl.＋ but ＋ S ＋ V ＋ O（＝ to 的 N.Ph.＋ N.Cl.）

例句(5)：Regarding the wrong shipment, we suggest that you may keep them for sale or
(you may) return them to us by sea freight at your convenience.

句　型：複合複雜句

分　析：副片＋ S ＋ V ＋ O（＝用 that 帶出的兩個名詞子句）＋兩個副片
Adv. Ph.(＝ that ＋ S ＋ V ＋ O or S ＋ V ＋ O ＋ Adv. Ph.＋ Adv. Ph.)

重點：只要在簡單句中加入連接詞及名詞／形容詞／副詞的片語和子句，句子就可
以無限擴大成所謂的複合句、複雜句或複合複雜句，但是要注意在商業寫作
上，一個句子不要帶太多的片語或子句，一個句子最好不要超過兩個以上的
片語和子句。

練習 5　複合複雜句練習

　　將下列中文句子翻譯成英文，需注意(1)英文排列順序；(2)單數名詞前加冠詞；(3)時間地點副詞放句尾；(4)形容詞片語子句放 N 後面；(5)動詞時態／主被動問題（第1.～6.題的第 2 句為英文排列順序）。

1. 客人　說　你們　上批出貨　所用的　材料　（是）　比較好
　　 1　　 2　　 4　　　 6　　　 5　　　 3　　 7　　 8

　　客人　說 the 材料（which 你們　使用　在上批出貨上）是　比較好

2. 這　不是　線路板上　所用的　焊點　的　尺寸
　 1　 2　　 7　　　 6　　 5　 4　 3

　　這　不是 the 尺寸 of　 the pad which 被用在　線路板上

3. 如附　是　客人　對你們　所提供的　樣品　的　報告
　　 1　 2　 3　　 7　　　 8　　　 6　 5　 4

　　如附　是 the 客人的報告 for　 the 樣品 which 你們　提供

4. 客人　正在檢查　你們　上週　所用的　包裝
　　 1　　 2　　 4　 6　　 5　　 3

　　客人　正在檢查 the 包裝 which 你們　使用　上週

5. 這　和　我　上個月　寄給　你們的　規格上的　顏色　（是）不一樣
　 1　 6　 9　 7　　 8　　 5　　　 4　　 2　　 3

　　這是　不一樣 from the 顏色 on the 規格 which 我寄給你們 上個月

6. 我們必須將所有未出貨的訂單上的　這些項目，在未來四週內　裝出
　　 1　 2　　　 5(open orders)　　　 4　　 6　　 3

　　我們　必須　裝出　這些項目 on all open orders in the next 四週

7. 客人　將不下　訂單　給　我們　直到　你們　確認　接受　他們的　價錢和交期
　　 1　　 2　　 3　 4　 5　　 6　　 7　　 8　 (to)9　 10　　 11

8. 我們　剛收到　一份　修正的　報價單，此報價單　從 7 月 1 日　開始生效
　　1　　2　　　3　　　4　　　　5　　　6(which)　　　　8　　　　　7

9. 如果　你們　無法　改進　訂單上的 9/15 的交期，請　將已完成的　5000 個　先
　　1　　2　　3　　4　　7　　　　6　　5　　8　　10　　　　11　　12
　出貨
　　9

10.請　和 3M　查看　是否　他們　有　現成的　產品，而此產品有　穩定的　背膠；
　　1　　3　　2　　4　　5　　6　　7　　　8　　　9(with)　　　10　　　11
　此背膠　　不會　在高溫下　脫落
　12(which)　13　　15　　　14

生字 / 片語

1. 材料 material，上批出貨 on last shipment，比較好 better
2. 線路板上 on the PCB (printed circuit board)，尺寸 size，焊點 pad
3. 如附 the enclosed，報告 report
4. 檢查 check，包裝 packing
5. 不一樣 different，規格 specification
6. 裝出 ship，所有未出貨的訂單上 on all open orders，這些項目 these items
7. 確認接受 confirm to accept，交期 delivery time
8. 修正的報價單 revised quotation，生效 will be effected
9. 改進 improve，已完成的 completed
10. 查看 check，是否 if，現成的 existed，有穩定背膠 with the stable back adhesive，將不會脫落 won't be taken off，在高溫下 under the high temperature

七、主要詞類功用／位置說明及練習(Parts of Speech)

I. 名詞(Noun)

種類：(1)普通名詞(Common Noun)：有名稱的人、地、事、物、看得見、
很多的、普遍的、可數名詞

例：sample, catalog, order…

(2)集合名詞(Collective Noun)：一個團體，可數名詞

例：company, manufacturer…

(3)專有名詞(Proper Noun)：人名、地名、國名、公司名、特定唯一之單數名
詞，注意第一個字母要大寫

例：Taipei, Mr. Johnson, ABC Co…

(4)抽象名詞(Abstract Noun)：看不見、觀念性的事物、不可數、單數名詞

例：commission, quality, quantity…

(5)物質名詞(Materials Noun)：材料、氣體、液體、金屬等單數名詞

例：furniture, rice, steel, beer…

注意：(1)可數名詞（普通名詞／集合名詞）要注意單／複數，單數名詞前面要加冠
詞 a/an/the，但是如果有所有格代名詞 our/your 或指示代名詞 this/that/
these/those 就不要加冠詞；複數後面加 s 或 es

例：This is a new sample.　This is our new sample.

　　We place you an order.　We place you two orders.

(2)物質名詞不可數，如要數，前面加單位來數

例：two pieces of furniture（兩件傢俱）；five tons of steel（五噸鋼鐵）

位置：放在主詞(S)、受詞(O)、主補(S.C.)、受補(O.C.)的位置和介系詞後面

用法：在 S/O/S.C./O.C.這四個名詞的位置，可以有三個變化：

①放一個名詞單字，叫名詞單字(Noun word)

②放兩個字以上的片語，就叫名詞片語(Noun Phrase)

③放一個完整的句子（基本句型），就叫名詞子句(Noun Clause)

例：(1)We received your sample.（受詞是一個名詞單字）

(2)To ship the goods on time is important.（主詞是個名詞片語）

(3)We know that you import toys.（受詞是一個名詞子句）

練習 6　名詞練習

將下列 1～10 中文句子翻譯成英文；11～15 為選擇題

1. 這個　客人　明天　將去　你們的　公司及工廠　（用 office/factory）
　　1　　2　　6　　3　　4　　　5

2. 我　在這家公司　上班　（用 company）
　　1　　3　　　2

3. 我們　是　（一家）　工廠　（用 manufacturer）
　　1　　2　　3　　4

4. 你　可以　給　我們　多少佣金？　(how much commission)
　　3　　2　　4　　5　　1

5. 你　要　買　多少數量？　(how much quantity)
　　2　3　4　　1　　（注意：兩個 V 中間用 to 分開，主詞前補一個 do）

6. 我們的　品質　（是）　很好　（quality 用單數 V）
　　1　　2　　3　　4

7. 請　寄　兩個　樣品　給　我們　（sample 注意用複數）
　　1　2　3　4　5　6

8. 我們　需要　新樣品　來更換舊的　S＋V＋O＋O.C.（＝名片 to replace the old one）
　　1　　2　　3　　4　　（注意：sample 如為單數，要加冠詞；複數加 s）

9. 我們　認為　你們的　價錢（是）　太貴了 S＋V＋O（＝ that 的名詞子句）
　　1　　2　　3　　4　　5　　6

10. 我們　知道　你們　在大陸　有　一家　工廠　S＋V＋O（＝ that 的名詞子句）
　　1　　2　　3　　7　　4　5　　6　　（工廠：factory）

11. (　)　The shipment of your O/No.001 will be ready for _____.　(a)ship　(b)shipment　(c)deliver　(d)shipping

12. (　)　Please send us a new _____ for test.　(a)sample　(b)samples　(c)catalogs　(d)prices

13. (　)　The new model has several _____ which were developed by our R&D department.　(a)feature　(b)features　(c)advance　(d) function

14. (　)　Please send us your shipping _____ as soon as you ship the goods.　(a) advise　(b)advice　(c)adviser　(d)advisee

15. (　)　We will settle the outstanding payment by bank _____ next Monday.　(a)transfer　(b)transport　(c)transit　(d)transaction.

2. 代名詞(Pronoun)

位置：(1)代名詞(Pronoun)代替名詞(Noun)，因此位置和名詞相同，放在主詞(S)，受詞(O)，主補(S.C.)，受補(O.C.)

(2)所有格代名詞(Possessive Pronoun)＝形容詞(Adj.)，放名詞(N)前面

種類：(1)人稱代名詞(Person)

人稱(Person)：第一人稱(First Person)：I, we

第二人稱(Second Person)：you

第三人稱(Third Person)：he, she, it, they

位格(Case)：

人稱 (Person)	主格(s) (Nominative)	受格(O) (Objective)	所有格 / 所有代名詞 (Possessive)	反身代名詞 (Reflexive)
第一（單數）	我　I	me	my/mine	myself
第一（複數）	我們　we	us	our/ours	ourselves
第二（單數）	你　you	you	your/yours	yourself
第二（複數）	你們　you	you	your/yours	yourselves
第三（單數）	他　he 她　she 它　it	him her it	his/his her/hers its/its	himself herself itself
第三（複數）	他們　they	them	their/theirs	themselves

注意：①人稱代名詞放在不同的位格（主格／受格／所有格）要跟著變化

②所有格代名詞(my, our, your…)等於形容詞單字，放在 N 前面

③所有代名詞(mine, ours, yours…)等於所有格的代名詞＋名詞

(my ＋ N ＝ mine, your ＋ N ＝ yours, our ＋ N ＝ ours…)，當名詞用，　放在 S, O, S.C., O.C.的位置，例如：This sample is ours.

④反身代名詞(myself, yourself…)是加強語氣的代名詞(emphatic pronoun)，一樣放在 S, O, S.C., O.C.的位置，例如：I hurt myself. 或介系詞 by 後面，例如：We do it by ourselves.或當同位語：We ourselves made this prototype sample.

⑤中文中的本公司／敝公司，在翻譯成英文時用 We 表示；本公司的＝Our，貴公司＝You，貴公司的＝Your

例：We like your products. （本公司喜歡貴公司的產品）

(2)指示代名詞(Indication Pronoun)

常用的有：this, these, that, those, same, the same as, such as ＝ like

例：<u>Those</u> are our new samples.（those 當主詞）

Your style is <u>the same as</u> our design.（same 當主補）

We have many customers <u>such as</u> IBM, Motorola…

（such as 當受詞 customer 的同位語）

(3)不定代名詞(Indefinite Pronoun)──Some, Any

例：If you want the sample of this style, we still have <u>some</u>.

If you have <u>any</u>, please give one to me.

(some 和 any 當受詞)

(4)疑問／關係代名詞──who, whom, whose, which, what, when, that

例：<u>Who</u> is your boss？（疑問代名詞 who 當主詞）

<u>Whom</u> are you talking to？（疑問代名詞 whom 當受詞）

The sample <u>which</u> we sent you is new.（which 當關係代名詞）

注意：疑問代名詞＋不定詞＝名詞片語（當 S, O, S.C., O.C.）

例：what to do, which to take, how to make it, when to leave,
where to go

例：(1)Please advise us <u>how to make it</u>.（當受詞）

(2)Please tell him <u>which to take</u>.（當受詞）

(5)其他形容代名詞──one, ones, none (＝ no one), each, every, both, all, either, neither, many, much, few, a few, little, a little

例：<u>Both</u> are made in Taiwan.（both 兩者，當主詞）

<u>Either</u> will do.（either 隨便哪一個，當主詞）

<u>Neither</u> will do.（neither 二者都不，當主詞）

<u>Many</u> think this product is useful.（many 多數，當主詞）

Is there <u>much</u>？（much 多量，當主詞）

3. 形容詞(Adjective)

　　功用：修飾名詞(N)、代名詞(Pron.)

　　位置：放在所要修飾的名詞(N)的前面或後面。形容詞單字放名詞前面，稱前修飾(Pre-Modification)，形容詞片語／子句放名詞後面，稱後修飾(Post-Modification)

> 公式： | 形容詞單字
Adj. Word | ＋ N ＋ | 形容詞片語／子句
Adj. Ph./Adj.Cl. |

　　注意：(1)中文全部前修飾，不論是形容詞單字／片語／子句，加一個的，全部放名詞前面：形容詞單字／片語／子句　＋的＋N；（英文要分前後修飾）

　　　例：①這是　一張　大的　新的　紅色的　海報

　　　　　This is　a　big　new　red　poster.

　　　　　（Adj.單字全部放名詞 poster 前）

　　　　②樣品室裡的　樣品　是新的

　　　　　The samples in the showroom　are new.

　　　　　　　　N　　＋　Adj.　Ph.

　　　　　（Adj 片語，中文在 N 前，英文要放 N 後）

　　　　③我們上週寄的　樣品　是新的

　　　　　The sample which we sent last week　is new.

　　　　　　　　N　　＋　　　Adj.　Cl.

　　　　　（Adj 子句，中文放 N 前，英文要放 N 後）

　　　(2)中文常將「名詞（物）＋的」當形容詞用，放名詞前，英文要倒過來翻：

　　　　中文：N1 的 N2→英文：N2 of/for/in N1

　　　例：產品的品質　　　→the quality of the product

　　　　　鞋子的出口商　　→the exporter of Shoes

　　　　　100 元的支票　　→the check for 100 dollars

　　　　　1000 個的價錢　　→the price for 1000 pcs

　　　　　台北的交通　　　→the traffic in Taipei

　　　　　（N1 和 N2 為同一物，的＝ of；N1 和 N2 不同物，的＝ for）

(3)形容詞單字排列順序：代名形容詞＋數量形容詞＋性狀形容詞＋名詞

代名形容詞		數量形容詞		性狀形容詞						
1	2	3	4	5	6	7	8	9	10	
前置冠詞的形容詞	冠詞指示 Adj. 所有 Adj. 不定 Adj.	序數	基數	性質狀態	大小長短形狀	新舊溫度	顏色	國籍	物質材質	名詞 N
all	a/an/the	first	one	kind	large	new	red	Chinese	iron	
both	this/that	second	two	fine	small	old	blue	English	steel	
such	our/your	third	three	good	big	hot	green	Japanese	stone	
	some/any	fifth	five	bad	round	cold				

(4)形容詞片語／子句排列順序

　　中文：形片／子句(1)＋的＋形片／子句(2)＋的＋ N

　　英文：N ＋形片／子句(2)＋形片／子句(1)

　　例：我們昨天傳真給你的　圖面上的　尺寸　是錯的

　　　　　Adj. Cl. (1)　　　　+ Adj. Ph.(2)+ N

　　The size　on the drawing　which we faxed you yesterday　is wrong.

　　　　N　　　 + Adj. Ph.(2)　 + Adj. Cl. (1)

形容詞的比較(Degrees)

① 一般級(Positive Degree)

　　例如：large, small, good, new, old, heavy, competitive

② 比較級(Comparative Degree)

　　一個音節或兩個音節的形容詞，在字尾加 er，例如：larger, heavier

　　兩個音節以上，在形容詞前加 more，例如：more competitive

③ 最高級(Superlative Degree)

　　一個音節或兩個音節的形容詞，字尾加 est，前面要加 the，例如：the largest, the newest

　　兩個音節以上，在形容詞前加 the most，例如：the most competitive

　　注意：(1)最高級的形容詞前，要加 the。(2)比較級前面只能加 much，例如 much lighter, much more competitive.

練習 7　形容詞片語／子句練習

1. 中文：N1 的 N2→英文 N2 ＋ of/for/in ＋ N1

(1)請　報　1000 個　的　價錢　（的＝ for）
　　1　　2　　　5　　　　4　　　3

(2)你們　可以　給　我們　這兩個型式　的　報價及樣品　嗎？（的＝ for）
　　2　　　1　　3　　4　　　7　　　　6　　　　5　　（注意：N ＝複數＋ s）

(3)我們　是　化學產品(chemical products)　的　主要的製造商　（的＝ of）
　　1　　2　　　　5　　　　　　　　　　　4　　　3（注意：N 單數前＋ a）

(4)我們的產品　的　品質　（是）　很優良的(excellent)（的＝ of）
　　　3　　　2　　1　　　4　　　　　5

(5)樣品室裡　的　許多樣品　是　我們的新設計(design)（的＝ in）
　　　3　　2　　　1　　　4　　　　5

2. 形容詞片語／子句，中文加的，放 N 前→英文：N ＋ Adj. Ph./Adj. Cl.

注意：單數 N 前面，要記得加冠詞；複數名詞後面，要加 s/es

(1)我們　接受　你們（的）　4/20 的　報價單上的　價錢
　　1　　2　　　3　　　　6　　　5　　　4

接受：accept；報價單上：on the quotation；4/20 的：of 4/20

(2)請　告訴　我們　你要買的　數量
　　1　　2　　3　　　5　　　4

你要買的：which you want to buy；數量：quantity

(3)具有良好品質的　貨　可以　符合　你們的　需求
　　　2　　　　1　　3　　4　　5　　　6

具有良好品質的：with good quality；符合：meet；需求：requirement

(4)我們所報的　價錢　不含　你的　佣金
　　　2　　　1　　3　　4　　5

我們所報的：which we offered；佣金：commission

(5)請　報　如附目錄上的　每種產品的　最低價
　　1　　2　　　5　　　　4　　　　3

如附目錄上的：on the enclosed catalog；每種產品的：for every product

(6)謝謝　你們　在降價方面的　支持　（thanks/thank you 後面要加 for）

　　　1　　2　　　4　　　3

　　在降價方面：in price reduction；支持：support

(7)我們可以給你的　最好的　價錢　是　每個美金 10 元(US$10/pc)

　　　　3　　　　1　　2　4　　　5

3. 選擇

(1) (　) Our new designed model A101 is _____ than the old one A100.

　　　　(a)much lighter　(b)more cheaper　(c)more better　(d)more function

(2) (　) We will place you _____ orders if your price is competitive.

　　　　(a)regularly　(b)regular　(c)firmly　(d)formally

(3) (　) The _____ is our new catalog for your reference.

　　　　(a)enclose　(b)enclosing　(c)enclosed　(d)attach

(4) (　) Your _____ comments to our items will be highly appreciated.

　　　　(a)value　(b)valuable　(c)valuably　(d) valuables

(5) (　) Due to the _____ variety of our trade, we have both import and export departments.

　　　　(a)highly　(b)high　(c)lengthily　(d)widely

(6) (　) With regard to the _____ order, please send us a _____ P/I.　(a)additional; revised　(b)addition; revision　(c) additionally; revised　(d)additional; revising

(7) (　) It is _____ for the users to play many games with our video games VG-01.

　　　　(a)usefully　(b)easily　(c)freely　(d)easy

(8) (　) To be _____ in the market, we need to have 10% discount from you.

　　　　(a)compete　(b)competitive　(c)competition　(d)competitively

(9) (　) Trusting the _____ price will invite you to include this item into your line.

　　　　(a)attract　(b)attraction　(c)attractive　(d)attractively

(10) (　) The _____ sample will be dispatched upon receiving your sample charge.

　　　　(a)relevant　(b)relate　(c)relatively　(d)relevantly

4.副詞(Adverb)

功用：修飾動詞(V)、形容詞(Adj.)、副詞本身（Adv.）、子句和句子

位置：放在所修飾的字之前或之後

公式： | 副詞單字 | ＋ V/Adj./Adv. / 子句 / 句子＋ | 副詞片語 / 子句 |

種類：表示時間／地點／目的／方法／原因／條件／態度／結果／讓步／次數／程度／肯定／否定

注意：中文位置大部分放在 S 和 V 中間，英文則大多放在句尾，排列順序如下：(1)和……人＋(2)方法＋(3)地點＋(4)時間（小→大）＋(5)目的

級數：亦分原級、比較級和最高級（造法和形容詞相同），例如：soon, sooner, the soonest

常用副詞呈現方式

種類	副詞單字	副詞片語	副詞子句
時間	now, then, soon, today, before, yesterday, tomorrow, sometime, early, late, already, yet, still, finally, recently, immediately	next week, by return, as soon as possible, right away, at once, on Jan. 1, 1999	when ＋子句 before ＋子句 after ＋子句 since ＋子句
地點	here, there, far, near, abroad, worldwide, everywhere	at your hotel, in Taipei	where ＋子句 in which ＋子句
目的		for your reference for your approval for your evaluation for business	so that ＋子句
方法		by L/C, by air	
原因		due to/owing to ＋ N because of ＋ N	because ＋子句 since/as ＋子句
結果			therefore ＋子句 so/hence ＋子句
條件		subject to ＋ N	if ＋子句
讓步			although ＋子句 but ＋子句
態度	well, hard, quickly, kindly carefully, sincerely	with thanks with regret	
次數	always, usually, sometimes, often, seldom, never, ever	twice a year	
程度	very, much, enough, only, simply, little, certainly, just, more, most, too, so, almost		
肯定／否定	yes, no, not, never, indeed	of course	

應用例句

(1)Please arrange the shipment <u>immediately</u>.（副單字：表時間）

(2)The customer will go to your factory <u>next Monday.</u>（副片：時間）

(3)Please reply to our letter <u>as soon as possible</u>.（副片：時間）

(4)We will give you a new sample <u>when you come here</u>.（副子句：時間）

(5)Our products are welcomed by customers <u>worldwide</u>.（副單字：地點）

(6)We will pick you up <u>at the airport</u>.（副片：地點）

(7)We sent you a parcel <u>in which you will find our profile</u>.（副子句：地點）

(8)We sent you our catalog <u>for your reference</u>.（副片：目的）

(9)Please advise us the quantity, <u>so that we can offer you</u>.（副子句：目的）

(10)Please ship the goods <u>by air freight</u>.（副片：方法）

(11)<u>Due to the typhoon</u>, the shipment will be delayed.（副片：原因）

(12)We can't ship the goods now <u>because we have no stock</u>.（副子句：原因）

(13)Our quality is excellent, <u>so our product is selling very well</u>.（副子句：結果）

(14)<u>If you can give us 5% discount</u>, we will place you an order.（副子句：條件）

(15)<u>Although your quality is good</u>, your price is too high.（副子句：讓步）

(16)We <u>sincerely</u> recommend you this new model.（副單字：態度）

(17)We received <u>with thanks</u> your letter dated Oct. 1.（副片：態度）

(18)We <u>always</u> spend a lot of money on R&D.（副單字：次數）

(19)This customer places us the order <u>twice a year</u>.（副片：次數）

(20)Your quality is not good <u>enough</u>.（副單字：程度）

(21)<u>Of course</u>, we can specially make it for you.（副片：肯定）

注意：(1)副詞單字，放在所要修飾的 V, Adj., Adv.或句子的前面；副詞片語，放在後面；如不強調，則都放句尾，所以是移動性的

例：① We know <u>from the advertisement</u> that you import computers.（強調 V）

② We know that you import computers <u>from the advertisement</u>.（不強調）

③ <u>From the advertisement</u>, we know that you import computers.（強調句子）

(2)副詞子句（when, if, because, so, but…帶出的句子），不可單獨用，句尾不可打句點，要和主要子句一起用。如放主要子句前面，要打逗點；放後面則可省略。

例：If you order 10,000pcs, we will give you 10% discount.

＝ We will give you 10% discount if you order 10,000pcs.

(3)中文位置：S＋時間（大→小）／地點／方法／原因／根據…／和…人＋V

英文位置：S＋V＋O＋和…人／方法／地點／時間（小→大）／目的／

根據…

注意：要將副詞從中文的 S 和 V 中間拿出，放到每個句型的句尾）

例：我老闆　明天早上　會　到機場　接您。

My boss will pick you up at the airport tomorrow morning.

(4)中文的句子表達方式：雖然…但是…；因為…所以…兩個副詞連接詞都會
出現；但是要注意英文句子，如果寫了 although/though，but 就不可以寫出；
寫了 but，although/though 就要去除：同樣寫了 because，so/therefore 就不
可以寫出來；寫了 so/therefore，because 就不要寫。

例 1：雖然你們的品質不錯，但是價錢太高了。

Although your quality is good, your price is too high.

Your quality is good, but your price is too high.

例 2：因為你們延遲交貨，所以客人決定取消訂單。

Because you delayed the shipment, the customer decided to cancel the
order.

You delayed the shipment, so the customer decided to cancel the order.

練習 8　副詞片語／子句練習

注意：翻譯時，要注意中文副詞位置和英文不同，不可照中文順序翻，要將副詞從
中文的 S 和 V 中間拿出，大多放句尾：

中文：S ＋副詞＋ V ＋ O → 英文：S ＋ V ＋ O ＋副詞

1. 副詞片語練習

(1)我　　明天下午 2：00　　在飯店　　和你　　見面（未來式）

　　1　　　　　5　　　　　　4　　　　3　　　　2

在飯店：at the hotel；明天下午 2：00 ＝ at 2：00 p.m. tomorrow

(2)我們　　可以　　在一週內　　開出　　L/C.

　　1　　　2　　　　5　　　　3　　　4

在一週內：within one week

(3)我們　　希望　　這週五前　　收到　　樣品（注意：V to V，名詞單複數）

　　1　　　2　　　　5　　　　3　　　4

這週五前：before this Friday

(4)我們　　從廣告上　　知道　　你們　　生產　　射出成型機(Injection Machine)

　　1　　　　3　　　　2　　　4　　　5　　　　6

從廣告上：from the advertisement

(5)我們　　同意　　按照你們的價錢　　接受　　這張　　訂單（注意：V to V）

　　1　　　2　　　　6　　　　　3　　　4　　　5

按照你們的價錢：according to your price

(6)請　　在下月底前　　決定　　你們的　　訂單

　　1　　　5　　　　2　　　3　　　4

下月底前：before the end of next month

(7)我　今天晚上　和另一家公司　有　一個　約會(appointment)
　　1　　6　　　　5　　　　2　　3　　4
今晚：tonight

(8)請　根據 10 台和 100 台　報　最好的　價錢　（注意 N＋s）
　　1　　　5　　　　　　2　　3　　4
根據 10 台和 100 台：based on 10 sets and 100 sets

2. 副詞子句練習

(1)若經索取（＝如果你要求），本公司　將寄上　最新的　價目表　給貴方(if)
　　　　　1　2　3　　，　1　　2　　　3　　4　　5
要求：request/ask；價目表：price list

(2)如果　你　同意，請　寄　三個　免費樣品　給我們(if)
　　1　　2　　3　，1　2　3　　4　　　5
同意：agree；免費樣品：free samples 或 samples free of charge

(3)如果　你們　可以　給　我們　5%折扣，我們　將下　訂單　給你們(if)
　　1　　2　　3　4　5　　6　　，1　　2　　3　　4
折扣：discount；下訂單給：place the order with

(4)因為　這　是　一張　大訂單，　我們　希望　貴方　重報　較好的　　價錢(as)
　　1　　2　3　4　　5　，　1　　2　　3　　4　　5　　　6
重報：re-offer 或 re-quote；較好的：better

(5)當　我們　一收到　L/C，我們　就出　貨(when)
　　1　　2　　3　　4　，1　　2　　3
出貨：ship the goods 或 make the shipment

(6)當　你們　出　貨時，請　告知　出貨明細(when)
　　1　　2　　3　4　，1　2　　　3
出貨明細：shipping details

(7)請　指出　你們喜歡的　型號，以便　我們　能　報　價(so that)

　　1　　2　　　4　　　　3　　5　　6　　7　8　　9

　　型號：model number/style number

　　指出：point out/indicate 或 specify

　　報價：quote the price/make the offer 或 give the quotation

(8)你們的　品質　（是）　不錯，但是　價錢　（是）　太貴了(but)

　　1　　2　　3　　4　　5　　6　　7　　8

(9)客人　在收到你們的樣品前　將不決定　他們的訂單　(before)

　　1　　　　　4　　　　　　2　　　　3

　　在收到你們的樣品前：before receiving your sample

　　　　　　　　　　或 before the receipt of your sample

　　　　　　　　　　或 before they receive your sample

(10)我們的　品質（是）很好，因此　我們的　產品　很暢銷（＝賣的很好）(so)

　　1　　2　3　4　　5　　6　7　8

　　很暢銷：is/are selling very well; has/have been selling well

3. 選擇

(1) (　) Please _____ send us the catalogues and price list for the machines you produce.
(a)kind　(b)kindly　(c)quick　(d)urgent

(2) (　) This year we _____ developed many new stylish items and _____ invite you to visit our booth no.123 at Taipei Computer Exhibition.　(a)special; sincere　(b)specially; sincere　(c)specially; sincerely　(d)special; sincerely

(3) (　) We _____ need 10,000pcs of P/N 700 in house by this Friday. It is _____ . Please help.　(a)urgent; urgently　(b)urgently; urgently　(c)urgent; urgent　(d)urgently; urgent

(4) (　) We are waiting for your quotation to the above items _____ .　(a) as soon　(b)as soon as　(c)soon　(d)sooner

(5) (　) Your early reply will be _____ appreciated.　(a)highly　(b)most　(c)more　(d) high

5. 連接詞(Conjunction)

　功用：連接兩個同等的單字、片語或子句

　注意：對等，即所連接的詞性前後要一致，例如：V and V, N and N, Adj. or Adj.，片
　　　　語 and 片語，子句 but 子句

　種類：(1)簡單連接詞：and, or, nor, but, so, therefore, yet, still, however, etc.

　　　　例：① We enclose our catalog and price list.(N and N)

　　　　　　② Our price is lower but our quality is better. （句子 but 句子）

　　　　　　③ We worked hard, yet we lost the order. （句子 yet 句子）

　　　　　　④ Your price is too high, so we cannot get the order. （句子 so 句子）

　　　　　　　注意錯誤連接：I have a book and good. （N and Adj.不對等）

　　　　(2)片語連接詞：as soon as （一…馬上…） ，as long as, so long as （只要）

　　　　例：① As soon as we receive your L/C, we will ship the goods.

　　　　　　② As long as your price is competitive, we will buy from you.

　　　　　　③ So long as you keep good quality, we will buy from you regularly.

　　　　(3)相關連接詞：so…that, either…or, neither…nor, not only…but also

　　　　例：① Your price is so high that we couldn't get the order.

　　　　　　② You can pay us either by L/C or by T/T.

　　　　　　③ We received neither your order nor your reply.

　　　　　　④ We not only have good products but also offer good service.

　　　　(4)副詞連接詞：when, if, after, before

　　　　注意：副詞子句和主要子句的主詞相同時，才可省略，改為分詞片語。

　　　　例：① After we receive your L/C, we will ship the goods.

　　　　　　　→After receiving your L/C, we will ship the goods.

　　　　　　② When we receive your L/C, we will ship the goods.

　　　　　　　→When receiving your L/C, we will ship the goods.

　　　　注意：①兩個子句中的主詞不同時，不可改成分詞片語。

　　　　　　　例：When you open the L/C, we will start the production.

　　　　　　　　　（主詞不同）

→When opening the L/C, we will start the production.

（錯誤）

②用 and 連接時，特別注意對等：片語，片語 and 片語。

例：Due to excellent quality, the best price and prompt shipment, our products are welcomed by consumers.

6.介系詞(Preposition)

功用：介系詞＋(the)＋ N/V-ing ＝介系詞片語

位置：當名詞片語、形容詞片語或副詞片語用

種類：(1)常用介系詞：in, on, of, for, at, with, by, from, to, before, after, without, regarding 的用法，請參照下列附表。

(2)其他介系詞：about（大約／在附近），above（在上），across（在…另一邊），against（面對／靠／反對），along（沿著），among（在三者之間），around（在周圍），behind（在…後面），below（在下），beneath（緊貼在下面），beside（在旁邊），besides（於…之外），between（在兩者之間），beyond（越過），but（除…之外），down（向下），during（在…期間），except（除外），into（入），like（像），off（離開），over（超過／越過），round（周圍），since（自從），through（穿過），till/until（直到），toward（向），under（下面），upon（＝ on），up（向上），within（在…期間）。

(3)片語介系詞(Phrase Preposition)：according to（按照），as for（至於），as to（關於），at home in（精通／熟悉），because of（因為／由於），by means of（以／藉由），for the purpose of（為了…目的），for the sake of（為…緣故），for want of（以缺乏…之故），in accordance with（按照），in answer to（回答），in case of（假如），in charge of（負責／管理），in front of（在…前面），in memory of（紀念），in place of（代替），in need of（需要），in (with) regard to（關於），in spite of（雖然／不顧），instead of（代替），on account of（由於），owing to/due to（由於），thanks to（幸虧）等等。

常用介系詞用法	中文例句	英文例句
in		
(1)在…裡面	在台北／在我們的樣品室裡	in Taipei／in our showroom
(2)在…方面	謝謝你在降價方面的支持	Thank you for your support in price reduction.
	我們在處理小金額出貨方面有困難	We have the difficulty in handling the small amouth shipment.
(3)在…行業	我們是電腦行業的主要工廠	We are the leading manufacturer in the Computer field.
(4)年／月	我們公司在 1990 年 7 月成立	Our company was established in July, 1990.
(5)用…顏色	我們的目錄是用紅色印刷的	Our catalog was printed in red color.
(6)用…語言	我們的目錄是用英文印刷的	Our catalog was printed in English.
(7)用…幣別	請用美金報價	Please quote the price in U.S. dollar.
on		
(1)在…上面	在桌上	on the table
（電話／電視／收音機／網站上）	我們在每批出貨上做 100%測試	We do 100% test on every shipment.
	我們老板正在電話線上（他正忙線中）	My boss is on the line.
(2)日期	我們在 7 月 1 日收到你們的 L/C	We received your L/C on July 1.
(3)星期	客人週五來我們公司	The customer will come to our office on Friday.
of		
(1)屬於…的	我們收到你們 7 月 1 日的信	We received your letter of July 1 (= dated July 1).
(2)N1 的 N2→N2 of N1	產品的品質	the quality of the product
	本公司是電腦產品的出口商	We are the exporter of computer products.
for		
(1)為了…目的	我來這裡作生意的	I come here for business.
(2)N1 的 N2→N2 of N1	這是 1,000 元的支票	This is a check for $1,000.
	請報 10,000pcs 的價錢	Please offer the price for 10,000pcs.
(3)以便於	請告知數量以便於報價	Please advise the quantity for quoting.

常用介系詞用法	中文例句	英文例句
(4)供…	供參考／評估／確認／測試／考慮	for reference/for evaluation/for approval/for test/for consideration
(5)完成一段時間（前面）	我們這家公司工作 3 年了	I have been working in this company for 3 years.
(6)給（送給／帶給）	和我們合作將帶給你們大利潤	To cooperate with us will bring a big profit for you.
at		
(1)特定的地點（前面）	在飯店／在機場	at the hotel/at the airport
(2)幾點幾分（前面）	客人下午 3 點到我們公司	The customer will come to our office at 3:00 p.m.
(3)價錢條件（前面）	我們的報價是 FOB TAIWAN	Our price is at FOB Taiwan.(= based on FOB Taiwan)
(4)門牌號碼（前面）	信義路 4 段 30 巷 20 弄 10 號	at No. 10, Alley 20, Lane 30, Sec. 4, Hsin-Yi Road
with		
(1)和…人	我將和老闆去接你	I will pick you up with my boss.
(2)跟隨著	我們寄給你一個附檢驗報告的樣品	We sent you a sample together with a test report.
(3)具有	在 R&D 方面具有多年經驗，我們在電腦業很有名	With many years' experience in R&D, we won a good reputation in the Computer field.
(4)拿	讓我幫你拿行李	Let me help you with your luggage.
(5)給（下訂單給）	我們將將下訂單給你們	We will place the order with you.
by		
(1)在…旁邊	我們的工廠在河邊	Our factory is by the river.
(2)用…方法	我們的付款是用 L/C	Our payment is by L/C.
(3)乘坐…交通工具	我們將開車去接你	We will pick you up by car.
(4)= before 在（日期）前	我們 7/20 前會出貨	We will ship the goods by 7/20.
(5)用空運／海運／郵包	請用空運／海運／郵包出貨	Please ship the goods by air/by sea/by air parcel.
from		
(1)從（A 到 B）	從 A 到 B 很遠	It is far from A to B
(2)從（日期）起	我們的新報價單從 8/1 開始生效	Our new quotation will be effected from Aug. 1.

MODERN BUSINESS ENGLISH 現代商用英文

常用介系詞用法	中文例句	英文例句
(3)從（地點）	我們從廣告上得知你的名字	We know your name from the advertisement.
to		
(1)（從 A）到(B)	從我們公司到工廠不遠	It is not far from our office to the factory.
(2)(give/send/open)…給	請寄新目錄給我們	Please send the new catalog to us.
	請開信用信狀給我們	Please open the L/C to us.
	請下訂單給我們	Please issue an order to us.
(3)當不定詞 V to V	我們喜歡和你們作生意	We like to do business with you.
(4)固定片語		
look forward to（介）	我們盼望盡快收到你們的回音	We look forward to receiving your reply soon.
due to/owing to（介）	由於原料短缺，我們無法準時出貨	Due to material shortage, we can't ship on time.
with a view to（介）	考慮到和你們的良好關係，我們將給你們 10%折扣	With a view to the good relationship with you, we would give you 10% discount.
in order to（不定詞）	為了省錢，請一批出貨	In order to save money, please ship by one lot.
before		
(1)當副詞＝從前	我從前見過你	I met you before.
(2)當連接詞＋子句	我們在收到 L/C 前不出貨	We won't ship the goods before we receive the L/C.
(3)當介系詞＋N/V-ing		We won't ship the goods before receiving the L/C.
after		
在…後	交期是收到 L/C 後 30 天	Delivery time is 30 days after we receive the L/C.
without		
不要	請不要遲疑盡快告知	Please advise us without hesitation.
	請不要延遲出貨	Please ship the goods without delay.
regarding		
有關	有關訂單 001，請增加 500 個	Regarding O/No. 001, please add 500pcs.

練習 9　介系詞練習

在下列空格內填入正確的介系詞

1. Thank you for your support _____ price reduction.

2. We do 100% test _____ every shipment.

3. We received your letter _____ December 20.

4. I came here _____ business.

5. We are the leading exporter _____ sport shoes.

6. Please quote the price _____ 10,000pcs.

7. Please send us your sample _____ approval.

8. To cooperate with us will bring a big profit _____ you.

9. _____ many years' experience, we are a reputable exporter _____ this field.

10. _____ sample approval, we will place the order _____ you.

11. Our price is _____ CIF New York.

12. Our Payment is _____ L/C.

13. Please ship our goods _____ sea freight.

14. Our new quotation will be effected _____ January 1.

15. Please send your new catalog and price list _____ us.

16. We will not ship the goods _____ receiving your L/C.

17. Our best delivery time is 45 days _____ receiving your order.

18. We know your name _____ the advertisement.

19. I have been working _____ this company _____ two years.

20. The quality _____ our products is excellent.

21. Our catalog was printed _____ English.

22. Please quote us the price _____ U.S. dollar.

23. We will pick you up _____ my boss _____ your hotel _____ 3:00 p.m.

24. We sent you a sample together _____ an inspection report.

25. Let me help you _____ your luggage.

26. Please let us know your comments _____ hesitation.

27. _____ _____ the labor shortage, the shipment will be delayed.

28. _____ O/No. 52A-001, it will be shipped next week.

7. 動詞(Verb)

位置：放在 V 位置

種類：分主要動詞(Main Verb)包含動作動詞(Function Verb)及聯綴動詞(Linking Verb)和助動詞(Auxiliary Verb)。

(1)動作動詞(FV)：buy, take, send, receive, open…

(2)聯綴動詞(LV)：主要是 Be 動詞和感官動詞：be, keep, feel, look, seem, sound, taste, smell, become, appear, get, turn, grow

(3)助動詞(Aux.)：will, would, shall, should, can, could, may, must…

注意助動詞不可單獨放 V 位置，一定要和原形動詞一起用：助動＋原形動詞(Aux.＋ V-root)；此外，注意在被動式(Be ＋ p.p.)裡的 Be 動詞及完成式(have ＋ p.p.)裡的 have 也屬助動詞

注意：(1)在動詞(V)位置，最少放 1 個字，最多可以放到 5 個字

(2)寫作時，要注意時態，主動（12 種）、被動（9 種）及語氣（5 種）的 26 種變化，要依文法規定，參照下面時態表來寫，不可自創一態，因為中文動詞沒有時態語氣變化，主被動也不清楚

(3)實際應用時，常用的變化只有 12 種左右，主動 6 種：簡單現在式、簡單過去式、簡單未來式、現在進行式、現在完成式及現在完成進行式；被動 5 種：簡單現在／過去／未來及現在進行／現在完成的被動式；假設語氣 1 種：可能的未來，所以至少要牢記這 12 種動詞變化

(4)過去式(p.t.)和過去分詞(p.p.)使用時，要注意規則(V ＋ ed)及不規則(go-went-gone, come-came-come, buy-bought-bought)的變化

(5)動詞亦分及物 Vt(Transitive)和不及物 Vi(Intransitive)

　①及物動詞(Vt)：動詞後面有受詞就稱及物動詞

　　‧有受詞，句子即完整，可打句點，稱完全及物(Complete Transitive)，例如句型 3→S ＋ V ＋ O 和句型 4→S ＋ V ＋ O ＋ O

　　‧有受詞，仍須補語補充說明的動詞，稱不完全及物(Incomplete Transitive)，例如句型 5→S ＋ V ＋ O ＋ O.C.

　　‧主要的不完全 Vt：make, call, keep, find, think, consider, wish, choose, elect, make, name, see, hear, feel, have, get, let…

　②不及物動物(Vi)：動詞後面沒有受詞，就稱不及動物詞

‧沒有受詞，即完整，可打句點，稱完全不及物(complete intransitive)，例如句型 1→S ＋ V

‧沒有受詞，仍須補語補充說明的動詞，稱不完全及物(incomplete intransitive)，例如句型 6→S ＋ LV ＋ S.C.

形式：動詞的形式有 5 種，如下表所示

原式 V-root	現在式 Present	過去式 Past (p.t.)	過去分詞 p.p.	現在分詞 V-ing
be	is/am/are	was/were	been	being
do	do/does	did	done	doing
have	have/has	had	had	having

注意：(1)只有現在式和過去式可以一個字放動詞(V)位置。

(2)原形／過去分詞／現在分詞不可單獨放動詞(V)位置。

①原形動詞放 3 個位置：

（Please 後，助動詞後及不定詞 to 後面）

Please ＋ V-root

Aux.＋ V-root

To ＋ V-root

②過去分詞放 2 個位置：

have ＋ p.p.（＝完成式）

be ＋ p.p.（＝被動式）

③現在分詞放 2 個位置：

be ＋ V-ing（＝進行式）

V ＋ V-ing（＝動名詞）

(3)使用現在式動詞，句中大都有下列習慣性副詞：usually, normally, regularly, often, frequently, sometimes, rarely, always, never, every day, week, month, year, once a week, twice a month 等。

(4)下列動詞不能用進行式：like, dislike, prefer, think, believe, know, mean, see, hear, feel, seem, have, need, own, want。

(5)注意 be, do, have 這三個動詞，也可當助動詞，用法如下：

	當動詞 中文意思	當助動詞 中文意思
be	是、在、到 例： We are the exporter.（是） The sample is here.（在） I have been to H.K.（到過）	be ＋ V-ing（進行式＝正在） be ＋ p.p.（被動式＝被） 例： We are making your sample.（正在） The sample was received.（被收到）
have has	有、吃 例： We have 10 sales.（有） I had my lunch.（吃）	have ＋ p.p.（已經，曾經） 例： We have sent our sample.（已經寄） I have been to USA.（曾經到過）
do does did	做 例： Please do 100% test.（做）	幫助造否定句／疑問句 例： We don't produce this item.（否定） Do you produce this item?（疑問句）

時態：動詞 12 種主動＋9 種被動變化

		主動	被動
(1)簡單式：現在 (2)　　　過去 (3)　　　未來		1 個字／用現在式 V 1 個字／用過去式 V will/shall ＋ V-root	is/are/am ＋ p.p. was/were ＋ p.p. will/shall ＋ be ＋ p.p.
(4)進行式：現在 (5)　　　過去 (6)　　　未來		is/are/am ＋ V-ing was/were ＋ V-ing will/shall ＋ be ＋ V-ing	is/are/am ＋ being ＋ p.p. was/were ＋ being ＋ p.p. will/shall ＋ be ＋ being ＋ p.p.
(7)完成式：現在 (8)　　　過去 (9)　　　未來		have/has ＋ p.p. had ＋ p.p. will/shall ＋ have ＋ p.p.	have/has ＋ been ＋ p.p. had ＋ been ＋ p.p. will/shall ＋ have ＋ been ＋ p.p.
(10)完成進行式：現在 (11)　　　　過去 (12)　　　　未來		have/has ＋ been ＋ V-ing had ＋ been ＋ V-ing will/shall have ＋ been ＋ V-ing	— — —

時態用法

(1)簡單現在式：事實、真理、習慣（句中有 usually, always, everyday…）

(2)簡單過去式：句中有過去時間：yesterday, just, last, week, before…

(3)簡單未來式：句中有未來時間：tomorrow, next month, soon…

(4)現在進行式：中文有「現在正在」或英文句中有 "now, currently, at the moment" 時

(7)現在完成式：剛完成或已完成的動作，中文有「已經、曾經、尚未、未曾」或英文句中有 "already, ever, never"，是過去式加強語氣

　　例：我們昨天已經寄出我們的樣品了　（用現在完成式）

　　　　We have sent our sample yesterday.　（不可用過去完成）

(10)現在完成進行式：從過去到現在，已完成部分但仍在繼續的動作

(5)(6)(8)(9)(11)(12)過去進行／完成／完成進行，未來進行／完成／完成進行：這 6 種時態要有兩個動作時間做比較時才用，所以使用機會不多。

例：當你昨天來時，

　　(1)我們那個時候正在做你們的訂單（過去進行）

　　(2)我們已經完成了你們的訂單（過去完成）

　　(3)我們已經完成一半訂單，仍在繼續（過去完成進行）

　　When you came here yesterday,

　　(1)We were producing your order.

　　(2)We had completed your order.

　　(3)We had been producing half of your order.

例：當你明天來時，

　　(1)我們將正在生產你們的訂單（未來進行）

　　(2)我們將已經完成你們的訂單（未來完成）

　　(3)我們將已完成一半你們的訂單（未來完成進行）

　　When you come here tomorrow,

　　(1)We will be producing your order.

　　(2)We will have completed your order.

　　(3)We will have been producing half of your order.

被動用法

　　一般動詞，人當主詞時，用主動：人(S)＋主動；東西／事情當主詞時，用被動：事／物(S)＋被動

　注意：(1)中文的被動不清楚，如果是事／物開頭，記得動詞要用被動

例：①樣品正要寄（現在進行被動式）

The sample is being sent.

②目錄下週寄（未來被動）

The catalog will be sent next week.

③ L/C 已經開了（現在完成被動）

The L/C has been opened.

(2)使役動詞例外，人(S)＋被動＋介系詞＋物→物(S)＋主動＋人

常用的有satisfy（使…滿意），interest（使…感興趣），surprise（使…驚訝）

注意：不同的使役動詞跟不同的介系詞，主動不加介系詞

例：① We are satisfied with your quality.（人＋被動）

= Your quality satisfies us.（物＋主動）

② We are interested in your products.

= Your products interest us.

③ We are surprised at your high price.

= Your high price surprises us.

語氣(Mood)3 種

(1)直説法(Indicate Mood)：肯定句／否定句／疑問句

例：We are the manufacturer.（肯定句）

We are not the manufacturer.（否定句）

Are you the manufacturer?（yes/no 疑問句）

What are your main products?（WH 疑問句）

(2)祈使法(Imperative Mood)：命令／請求／勸告的句子，用原形動詞開頭，省略 you

例：Come to my office right now.（命令）

Please send one free sample to us soon.（請求）

Don't make the same mistake again.（勸告）

(3)假設語氣(Subjunctive Mood)：5 種，注意動詞寫法和直說法不同

①和現在事實相反的假設語氣

　　If 子句用過去式動詞（注意 Be 動詞只能用 were），主要子句用過去式助動詞＋原形動詞

公式：If ＋ S ＋過去式 V/were, S ＋ would/could/should/might ＋ V-root

　　例：If I were you, I would take this business.

　　　　（如果我是你，我將接受此筆生意）

　　　　If I had enough money, I would buy this computer.

　　　　（要是我現在有足夠的錢，我將會買這台電腦）

②和過去事實相反的假設語氣（兩個子句的動詞都用完成式）

　　多用在抱怨的語氣；If 子句用過去完成式，主要子句用過去式助動詞＋完成式

公式：If ＋ S ＋ had ＋ p.p., S ＋ would/could/should/might ＋ have ＋ p.p.

　　例：If you had opened the L/C earlier, we wouldn't have delayed the shipment.

　　　　（要是你早點開信用狀，我們就不會延誤出貨了）

③可能的未來

　　If 子句用現在式代替未來式，主要子句用簡單未來式

公式：If ＋ S ＋現在式 V, S ＋ will/shall ＋ V-root

　　例：If you can give us 5% discount, we will buy from you.

　　　　（如果你們能給我們 5%折扣，我們將跟你們購買）

④不可能的未來（和未來事實相反的假設語氣）

　　If 子句用 were to ＋原形 V，主要子句用過去式助動＋原形 V

公式：If ＋ S ＋ were to ＋ V-root, S ＋ would/could/should/might ＋ V-root

---

　　例：If I were to be the President next year, I would be very happy.

　　　　（如果我明年當選總統，我將非常高興）

⑤一半可能的未來（不確定的未來）

If 子句用過去式助動詞＋原形動詞，主要子句用簡單未來式

公式：If ＋ S ＋ would/could/should/might ＋ V-root，S ＋ will/shall
　　　＋ V-root

例：If he would come here next week, we will give him new samples.

（如果他下週來這兒，我們將給他新樣品）──不確定他來不來

例外動詞

英文句子嚴格要求一個句子一個動詞，如有第二個動詞出現，必須用不定詞to隔開，或將下一個動詞改為動名詞V-ing，不可以將動詞連著出現在同一個句子中。一般動詞後兩種用法皆可（V to V 或 V ＋ V-ing），如以 to 分開（即第二個動詞用不定詞片語出現）則表示未來動作或即將要做的動作；如使用V-ing（即第二個動詞用現在分詞片語出現）則表示一件事或過去動作。但下列動詞則為例外，要特別注意：（此亦為各種英文考試常考部分）

1. V to V（下列動詞後一定要用不定詞 to 接下一個動詞）

afford（力足以／供給），agree（答應／同意），decide（決定），desire（想要），determine（下決心），expect（期待），fail（未能成功），hope/wish（希望），learn（學習），manage（處理／支配），plan（計畫），pretend（假裝），promise（答應），refuse（拒絕），swear（發誓），want（要）

例：We expect to hear from you soon.

2. V ＋ V-ing（下列動詞後一定要用現在分詞 V-ing 接下一個動詞）

admit（承認／容許），avoid（避免），consider（考慮），deny（否認），dislike（討厭），enjoy（喜愛），envy（羨慕／忌妒），escape（逃脫／免除），finish（完成），imagine（想像），keep（保持），mind（介意），miss（錯過），postpone（展緩／延擱），practice（練習），quit（停止），spend（花費），suggest（建議）

例：I spent US$1,000 buying this computer.

3. go ＋ V-ing（如表示要去做某事時，go 後面要跟 V-ing）

go swimming（去游泳），go shopping（去逛街），go hiking（去健行），go mountain climbing（去爬山），go biking（去騎腳踏車），go fishing（去釣魚），go camping（去露營）

例：I will <u>go hiking</u> this Saturday.

4. V ＋ 人 ＋ Vrt（下列動詞後 N 如爲人，一定要跟原型動詞）；V ＋ 物 ＋ p.p.

let（讓），help（幫忙），make（使得），have（使得）

例：<u>Let</u> me <u>help</u> you <u>wash</u> your car.

　　He <u>had</u> his house <u>pained</u> in green.

5. V ＋ Vrt/V-ing/p.p.（下列感官動詞後受詞爲人時，可以跟原型動詞／動名詞；受詞爲物時，則跟過去分詞＝感官動詞＋人＋ Vrt/V-ing；感官動詞＋物＋ p.p.）

see/watch/look（看／看起來），hear/sound（聽／聽起來），smell（聞起來），feel（覺得），taste（嘗起來）

例：I <u>saw</u> a young man <u>washing</u> the car.

　　I <u>saw</u> the car <u>washed</u> by a young man.

6. 情緒動詞／使役動詞

情緒動詞／使役動詞的使用，和一般動詞剛好相反。一般動詞使用時：人當主詞時用主動，事／物當主詞時用被動；但使用情緒動詞／使役動詞時：人當主詞時用被動（注意被動的介系詞都不相同），事／物當主詞時用主動：

> 公式：人＋ be ＋ p.p.＋介系詞＋事／物（用被動）
>
> 　　　事／物＋ V ＋人（用主動）

例：We <u>are interested</u> in you products.（人＋被動 be ＋ p.p.＋介系詞 in ＋物）

　　（我們對你們的產品感興趣）

　　Your products <u>interest</u> us.（物＋主動＋人）

　　（你們的產品使我們感興趣）

　　This is an <u>interesting</u> book.（interesting 當形容詞單字用；形容物）

　　（這是一本有趣的書）

　　This is an <u>interested</u> man.（interested 當形容詞單字用；形容人）

　　（這是一個有趣的人）

常用情緒動詞／使役動詞如下：

Vrt 原型	V-ing/adj 現在分詞	p.p.過去分詞	（被動後跟的 介系詞）	（中文）
(1)bore	boring	bored	(with)	（厭煩）
(2)disappoint	disappointing	disappointed	(in)	（使…失望）
(3)disgust	disgusting	disgusted	(at/with)	（使…厭惡）
(4)embarrass	embarrassing	embarrassed	(about)	（使…困窘）
(5)excite	exciting	excited	(at/about/by)	（使…興奮）
(6)exhaust	exhausting	exhausted	(in)	（精疲大盡）
(7)frighten	frightening	frightened	(by)	（使…驚嚇）
(8)interest	interesting	interested	(in)	（使…感興趣）
(9)impress	impressing	impressed	(with/by)	（使…印象深刻）
(10)please	pleasing/pleasant	pleased	(with)	（使…高興）
(11)satisfy	satisfying/satisfactory	satisfied	(with)	（使…滿意）
(12)surprise	supprising	surprised	(at)	（使…驚訝）
(13)tire	tiring	tired	(with)	（使…疲倦）
(14)worry	worrying	worried	(about)	（使…擔心）
(15)confuse	confusing	confused		（使…迷惑）

選擇（例外動詞練習）

1.( ) We _____ in the products which you displayed at your booth.   (a) interest   (b) are interested   (c) interesting   (d) are interesting

2.( ) Please help us ____ this kind of product.   (a)find   (b)to find   (c)finding   (d) found

3.( ) We spent one week _____ this sample for you.   (a)to make   (b)making   (c) made (d)is making

4.( ) We will consider _____ from you if you can offer us 10% off.   (a)to buy   (b)buying (c)bought   (d)to be bought

5.( ) We decided _____ importing from you.   (a)stop   (b)to stop   (c)stopping   (d) stopped

練習 10　動詞時態練習

1. 簡單現在式（注意第 3 人稱單數主詞，V ＋ s）

   (1)我們的工廠　目前　生產　這種產品　（現在時間）
   　　　　1　　　　4　　2　　　3

   (2)這　是　我們的　最好價錢　（事實）
   　1　2　　3　　　　4

   (3)品質　（是）　很重要　（真理）
   　1　　2　　　3

   (4)這個客人　每年　來　台灣　兩次　（習慣）　(twice a year)
   　　1　　　5　　2　　3　　4

2. 簡單過去式（句中有明確過去時間 yesterday, last month, just）

   （主動 1 個字，注意不規則變化；被動：was/were ＋ p.p.）

   (1)我們　上個月　寄了　兩個樣品　給　你們
   　1　　6　　　2　　　3　　　4　5

   (2)我們　剛收到　你們寄來的　目錄(the catalog)
   　1　　2　　4(Adj. Cl.)　　　3

   (3)訂單號碼 001 的　貨　昨天　裝了　（被動）　(of O/No.001)
   　2(Adj. Ph.)　　1　4　　3

3. 簡單未來式（句中有明確未來時間 tomorrow, next week, soon）will ＋原 V

   (1)這批貨　將於　下週　裝出　(will be shipped)
   　1　　　3　2

   (2)客人　將於　下月初　拜訪　你們的工廠　(early next month)
   　1　　　4　　2　　3　　　　(in the early of next month)

4. 現在完成式（句中有「已經、曾經、尚未、未曾」）have/has ＋ p.p.

   (1)我們　已經完成　你們的訂單
   　1　　2　　　3

   (2)信用狀(L/C)　上週　已經開了　（被動）(have/has been ＋ p.p.)
   　1　　　3　　2

5. 現在進行式（中文有「現在正在」）is/are/am + V-ing

(1)我們　現在　正在做　你們的訂單
　　1　　4　　2　　　　3

(2)我們　正要　用郵包　寄　我們的樣品　給　你們(by air parcel post)
　　1　　6　　　2　　3　　　　4　　5

6. 現在完成進行式 have/has been + V-ing

(1)我　已經　在這個公司　工作　3年了　(for three years)
　　1　　3　　　　　2　　4

(2)有關 P/O No.001，我們　已經生產了　20,000 pcs 了
　　1　　　2　　　3　　　　4　　　5

下列 6 種時態要有兩個動作時間做比較時，才使用

7. 過去進行式（過去正在）was/were + V-ing

(1)你　上週　來　我們工廠　時，我們　正在生產　這種產品
　2　　5　　3　　4　　1，1　　2　　3

(2)我們　昨天　離開　你們公司　時，天　正在下雨（時＝ when）
　2　　5　　3　　4　　1，1　　2

8. 過去完成式 had + p.p.

(1)當　我們　收到　你們的報價單時，我們　已經下　訂單　給別人了
　1　　2　　3　　4　　，1　　2　　3　　4

(2)當　你　上週　來　台北時，我們　就已經給　你　新樣品了
　1　2　　5　　3　　4　，1　　2　　3　　4

9. 過去完成進行式 had been + V-ing

當　我們　收到　你們的信用狀時，我們　已經生產了 5000 pcs 了
1　　2　　3　　4　　，1　　2　　3

10.未來進行式 will be ＋ V-ing

(1)當　你　明天　到　我們工廠時，我們　將正在生產　你們的訂單
　　1　2　5　3　　　　4　，1　　　2　　　　　3

(2)當　你　下週　到　我們工廠時，我們　將正在做　你們的樣品
　　1　2　5　3　　　　4　，1　　　2　　　　3

11. 未來完成式 will have ＋ p.p.

當　我們　10月底　到台北時，我們　希望　你們將已經完成我們的訂單
1　2　　4　　　3　，1　　2　　3　　4　　5

12.未來完成進行式 will have been ＋ V-ing

到下個月底，我們　將已經完成　這張訂單的 1/2　(1/2 of this order)
　　　　　　1　　　2　　　　3　　　4

13.選擇

(1) (　) We _____ our latest samples to you for a week.

(a)have send　(b)are sending　(c)have been sent　(d)have sent

(2) (　) Thank you for your attention to the above and look forward to _____ from you soon.

(a)hear　(b)heard　(c)hearing　(d)heart

(3) (　) We not only supply quality products, but also _____ good after sales service.

(a)render　(b)rendering　(c)rendered　(d)have rendered

(4) (　) Our factory _____ been established for 10 years.

(a)have　(b)had　(c)has　(d)was

(5) (　) Few makers _____ this kind of product at present.

(a)produce　(b)producing　(c)produced　(d)produces

(6) (　) The enclosed _____ our latest catalogue and quotation for reference.

(a)is　(b)are　(c)has　(d)have

(7) (　) We will release the shipment as soon as we _____ your L/C.

(a)received　(b)receiving　(c)receive　(d)receipt

(8) (　) If you _____ the L/C earlier, we would have shipped the goods on time.

(a)opened　(b)have opened　(c)had opened　(d)were opening

(9) (　) I would accept this order if I _____ you.

(a)am　(b)was　(c)were　(d) been

(10) (　) Our best delivery time is 30 days after order _____.

(a)confirm　(b)confirmed　(c)confirming　(d)confirmation

練習 11　動詞練習（注意時態、語氣、主被動、形容詞片語／子句放 N 後）

1. <u>貴公司的訂單號碼 52N-025 的</u>　貨　已於　昨日　用空運　裝出（現在完成式被動）
　　　　　　　　2　　　　　　　1　　　5　　4　　3

2. <u>貴公司索取的</u>　目錄　很快　會寄　去　貴公司　（簡單未來式被動）
　　　　2　　　　1　　6　　3　　4　　5

3. <u>你們寄來的</u>　樣品　昨天　收到了　（簡單過去式被動）
　　　2　　　1　　4　　3

4. 如果　你們　有　庫存(stock)，請　即刻　空運　給　我們（用可能的未來，主動）
　　1　　2　　3　　4　　　1　　5　　3　　4

5. 如果　你　下週　來　台灣，我們　將給　你　新樣品（用可能的未來，主動）
　　1　　2　5　　3　　4　　1　　2　　3　　4

6. 如果　信用狀　下週　能被收到，貨　就可　<u>於月底</u>　被裝出（可能未來，被動）
　　1　　2　　4　　3　　　1　　2　　4　　3

7. 如果　你們　能　<u>於30天內</u>　出貨，我們　就下　訂單　給　你們（可能未來，主動）
　　1　　2　4　　　3　　1　　2　　3　4　5

8. 如果　上週　沒有　颱風，我們　就會　準時　出貨（過去相反的假設語氣）
　　1　　4　2　　3　　1　　2　　4　　3

颱風：typhoon；出貨：ship the goods, make the shipment

9. 如果　<u>不是</u>　你們　給　錯誤的規格，我們　就不會做　錯貨（過去相反假設語氣，否定）
　　1　　2　　3　4　　　1　　2　　3

錯誤的規格：wrong specification；做錯貨：make the wrong shipment

10.如果　箱子　沒有破損，我們　就不會要求　賠償（過去相反假設語氣，否定）
　　1　　2　　3　　1　　2　　3

要求賠償：ask for compensation

八、英文文法寫作易錯地方自我檢查(Mistake Checking)

1. 普通名詞前要加冠詞 the/a/an 或所有格代名詞 our, your, their

   例：(1)We are a manufacturer.

   (2)Please tell us the model number you want.

   (3)We sent you our new sample.

   (4)We sent you a new sample.

2. 介系詞後要有定冠詞 the，加名詞或直接跟動名詞＝介＋ the ＋ Noun 或介＋ V-ing

   例：on the table, in doing R&D

3. 注意名詞複數問題，可數名詞複數時，要加 s 或 es

   例：(1)We sent you two samples yesterday.

   (2)We have many kinds of models of computers.

4. 主詞單數＋單數動詞；主詞複數＋複數動詞或 are/were, have（要一致）

   第三人稱單數現在式主詞，動詞要加 s 或 es 或用單數動詞 is/was, has/does

   例：(1)Our company has been established for 10 years.

   (2)Our company produces this kind of product now.

   (3)The customer has been waiting for you here since 10:00 a.m.

   (4)Few makers produce this product in Taiwan now.

   (5)Enclosed are our new catalog and samples.

   (6)Enclosed is our new quotation.

5. 過去動作記得用過去式動詞

   例：(1)我們剛收到你們寄來的樣品

   We just received the sample you sent.

   (2)你報的價錢太貴了

   The price which you quoted is too high.

6. 注意形容詞片語（兩個字以上）和子句，放名詞後（記得要緊跟在所形容的名詞後）

　　例：(1)Our payment is by irrevocable L/C at sight.

　　　　(2)Please send us the catalog of your computer products.

　　　　(3)We received the sample which you sent on 7/1.

　　　　　＝ The sample which you sent on 7/1 was received.

7. 副詞（時間、地點、目的、方法、原因、根據…、和…人）記得放句尾＝介系詞片語（放句尾）

　　例：(1)我明天和我老闆開車去飯店接你

　　　　I will pick you up with my boss by car at your hotel tomorrow.

　　　　(2)根據此圖面，請報出你最好的價格

　　　　Please quote your best price according to this drawing.

8. 連接詞注意對等：片語 and 片語；子句 and 子句

　　例：Due to good price and excellent quality, our product is popular.

　　　　As our prices are good and quality is excellent, our products are popular.

9. 注意一個句子，只能有一個動詞；第二個動詞要用 to 隔開或改成動名詞

　　V + to + V（表未來動作）；V + V-ing（表過去動作或一件事）

　　例：(1)We would like to establish business relationship with you.

　　　　(2)We like doing business with you.

　　　　(3)We decided to stop importing your products.

10. 注意原形動詞的三個位置

　　(1) Please 後面加原形 V

　　　　例：Please send us your sample.

　　(2) 不定詞 to 後，加原形

　　　　例：We hope to do business with you.

　　(3) 助動詞 will, shall, can, could, must, may, do, does, did 後，加原形 V

11. 注意完成式 Have ＋ p.p.（過去分詞）

例：We have sent our catalog to you.

12. 副詞子句用現在式動詞代替未來式，主要子句用簡單未來式

If/when/before/after/as soon as ＋ S ＋現在式 V（代未來），S ＋ will ＋原形 V

例：(1)If you accept our price, we will place an order with you.

(2)When we receive your L/C, we will ship the goods.

13. 注意：because, as, since ＋子句

because of, due to, owing to ＋片語

14. 注意使役動詞 interest (in), satisfy (with), surprise (at)的用法

人＋被動＋介系詞＋事／物＝事／物＋主動＋人

例：(1)We are interested in your products.＝ Your products interest us.

(2)We are satisfied with your quality.＝ Your quality satisfies us.

(3)We are surprised at your claim.＝ Your claim surprises us.

15. 否定句：在 Be 動詞或助動詞 will, can, have, has, do, does 後加 not 或名詞前加 no

例：(1)We have no sample on hand now.

(2)We are not the manufacturer.

(3)We can not send you free samples.

16. 注意：形容詞形容名詞：We will place you regular orders.

副詞形容動詞：We will order from you regularly.

17. 組合片語結構只有 4 種：

(1) 介＋ N/V-ing（介系詞開頭的介系詞片語）

(2) To ＋原形 V（不定詞 To 開頭的不定詞片語）

(3) V-ing ＋ N（V-ing 開頭的現在分詞片語）

(4) p.p.過去分詞片語（p.p.開頭的過去分詞片語）

18. 注意句型結構，先完成主詞、動詞及受詞或補語，副詞要放句尾／句首或動詞後，
    形容詞單字放名詞前，片語／子句放名詞後

19. 動名詞後面直接跟受詞：V-ing + O
    名詞和名詞中間要用 of 分開：N of N
    例：(1)Upon receiving your L/C, we will ship the goods.
       (2)Upon the receipt of your L/C, we will ship the goods.

20. Be 動詞後面跟形容詞：Be + Adj.(S.C.)
    動作動詞後面跟名詞：FV + N(O)
    例：(1)We are confident that our quality can satisfy you.
       (2)We have confidence that our quality can satisfy you.

21. Be 動詞當動詞用時，中文只有三個意思：是、在、到
    當助動詞用時，注意後面的動詞一定要用 V-ing 或 p.p.
    Be + N/Adj.(S.C.)   （當動詞）
    Be + V-ing   （進行式－當助動）
    Be + p.p.   （被動式－當助動）
    例：(1)This is our best price.（是）＝動詞
       (2)We are making your order.（正在）＝助動詞
       (3)The sample was received.（被）＝助動詞

22. 副詞單字如果要補充說明動詞，放在 Be 動詞／助動詞後面，動作動詞(FV)前面，但
    是 FV 後面如果有受詞(O)時，則搬到 FV 前面
    S + FV + Adv.   →   Please reply immediately.
    S + Adv.＋ FV + O   →   Please immediately reply our letter.

23. 我們收到 9/13 的信：We received the letter of 9/13.（形容 N，的用 of）
    我們在 9/13 收到信：We received the letter on 9/13.（形容 V，在用 on）

24.年／月的介系詞用 in，日期／星期用 on，月初／月中／月底用 in 或 at

例：in 1999

in July

on July 1, 1999

on Monday

in the end of July（＝ 7/20～7/31 一段時間）

at the end of July（＝ on 7/31 特定的點）

25.As 的用法

(1) As ＝ being（作為／是）：As a leading importer（作為一個主要的進口商）

(2) As（如同）：As you know（如你所知），As requested（如所要求）

(3) As per ＝ according to（按照）：As per your request（按照你們的要求）

(4) such as（例如）：We sell many products, such as shoes, clothes…

(5) as follows ＝ as below（如下）：We quote you our best prices as follows.

(6) as soon as（一…馬上…）（後加子句；和 upon 意思相同，但 upon ＋ N）：

We will confirm our order as soon as your sample is approved.

（＝ We will confirm our order upon your sample approval.）

(7) As ＝ because/since（因為）：As our quality is good, our products are selling well.

26.reach（抵達）、arrive（到達）、happen（發生）、rise（上升）這幾個動詞，注意用

主動：物(S)＋主動

例：(1)The goods arrived(reached) here in good condition.（貨完好抵此）

(2)An accident happened to me.（我發生了一個意外）

(3)The price will rise soon.（價錢很快將上漲）

27.after/before/when 當介系詞用時，後面跟 N 或 V-ing，不可跟過去式 V

例：After receiving your sample or After the receipt of your sample

After received your sample（錯誤）

28.look forward to（盼望），due to/owing to（由於），with a view to（爲了）爲介系詞，

後面＋ N/V-ing；介系詞 without, by, during ＋ N/V-ing；

used to（習慣於）＋原 V；be used to ＋ V-ing

例：We look forward to <u>hearing</u> from you soon.

Please advise us your comments without <u>hesitation</u>.

We used to <u>buy</u> from you based on FOB Keelung.

We were used to <u>buying</u> from you based on FOB Keelung.

29.let/help/make/have ＋ N（人）＋原形 V

let/help/make/have ＋ N（物）＋ p.p.

enjoy/ spend/ practice/finish/mind/ avoid ＋ V-ing

例：Please <u>help</u> us <u>find</u> this kind of product.

We <u>spent</u> one week <u>making</u> this sample for you.

They <u>had</u> the floor <u>cleaned</u>.

30.注意句子有 although 或 though，就不要寫 but；有 but 就不寫 although/though；相同

的，有 because 就不寫 so/therefore；有 so/therefore，就不寫 because

31.常用假設語氣公式

可能未來：If ＋ S ＋現在式 V， S ＋未來式 V (will ＋ Vrt)

和現在事實相反：If ＋ S ＋過去式 V，

S ＋ would/could/should/might ＋原形 V

和過去事實相反：If ＋ S ＋ had ＋ p.p.

S ＋ would/should/could/might ＋ have ＋ p.p.

32.when/before/after/as soon as ＋ S（主詞）＋ V（用現在式動詞代替未來式）

when/before/after/as soon as ＋ Ving（如去除主詞，動詞要使用動名詞）

33.soon/as soon as possible（盡快）放句尾；句尾不可放 as soon（此爲錯誤片語）

或 as soon as（一…馬上…此片語只可放句首或句中）

九、常用字彙表 (Useful Vocabulary)

注意下列可替換使用的字的用法及詞性變化。

1. 目錄

catalogue（有圖片文字的目錄，3-10 頁）

leaflet（單頁的目錄）

brochure（小冊子、整本、很詳細的目錄）

literature（純文字說明）

circular（傳單）

2. 提供、供給、寄送（樣品／目錄／報價單／貨物）

send（寄、送、發 fax、派）⎱

deliver（出貨）

dispatch（遞送）

supply（提供）

submit（提供）　　　　　　＋物

forward（轉交）

provide（供給）

furnish（供給）

present（呈送）⎰

offer（提供：物或服務）

render（給與：服務、幫助）

3. 最好的（價錢）

the most competitive（最有競爭性的）⎱

the best（最好的）

the lowest（最低的）

sensitive/sensible（最敏感的）　＋ prices

rock-bottom（最底的）

attractive（吸引人的）

keen（敏銳的）

breath-taking（令人窒息的）⎰

4. 大的（訂單／數量）

big/large/considerable/substantial/comprehensive/potential orders（大訂單）/heavy demand（大的需求）

5. 最新的（目錄／報價單／價目表／樣品／設計）

the newest/latest/current/updated

6. 競爭 compete (V) competition (N) competitive (Adj.)

競爭者／同行 competitor (N)

7. 關於／有關

regarding/concerning/as to/as for/with regard to/with reference to/in respect of（放句首）

8. 馬上／立刻／儘快（副詞，放句尾）

immediately/right away/at once/soon/as soon as possible/promptly/rapidly/by return/ in the near future/without hesitation（不要遲疑）

9. 按照／依照

according to/in accordance with/as per/in compliance with/with reference to/referring to （介系詞，後面＋ N/V-ing）/follow (V)

10. 公司簡介　company profile

11. 操作手冊／說明書　Instruction Manual

12. 報價　offer the price/quote the price/give the quotation

13. 出貨　ship the goods/deliver the goods

　　　　make the shipment/effect the shipment

14. 出貨前　before shipment/prior to shipment

　　出貨後　after shipment

　　出貨同時　upon shipment

15. 分批出貨　partial shipment

16. 轉運　transshipment

17. 樣品確認後　after sample approval

18. 售後服務　after-sales service

19. 報價單　quotation/offer sheet；價目表　price list；估價單　estimate；詢價 inquiry；招標單　invitation；標單　bid；投標商　bidder；押標金　bid bond；公開 投標　public tender；開標　open tender；還價 counter offer；決標　award

20. 樣品費　sample charge；郵費　postage

21. 展覽　fair/exhibition/show；展示　display；示範　demonstration

22. 攤位　booth/stand

23. 不流行　out of fashion/out of time/out of date

　　不生產　out of production

　　淘汰　phase out

24. 原料　raw material

25. 手工　workmanship

26. 盼望　look forward to ＋ V-ing/N；期盼　expect

27. 試銷訂單／樣品訂單　trail order/initial order/first order/sample order

　　正式訂單　formal order

　　確定／肯定的訂單　firm order

　　重複（循環）訂單　repeat order

　　定期採購訂單　regular orders

　　最低訂購量　minimum order/minimum quantity/min. order quantity

　　尚未出貨的訂單 pending orders/open orders

28. 財力證明　bank reference

　　信用／財務狀況　financial standing, credit status

29. well ＋ p.p.＝ Adj.

　　組織良好的　well-organized

　　經驗良好的　well-experienced

　　表現良好的　well-performed

　　著名的　well-known

　　設備良好的　well-equipped

30. 如所指示　as (you) instructed

　　如所規定　as specified

　　如所要求　as requested

31. 打樣樣品　counter sample

　　原型樣品　prototype

　　代表性樣品　representative sample

　　出貨樣品　shipping sample

32. 品質管制　quality control

33.研究發展　research and development (R&D)

34.產能　production capacity；月產量　monthly output/monthly capacity

35.銷售管道　sales channel/sales outlet

36.貶值　devaluation(N), devaluate (V)

37.升值　appreciation(N), appreciate (V)

38.佣金　commission

39.淡季　off-peak season/calm season/flat season

40.旺季　peak season/hot season

41.樣品室　show room/sample room

42.營業額　sales turnover/sales amount

43.裝配線　assembly line；生產線　production line

44.報關行　customs broker；攬貨公司　forwarder；運輸公司　shipping company

45.庫存　stock；倉庫　warehouse

46.利潤　profit；最低的利潤　the smallest/lowest/narrowest profit

47.客戶　customer/client；長期客戶　regular customer

48.管銷費用　overheads

49.手續費　handling charge；啟動費　set up charge

50.專利　patent；商標／品牌　brand

51.付款條件　terms of payment/payment terms

52.包裝方式　packing method (way)；淨重　net weight (NW)；毛重　gross weight(GW)

53.散裝　bulk packing；小包裝　small packing；個別包裝　individual packing

54.交期　lead time/shipping date/shipment date/delivery time

55.相關的　relative/related/relevant

56.進一步的資料　further information

57.連鎖店　chain stores

58.緊急的　urgent (adj); urgently (adv)

59.價錢　price；淨價　net price；市場價　market price；零售價　retail price

60.手工做的　hand-made

61.工廠　manufacturer；出口商　exporter；進口商　importer；經銷商　distributor；大盤　wholesaler；小盤　dealer；零售商　retailer；分公司　branch office；用戶　end user

62.商會　Chamber of Commerce；商務辦事處　Commercial Office

63.免費　free of charge

64.整廠輸出　turn-key plant project；合資　joint venture

65.技術　technical know how

66.市場需求　market demand

67.用空運　by air；用海運　by sea (freight)；用郵包　by air parcel post；用快遞　by courier；印刷品　by printed matter

68.出貨嘜頭　shipping mark；出貨通知　shipping advice

69.出貨文件　shipping documents；發票　invoice；包裝表　packing list；保險單　Insurance Policy；海運提單　bill of lading；空運提單　air way bill；產地證明　certificate of origin

70.匯率浮動　fluctuation of the exchange rate

71.外匯　foreign exchange；匯差　the loss of the exchange rate/exchange rate loss

72.確認（訂單／價錢／條件）　confirm (V), confirmation (N)

73.材料成本　material cost；人工成本　labor cost

74.替代品　substitute（A 不生產，B 代替）；替換物　alternative（A 或 B 皆可，任選一）；賠貨　replacement（A 壞掉，重送新的 A 替換）

75.預付發票　proforma invoice

76.全電信用狀　full cable L/C；短電　short cable L/C；郵寄信用狀　mailed L/C

77.意見　comments/opinion

78.對帳單　statement；應收帳單　debit note；應付帳單　credit note

79.折扣　discount；降價　reduce the price/price reduction；讓步　concession

80.賠償　compensate (V), compensation (N)；索賠　ask for compensation from/claim to

81.代理商　agent；獨家代理　sole agent/exclusive agent；合約　agreement, contract

82.市場不景氣　market depression/market recession

83.電匯　T/T(telegram transfer/cable transfer)；信匯　M/T(mailed transfer)

84.不可撤消即期信用狀　irrevocable L/C at sight

85.運費已付　freight prepaid；運費對方付　freight collect

86.銀行本票　bank check；旅行支票　traveler's check

87.付款交單　document against payment (D/P)

88.承兌交單　document against acceptance (D/A)

89.信用狀修改書　L/C amendment

90.修改費　amendment charge

91.貨櫃　container；貨櫃船　container vessel；散裝船　bulk vessel

92.慢出貨／延遲出貨　delay in shipment/late shipment

93.取消訂單　cancel the order

94.結清會計年度　clear the fiscal account

95.現金周轉　cash flow

96.最後期限　deadline

97.方便　convenient (Adj.), convenience (N)；方便於　facilitate (V)

98.機器設備　equipment；所有設備（含軟應體管理等）　facility

99.未付款項　outstanding payment

100.付清／結清／解決　settle (V) , settlement (N)；餘額　the balance amount；差額　the difference；尚未出貨的訂單　pending orders/open orders

101.瑕疵文件　discrepant documents/defective documents；瑕疵費用　discrepancy fee

102.保險公司　insurance company

103.推薦　recommend (V), recommendation (N)

104.和規格有誤差（差異）　deviation from the specification

105.圖面　drawing；規格　specification

106.處理的方法　disposal/solution

107.生產　produce (V), production (N)；產品　product (N)

108.檢驗報告　inspection report/test report

109.商品檢驗局　Bureau of Commodity Inspection and Quarantine

110.最後出貨日　latest shipping date；有效期　expiry date (expiration)

111.供測試　for test；供確認樣品　for sample approval

112.供參考　for reference；供評估　for evaluation；供考慮　for consideration

113.準時　on time/punctually/without delay（不要延遲）

114.生產安排　arrange the production/production arrangement

115.進行生產　proceed the production；進行訂單　proceed the order

116.優點／特色　advantages/features/characteristic

117.如下　as follows/as below；下列的　following (Adj.)

118.事實上　in fact/as a mater of fact/actually

119.負責　take the responsibility

120.掛我們帳上／貴方付費　at our expense/for your account

121.低於成本　under the cost；低估價值　undervalue

122.進口稅　import duty

123.過失　fault/mistake/error (N)；錯誤的　wrong (Adj.)

124.困境　hard position/difficult position

125.同時　at the same time；此外　besides/meanwhile/in addition/furthermore

126.因此　so/therefore/hense/thus

127.然而　however/nevertheless；無論如何　anyway；換言之　on the other word

128.為了表示誠意　to show our faith/to express our sincerity

129.折讓　allowance

130.不良的樣品　defective samples

131.劣等品質　inferior quality/poor quality

132.抽驗　inspect the goods at random

133.可接受品質標準　Acceptable Quality Level (AQL)

134.主缺點　major defect；次缺點　minor defect

135.公平的／合理的　fair/reasonable

136.符合　match/meet

137.壽命期　life cycle

138.檢驗報告　inspection report/test report

139.付清欠款　settle the outstanding payment；付清　settlement (N)

140.狀況良好　in good condition/in perfect condition

141.免除不良　free from defects

142.建議　proposal

143.運輸途中　in transit

144.原始設備廠／原始委託製造　Original Equipment Manufacturer (OEM)

145.國際金融業務分行　Offshore Banking Unit (OBU)

146.記帳／月結　open account (O/A)

147.交貨付款　cash on delivery (COD)

148.下單付款　cash with order (CWO)

149.分期付款　installment

現代商用英文 MODERN BUSINESS ENGLISH

150.寄售　on consignment

151.船位　shipping space (S/S)；訂船位　book the shipping space

152.雜費　miscellaneous/petties

153.單邊貿易　unilateral trade

154.雙邊貿易　bilateral trade/two-way trade

155.三角貿易　triangular trade

156.多邊貿易　multilateral trade

157.直接貿易　direct trade

158.間接貿易　indirect trade

159.加工貿易　improvement trade

160.以物易物貿易　barter trade

161.轉口輸出　re-export

162.貿易順差　favorable trade balance；貿易逆差　unfavorable trade balance

163.配額　quota

164.傾銷問題　dumping problem

165.不准轉運　transshipment is not allowed (permitted)/prohibitted/forbidden；
　　可以轉運　transshipment is allowed (permitted)

166.船上交貨價（離港價）　Free on Board (FOB)

167.含運費在內價　Cost and Freight (CFR)

168.含運費保險在內價　Cost, Insurance and Freight (CIF)

169.目的地交貨價　Delivered at Place (DAP)；終點站交貨價　Delivered at Terminal (DAT)

170.離廠價　Ex-Factory/Ex-Works (EXW)
　　船邊交貨價　Free Alongside Ship (FAS)

171.信用狀　letter of credit (L/C)

172.不可撤銷即期信用狀　irrevocable letter of credit at sight

173.遠期信用狀　Usance L/C

174.可轉讓信用狀　transferable L/C

175.保兌信用狀　Confirmed L/C

176.憑轉信用狀　Back-to-Back L/C

177.擔保信用狀 Stand-by L/C ＝履約保證金 Performance bond

178.循環信用狀　Revolving L/C

179.履約保證信用狀　Performance Bond Credit

180.受益人　beneficiary

181.原始設計廠／原始委託設計　Original Designed Manufacturer (ODM)

182.除…以外　except（不含後面 N）；besides（包括後面 N）

183.修改　revise（圖面／規格／價錢），modify（模具），amend（信用狀）

184.賠貨／替代品　replace/replacement（換貨），substitute（代用品），alternative（另一可選用的同等品），exchange（交換／調換）

185.需求／要求　request（動作的要求），requirement（條件的要求），demand（數量的需求或請求）

186.因此／結果　so, therefore, hence, thus, for this reason, consequently, as a result, on the whole, in conclusion, finally, accordingly（如前所說）

187.在…廠內　in house

188.塑膠框　plastic housing

189.材料表　bill of materials (BOM)

190.第一張訂單　first release ＝ first order

191.第一次生產的樣品　first article sample

192.品管 Q.C. (Quality Control)；進料品管 IQC (Incoming Quality Control)；製程線上品管 IPQC (In Process Quality Control)；完成品品管 FQC(Final Quality Control)；全面品管 TQM (Total Quality Management)

193.品保 Q.A. (Quality Assurance)

194.標準檢驗程序 SIP (Standard Inspection Procedure)

195.標準作業程序 SOP (Standard Operation Procedure)

十、 NEW TOEIC 文法選擇試題範例

1. (　) The marketing department is adequately _____ to launch the perfume campaign, but they're concerned the budget may need to be increased.
(a)most prepared　(b)prepared　(c)preparedness　(d) more prepared

2. (　) To safeguard _____ investment before a disaster occurs, homeowners are advised to take out a comprehensive insurance policy.　(a)their　(b)they　(c)theirs　(d)them

3. (　) For more than 30 years, our firm _____ yearly analyses tracking changes in consumer demand habits.　(a)conducts　(b)conducted　(c)is conducting　(d)has conducted

4. (　) Our supplier from Singapore isn't here yet, but he said he'd arrive _____ 4:00 pm, at the latest.　(a)until　(b)from　(c)by　(d)to

5. (　) The recreational facilities have been _____ improved to include a tennis court and swimming pool.　(a) significant　(b)significance　(c)significantly　(d)signify

6. (　) _____ medical procedure contains a certain amount of risk, but in this case it's minimal.　(a)Most　(b)All　(c)Much　(d)Every

7. (　) Emesto Hernandez is the excellent sales representative _____ I mentioned when we were talking about our Latin American operations.
(a)whom　(b)whose　(c)which　(d)what

8. (　) One day next week, the head office _____ someone over here to assess our performance.　(a) sent　(b)will be sending　(c)to send　(d)has sent

9. (　) _____ properly contained, the contagious disease can quickly spread throughout the population.　(a)If　(b)Nevertheless　(c)Since　(d)Unless

10. (　) Whereas mutual funds are generally considered _____ than individual stocks, they can also carry a high degree of risk.　(a)safest　(b)safer　(c)safe　(d)unsafe

11. (　) It will be another five years before the modernization of the telecommunications network _____ .
(a)completes　(b)will complete　(c)was completed　(d)is completed

12. (　) Neither the city government _____ the power company has provided a reasonable explanation for yesterday's blackout.　(a)or　(b)either　(c)nor　(d)as well as

13. (　) The _____ a company gets, the easier it becomes for it to take over smaller rivals.
(a)largest　(b)largely　(c)larger　(d)large

14. (　) By the time the lawsuit is concluded, Grayson Chemicals will _____ more than £1 million on legal fees.

(a)spending　(b)spent　(c)have spent　(d)to spend

15. (　) _____ the amount of money spent on the advertising push, it still failed to help the firm increase its market share.

(a)Although　(b)Due to　(c)Because　(d)In spite of

16. (　) Collingstown is the city _____ this year's International Metal Expo will be held.

(a)when　(b)which　(c)where　(d) what

17. (　) _____ estimates for the construction of the bridge put the total cost at $125 million.

(a) Conservatively　(b)Conservation　(c)Conserve　(d)Conservative

18. (　) I've been a project coordinator here _____ two and a half years, and before that I was a design specialist.　(a)for　(b)since　(c)by　(d)in

19. (　) This year Zantek Inc. acquired rival firms in Argentina and Italy, _____ its ability to expand into those countries.

(a)maximized　(b)maximizes　(c)maximizing　(d)maximize

20. (　) The statement reiterated the firm's commitment to upholding high moral standards _____ protecting the interests of the shareholders.

(a)while　(b)so that　(c)because　(d)unless

21. (　) The new CEO _____ on improving the corporation's bottom line, since the firm has been losing money for years.

(a)to concentrate　(b)is concentrating　(c)concentrate　(d)concentrating

22. (　) _____ mounting debts and an inability to find new investors, the consultancy went out of business.　(a)Due to　(b)Since　(c)Because　(d)As a result

23. (　) Executives at that bank insist they don't know _____ precipitated the stock's nosedive.

(a)what　(b) how　(c)when　(d)where

24. (　) Since I've been working here for five years, and I've never taken a holiday, I'm eligible _____ a rather lengthy vacation.　(a)for　(b)to　(c)on　(d)under

25. (　) The emergency of a number of firms copying our original design was our _____ for developing a new model.　(a)failure　(b)catalyst　(c)competitor　(d)necessity

十一、國貿大會考歷屆英文考題範例

1. (　) Please send your sample for our quality _____ .

   (a)approve　(b)approval　(c)approved　(d)improve

2. (　) We will do everything possible to _____ that such a mistake will not happen again.

   (a)assure　(b)ensue　(c)ensure　(d)insure

3. (　) When the NT dollars appreciate against the US dollars, it means that the former has become _____ .

   (a)bigger　(b)smaller　(c)stronger　(d)weaker

4. (　) We are _____ your further confirmation.

   (a)wait　(b)waiting　(c)await　(d)awaiting

5. (　) Due to insufficient demand in the market, this item has been _____ .

   (a)phased in　(b)introduced　(c)continued　(d)discontinued

6. (　) _____ of this item will fill up a _____ container.

   (a)500 dozens; 20-foot　(b)500 dozen; 20-foot　(c)500 dozens; 20-feet　(d)500 dozen; 20-feet

7. (　) The cargo has _____ the destination port.

   (a) arrived　(b)been arrived　(c)reached　(d)been reached

8. (　) Our production will be _____ during the Lunar New Year holidays.

   (a) suspended　(b)surrendered　(c)resumed　(d)retrieved

9. (　) We are very interested in the micro projectors displayed at your _____ at the computer exhibition in Taipei last month.

   (a)seat　(b)company　(c)booth　(d)place

10. (　) Unless we receive the long overdue payment within 10 days, we would _____ .

    (a)give you a discount　(b)open the L/C to you　(b) take a legal action to collect it (d)urge you to place further orders

11. (　) We have learned from the British Embassy that your gloves are _____ natural materials.

    (a)made of　(b)made from　(c)extracted from　(d)made up for

12. (　) We hope that you can _____ the conditions detailed below.

    (a) fit　(b)meet　(c)suit　(d)conform

13. (　) We offer a very reasonable quotation for our watches _____ their good quality.

(a)in spite of　(b)as long as　(c)as far as　(d)no matter of

14. (　) We ask that you promptly open an irrevocable L/C in _____ favor, valid until November 30.

(a)our　(b)your　(c)better　(d)best

15. (　) We are pleased to _____ our acceptance as per the enclosed proforma invoice.

(a)keep　(b)take　(c)cover　(d)confirm

16. (　) We _____ to hear that this item is no longer available.

(a) regret　(b)are regret　(c)are regretted　(d)are regrettable

17. (　) Enclosed are two copies of sales confirmation. Please _____ and send back one copy.

(a)sign　(b)signature　(c)countersign　(d)counter-offer

18. (　) If you are prepared to increase your _____ to 15%, we shall be pleased to purchase the complete stock.

(a)discount　(b)sales volume　(c)price　(d)cost

19. (　) The issuing bank unpaid due to the _____ of the documents presented.

(a)damage　(b)defect　(c)discrepancy　(d)dishonor

20. (　) Please send back one copy of the agreement with your _____ .

(a)acceptance　(b)rejection　(c)signature　(d)support

21. (　) The _____ is now US$1.00: NT$33.50

(a)exchange market　(b)exchange rate　(c)exchange reserve　(d)foreign exchange

22. (　) Under FOB terms, we will ship the goods to you with freight _____ .

(a)collect　(b)collected　(c)prepay　(d)prepaid

23. (　) Your price is 10% higher than _____ of your competitor.

(a)that　(b)these　(c)which　(d)whose

24. (　) We are forced to _____ our price by the material cost increase.

(a)arise　(b)arouse　(c)raise　(d)rise

25. (　) The order is _____ to our sample approval. Please send us the samples immediately.

(a)apt　(b)due　(c)owing　(d)subject

26. (　) The B/L is to show "To order of shipper" as the _____ .

(a) carrier　(b)consignee　(c)notify party　(d)shipper

27. (　) A: What are your terms of payment? B: _____

(a)We usually ask our customers to issue an irrevocable sight L/C.

(b)We can deliver them within 20 days after receipt of your confirmation.

(c)Two thousand pieces are our minimum order.

(d)Marine insurance is to be effected by the seller.

28. (　) _____ will fill up a 20-foot container?

(a)How much quantity　(b)How many quantities　(c)What quantity　(d)What a

quantity

29. (　) We would _____ it if you could place your order immediately.

(a)appreciate　(b)be appreciated　(c)be appreciating　(d)be appreciative

30. (　) Please T./T payment to our bank account at its _____ .

(a)validity　(b)expiry　(c)maturity　(d)acceptance

31. (　) We forwarded three samples to you by DHL air courier service yesterday, _____

number 20098.

(a)tracing　(b)tracking　(c)trucking　(d)trafficking

32. (　) Our import costs have risen considerably due to the _____ of the US dollars.

(a)appreciation　(b)increase　(c)increment　(d)inflation

33. (　) We look forward to _____ .

(a)receive your reply soon.　(b)your earliest reply　(c)your reply soon　(d)receiving

your earliest reply soon.

34. (　) Shipment will be made _____ .

(a)after receiving your L/C within 60 days　(b) after receipt of your L/C within 60 days

(c)within 60 days after receipt your L/C　(d)within 60 days after receipt of your L/C

35. (　) Attached is a _____ for the sample charge we are to collect from you.

(a)Cover Note　(b)Credit Note　(c)Debit Note　(d)Promissory Note

36. (　) US$58,040.50 is to be written in words as:

(a)US dollar fifty eight thousand forty and fifty cent only.

(b)US dollars fifty eight thousand fourty and fifty cents only.

(c)US dollars fifty eight thousand forty and fifty cents only.

(d)US dollars fifty eight thousand forty and 50% only.

37. (　) A:Where are your main markets? B: _____

(a)Our main sales areas are America, Europe and Asia.

(b)We have Marketing Dept., Design Dept., and Sales Dept.

(c)Our sales amount is about USD3,000,000 yearly.

(d)We have 3 branches: USA, Czech and Saudi Arabia.

38. (　) A: I'm interested in your tennis shoes. What's the packing method? B: _____

(a)Each piece is packed in an inner box, 4 pieces to an export carton.

(b)Each pair is packed in an inner box, 4 pairs to an export carton.

(c)We only accept payment by irrevocable L/C at sight in our favor.

(d)There are 16 cartons broken in last shipment.

39. (　) A: What's your minimum order quantity? B: _____

(a)Our monthly capacity is 50,000 pcs.

(b)The 1,500pcs of bicycles you ordered will be ready for shipment.

(c)We require 3,000 sets for each item.

(d)We sell about 50,000 doz. per month for this item.

40. (　) A: How long is your lead time? B: _____

(a)Please arrange the shipment for O/No.168 as early as you can.

(b)Please e-mail us the shipping advice upon shipment.

(c)Our delivery time is within 45 days after receipt of your order.

(d)There is no vessel available this week.

41. (　) A: What are the advantages of your products? B: _____

(a)We have the patent and offer one year guarantee.

(b)We have good relationship with the manufacturers.

(c)We can supply 10,000pcs monthly.

(d)Item no. SP-1968 has been out of production for 3 months.

42. (　) A: Now that the quantity is acceptable. If you can allow us 10% discount, we will place an order with you. B: _____

(a)The contract is available for two years.　(b)I am sorry. This is our bottom price.　(c) I'll send you some samples　(d)We are greatly surprised to find that there is a shortage of 10 cartons.

43. (　) A: May I ask you to cancel my order for the 1000 sporting goods?

B: Why? You know we've already committed ourselves to manufacturing the products. I always place my order with the manufacturers in advance.

A: Persistent bad weather here has seriously affected sales. I made the request with the deepest regret. Could you reduce the number from 1000 to 500?

B: In the circumstances I think I have no choice but to help you.

(a)A asked for the cutting of the price.　(b)B strongly disagreed to A's request.　(c)A asked to cancel the order on account of strike.　(d)A asked to reduce the ordered quantity.

44. (　) Which one of the following answer is not suitable for the dialogue?

A:How much do you think you could bring the price down?

B: ＿＿＿＿＿＿＿＿＿

(a)Our profit margin is not that large.　(b)I suppose we could reduce it by 10% if you guarantee to double your order.　(c)Our price is lower than those of competitors.　(d) I believe there is nothing left to be discussed.

45. (　) Under the trade terms CIP, the ＿＿＿＿＿ must contract for the cargo transportation insurance.

(a)buyer　(b)seller　(c)consignee　(d)carrier

46. (　) ＿＿＿＿＿ the terms of payment as stipulated in the contract, please establish an irrevocable letter of credit in our favor.

(a)In fact　(b)In accordance with　(c)In contrast with　(d)As a matter of fact

47. (　) The goods are ready for ＿＿＿＿＿ .

(a)ship　(b)shipped　(c)shipment　(d)shipments

48. (　) Shipment should be made within 60 days ＿＿＿＿＿ .

(a)before accepting L/C　(b)after application of L/C　(c)before received L/C (d)after receipt of L/C

49. (　) The trade terms FOB should be followed by named ＿＿＿＿＿ .

(a)port of shipment　(b)place of dispatch　(c)port of destination　(d)place of destination

十二、國貿業務丙級學科基礎貿易英文題庫範例

1. ( ) We would be delighted to _____ business relations with you.
   (1)enter (2)establish (3)open (4)set

2. ( ) Please return the damaged goods. We will replace them free of _____ .
   (1)expense (2)charge (3)pay (4)payment

3. ( ) Provided you can offer favorable quotation, we will _____ regular orders with you.
   (1)make (2)take (3)place (4)fulfill

4. ( ) We can send you a replacement, or if you like, we can _____ your account.
   (1)charge (2)credit (3)debit (4)deduct

5. ( ) Thank you for your enquiry _____ October 12 concerning DVD players.
   (1)date (2)dating (3)of (4)on

6. ( ) The new model has several additional _____ which will appeal to customers.
   (1)dimensions (2)features (3)specialities (4)measurements

7. ( ) As the photocopier is still under _____ , we'll repair it for free.
   (1)warranty (2)standard (3)instruction (4)construction

8. ( ) We would like to know whether the firm is _____ in settling its accounts promptly.
   (1)reasonable (2)favorable (3)advisable (4)reliable

9. ( ) As the time of shipment is fast approaching, we must ask you to fax the L/C and
   shipping _____ immediately.
   (1)advice (2)documents (3)manual (4)instructions

10. ( ) In regard to your invoice No 23130 for $2,578, which we expected to be cleared two
    weeks ago, we still have not yet received your _____ .
    (1)remittance (2)transfer (3)pay (4)account

11. ( ) We trust that the _____ will reach you in perfect condition.
    (1)packing (2)shipping (3)consignment (4)assignment

12. ( ) We would be grateful if you would allow us an _____ of three months to pay this
    invoice.
    (1)extension (2)exception (3)intention (4)invention

13. ( ) The goods you inquired about are sold out, but we can offer you a _____ .
    (1)substitute (2)compensation (3)refund (4)replace

14. (　) We _____ to inform you that our customers find your prices too high.

　　　(1)dislike　(2)regret　(3)advise　(4)require

15. (　) Owing to a fire in our warehouse, we have to _____ the shipping date to August 15.

　　　(1)cancel　(2)schedule　(3)postpone　(4)forward

16. (　) At the fair, we will _____ some of our newly-developed products.

　　　(1)secure　(2)procure　(3)exhibit　(4)expand

17. (　) The package _____ the dinner plates appeared to be in good condition.

　　　(1)containing　(2)maintaining　(3)included　(4)excluded

18. (　) Payment will be made by bank _____ .

　　　(1)transport　(2)transaction　(3)transit　(4)transfer

19. (　) We have _____ from the Chamber of Commerce in Boston that you are a leading

　　　manufacturer of waterproof watches in Taiwan.

　　　(1)known　(2)learned　(3)told　(4)written

20. (　) If you are not already represented here, we should be interested in acting as your ____

　　　agent.

　　　(1)travel　(2)collection　(3)forwarding　(4)sole

21. (　) The agency we are offering will be on a _____ basis.

　　　(1)competition　(2)commission　(3)compensation　(4)conversation

22. (　) _____ our latest catalog and price list for your reference.

　　　(1)We are enclosed　(2)Enclosed are　(3)Enclosed is　(4)Enclose

23. (　) We have arranged with our bankers to open a letter of credit _____ .

　　　(1)for your benefit　(2)in your interest　(3)in your account　(4)in your favor

24. (　) We are manufacturers of high quality _____ .

　　　(1)office equipment　(2)fax machine　(3)furnitures　(4)kitchenwares

25. (　) _____ your confirmation, we will execute the order.

　　　(1)Upon receipt of　(2)After receive　(3)When we will receive　(4)As soon as receive

26. (　) _____ you can see from the enclosed catalogue, we offer a wide range of products.

　　　(1)While　(2)As　(3)If　(4)Unless

27. (　) _____ shipment has been effected, we will advise you by fax.

　　　(1)As long as　(2)As far as　(3)As soon as　(4)As for

28. (　) If the quality of the goods comes up to our expectations, we can probably let you have _____ orders.

(1)trial　(2)regular　(3)rare　(4)usual

29. (　) _____ if you could send some samples of the material.

(1)We would appreciate　(2)We would be appreciated　(3)It would be appreciated (4)We would grateful

30. (　) If you have any questions, please _____ .

(1)do not be polite.　(2)do not hesitate to let us know.　(3)do not forget to tell us. (4)do remember asking us.

31. (　) We look forward to _____ .

(1)hear from you soon　(2)hearing of you soon　(3)you promptly reply　(4)your prompt reply

32. (　) We have quoted our most _____ prices.

(1)favor　(2)favoring　(3)favorite　(4)favorable

33. (　) _____ , we are enclosing our catalogue and price list.

(1)As requested　(2)As request　(3)As requiring　(4)As requires

34. (　) Please confirm the order _____ email and send us the shipping information along with your invoice.

(1)by　(2)in　(3)on　(4)through

35. (　) May we suggest that you visit our showrooms in Los Angeles _____ you can see a wide range of units?

(1)that　(2)which　(3)which　(4)where

36. (　) Our prices are relatively low in comparison with _____ .

(1)they　(2)them　(3)their　(4)theirs

37. (　) Please open the relative _____ as soon as possible so we can arrange shipment without delay.

(1)B/L　(2)L/C　(3)P/I　(4)T/T

38. (　) Which of the following terms is not related to payment ?

(1)L/C　(2)D/P　(3)O/A　(4)FOB

39. (　) A: This clock comes with batteries, doesn't it? B: _____

(1)That's right. There is a ten percent service charge.

(2)No. I'm afraid they're sold separately.

(3)Yes. You'll save time if you do.

(4)Yes. There have been several reports of damage.

40. (　) Would you please _____ this matter and send our order without further delay.

    (1)look into　(2)investigate into　(3)deal in　(4)take care

41. (　) The new model is _____ than the old one.

    (1)more efficiently　(2)more better　(3)less cheaper　(4)much lighter

42. (　) The following is a list of our _____ products.

    (1)late-developed　(2)fast-grown　(3)best-selling　(4)most cheap

43. (　) Regarding the damaged goods, we have filed a _____ with the insurance company.

    (1)claim　(2)complaint　(3)compensation　(4)commission

44. (　) Our prices are considerably lower than _____ of our competitors for goods of similar quality.

    (1)which　(2)that　(3)those　(4)ones

45. (　) We sent you a fax on October 12 _____ some information about your notebook computers.

    (1)request　(2)requests　(3)requesting　(4)requested

46. (　) Under the circumstances, we have no choice _____ the order.

    (1)but cancel　(2)but will cancel　(3)but to cancel　(4)but canceling

47. (　) We will grant you a 3% discount if your order _____ is over £15,000 for one shipment.

    (1)value　(2)quantity　(3)quality　(4)worthy

48. (　) We have instructed our bankers to _____ the L/C.

    (1)correcting　(2)settle　(3)revised　(4)amend.

49. (　) Please _____ the overdue payments immediately.

    (1)solve　(2)pay　(3)settle　(4)exchange

50. (　) A: How would you like your coffee? B: _____

    (1)Well done, please.　(2)Very well, thank you.　(3)Not for me, thanks.　(4)Black, please.

Part 2

MODERN BUSINESS ENGLISH MODERN BUSINESS ENGLISH
MODERN BUSINESS ENGLISH MODERN BUSINESS ENGLISH

E-Mail 格式／書信格式及信封打法

E-Mail Form/Letter Form & Envelop

MODERN BUSINESS ENGLISH MODERN BUSINESS ENGLISH
MODERN BUSINESS ENGLISH MODERN BUSINESS ENGLISH

一、E-Mail 格式

在 e 世代的商業書信溝通及往來中，e-mail 的書寫與傳遞方式成了溝通上非常重要的一部分，除了書信聯絡外，任何文件及圖面檔案的傳遞，都可透過 e-mail 的方式來傳送。一封 e-mail 內容中的基本要素包括下列幾項：

From（寄件者)：這一行內填入 e-mail 寄信人電子郵件地址
To（收件者)：這一行內填入 e-mail 收信人電子郵件地址
CC（副本）：這一行內填入 e-mail 副本收信人電子郵件地址
BCC（密件副本）：這一行內填入 e-mail 隱名副本收信人電子郵件地址
Sent（寄件日期）：電子郵件寄出日期及時間於寄出後會自動出現
Subject（主旨）：這一行可簡單輸入此 e-mail 信息的大意
Attachment（附加檔案）：（一般會出現迴紋針圖案）此行區可貼上任何附加檔案
Message text area（e-mail 信息內容區）：此信息內容為本文內容，和一般書信的架構、
　　　　　　　　　　　　　　　　　　　寫法和編輯功能都相同。

〈樣本如下〉：

mary0105: 收件匣	上一封 ｜ 下一封 ｜ 返回 ｜ 另開視窗閱讀信件

寄件人： Aurore Lion < Aurore.Lion@long..com > 加入通訊錄　檢舉垃圾信

收信人： mary0105@ms33.hinet.net , sb.wow@msa.hinet.net

副本： Simon < Simon@long.com > , John < John@long.com >

日期： Wed, 22 Feb 2012 10:55:39 +0100

主旨： RE: Shipment of Repair tender: FP-0010086

附檔： Payment.pdf(27.6K)

Dear Mary,
Our accounting department has received the payment from your side, so we confirm
that the shipment will be made early next week.
Best regards,
Aurore

mary0105: 收件匣	上一封 ｜ 下一封 ｜ 返回 ｜ 另開視窗閱讀信件

回信　全部回信　轉寄　檢視信件原始檔　友善列印　下載信件　刪除

二、書信格式

一篇完整的英文書信，基本上包括下列 16 個部分，並說明如下。

---

1. From: Merrybest Int'l Co.

   F1.4, No, 5, Alley 15, Lane 10, Sec.4

   Hsin-Yi Road, Taipei, Taiwan.

2. To: Best Equipment Co.   4. Date: _____

   P. O. Box 720693 Houstom, TX 77272, U.S.A.

3. Attn: Mr. Fred Jones/Manager   5. Our Ref. No: SQ-001

                                                                      Your Ref. No. BS-04

6. Subject: O/No. 001

7. Dear Mr. Jones,

8. Many thanks for your O/No. 001 and we confirm the acceptance of this order and

   will certainly follow your instruction⋯⋯

9. Very truly yours,

10. Merrybest Int'l Co.

   ⋯⋯⋯⋯⋯⋯

11. Mary Huang

12. V. President   16. MH/jw

13. C.c.: Mr. Enslin/Q.C. Dept.

14. P.s.: Please send us your relevevant L/C copy soon.

15. Encl.: P/I No. 2013

1. 信頭(Letter Head)：

    注意：中文的地址，由小到大：先寫樓 F1. 4，再寫號碼 No. 5，弄 Alley 15，巷 Lane 10，段 Section 4，街名 xxx Street 或 xxx Road，市 Taipei，國名 Taiwan

2. 收信人的公司名稱和地址(Complimentary Address)

3. 特定的收信人名字及頭銜(Attention)

4. 發信日期(Date)

    一般寫法有下列幾種：(1)July 15, 2012；(2)15th of July, 2012；(3)07/15/12 或 07-15-2012；(4)15/07/12 或 15-07-2012

    注意：第(3)(4)種寫法最好用於日期為 13 日以上較妥，以免造成月日混淆不清，因為美式寫法為月／日／年，歐洲式：日／月／年，日本式：年／月／日

5. 參考編號(Reference No.)

    分為發信人的參考編號 Our Ref. No.及收信人的參考編號 Your Ref. No.兩種，具選擇性的，特別用於報價函

6. 主題（Subject 或 Re）

    列出此篇文章的主旨

7. 尊稱(Salutation)

    (1)如已知收信人名，男性用 Mr.、女性用 Mrs.（已婚）Ms.（未婚）

    (2)Messrs. 為公司尊稱，放公司名稱前，新式寫法多已省略

    (3)如不知收信人名，可寫 Dear Sirs, Gentlemen, Dear Madam

8. 本文(Body)

    文章內容分：(1)主題；(2)說明；(3)結論三部分來布局

    打字方式分：(1)齊頭式(Block Form)：每段開頭都從第一個字開始打

    　　　　　　(2)縮進式(Indented Form)：每段開頭都空 5 至 7 個空格，再開始打

9. 結尾語(Complimentary Colose)

    常用的結尾語有 Very truly yours, Sincerely Yours, Faithfully Yours 或 Best Regards（多用於 e-mail 的結尾語）

10. 發信人公司名稱

11. 簽名(Signature)

12. 頭銜(Title)

總經理 President 或 General Manager、副總經理 Vice President、業務經理 Sales Manager、董事 Director

13. 副本抄送(Carbon Copy)

14. 再啓／附記(Postscript)

本文中未提到或追加補充之事

15. 附件(Enclosure)

附在信函後的其他文件資料

16. 發信人（簽名者）及打字者（或撰文秘書）的名字縮寫

MH 為簽名者，jw 為打字者，如果發信人和打字者為同一人，則不需註明此點

注意：(1)以上第 5.、13.、14.、15.、16.項為選擇性，視需要再加上。

(2)第 11 項簽名處，如寫信人（經理／老闆）正好不在公司，需由打字者（或撰文秘書）代為簽名時，可在人名前面加一個 for，例如：Julie Wang for Mary Huang（此方式表示書信內容為 Mary Huang 的意思，Julie Wang 只是代為簽字而已）；另一種方式則是由打字者（或撰文秘書）直接簽署自己名字，但在信上註明 C.c.給經理／老闆，例如：簽名處直接打 Julie Wang，底下 C.c. Mary Huang（此方式表示書信內容為 Julie Wang 的意思，副本抄送知會 Mary Huang）。

書信樣本(Letter)

MERRYBEST INTERNATIONAL CO.

Fl. 4, No. 141, Sec. 4, Hsin-Yi Road, Taipei, Taiwan, R.O.C.

Tel: 886-2-27051608   Fax: 886-2-27056741

E-mail Address: merrybest@msa.hinet.net

To: BATA Crane Ltd.   Date: July 15, _____

P. O. Box 9370, Daytona Beach

32150, U.S.A.   Our Ref. No. MB-012

Attn: Export Manager

Subject: Inquiry for Water Hammer Arresters

Dear Sirs,

As the leading manufacturer of Valve & Piping Accessories here in Taiwan, we are looking for the Water Hammer Arrestrs now. From the Name List of U.S. Suppliers, we know you are the manufacturer of these products and hope that you can mail your relevant catalogs or brochure to us soon.

Besides, please kindly make your best quotation according to our enclosed drawing and send it to us by e-mail for our evaluation as soon as you can.

Many thanks for your attention to the above and look forward to starting the business cooperation with you in the near future.

Very truly yours,

Merrybest Int'l Co.

Mary Huang

V. President

Encl.: Drawing No. MB050

三、信封打法

　　西式信封的打法和中式不同，發信人的公司名稱及地址印或打在左上角，收信人的公司名稱地址打在正中央，郵票貼在右上方，其他附註字樣，例如印刷品、快遞、掛號等，則打在左或右下方空白處（參下圖）。

Form: Merrybest Int'l Co.

　　P. O. Box 44-6, Taipei, Taiwan

Stamp
貼郵處

　　　　　　To: BATA Crane Ltd.

　　　　　　　P. O. Box 9370, Daytona Beach

　　　　　　　F1 32150, U.S.A.

AIR MAIL
PRINTED MATTER
REGISTERED
　　　　　　Attn: Export Manager

信封上常見附註字樣表

1.	AIR MAIL 或 PAR AVION	航空
2.	PRINTED MATTER	印刷品
3.	REGISTERED	掛號
4.	EXPRESS	快遞
5.	SAMPLES OF NO COMMERCIAL VALUE	無商業價值樣品
6.	C/O, PLEASE FORWARD TO	煩轉交某某人
7.	PHOTO INSIDE	內有照片
8.	PRIVATE, PERSONAL	本人親啟
9.	CONFIDENTIAL	機密文件
10.	URGENT	緊急文件

四、商業信函摺疊方式

商業信函收入信封之前，最好用三摺式，將文章內容遮住，只露出收信人的公司名稱地址及人名，詳參下圖摺疊方式。

MERRYBEST INTERNATIONAL CO.

Fl. 4, No. 141, Sec. 4, Hsin-Yi Road, Taipei, Taiwan

Tel: 886-2-27051608   Fax: 886-2-27056741

E-mail Address: merrybest@msa.hinet.net

To: BATA Crane Ltd.   Date: July 15, _____

P. O. Box 9370, Daytona Beach

FL 32150, U.S.A.   Our Ref. No. MB-012

Attn: Export Manager

Su

Part 3

ODERN BUSINESS ENGLISH MODERN BUSINESS ENGLISH
ODERN BUSINESS ENGLISH MODERN BUSINESS ENGLISH

書信寫作技巧及文章
內容好壞對照範例

Effective Writing Skills of Business
Letters & Examples

一、書信寫作技巧

一篇好的商業書信，在文章寫作上必須注意下列幾個重點：

1. 一個段落，只寫一個主題或只談一件事（One paragraph one idea）

   例如這個段落談送樣品的事，付款的事就另起一個段落來寫，交期或其他事也是一件件分段敘述。

2. 一封信只談一個主題（One letter one subject）

   一封信裡的每個段落都要針對同一個主題來討論，例如這封信是討論某張訂單，所有段落都要針對此訂單內容作討論，不要扯到別的出貨、產品或其他主題上去。

3. 每篇文章布局嚴格要求分三部分

   (1) 主題 main idea (beginning)

   寫此信的目的，要告訴對方什麼事，或要對方做什麼配合，在第一段馬上直接告知。

   (2) 說明 examples/explanation to support the main idea (middle)

   為什麼要有此要求，理由為何？用第二段來說明及支持主題。

   (3) 結論 conclusion (ending)

   希望對方如何配合，或怎麼合作。

   舉例來說，如果有一批出貨因為颱風來襲要延期，寫給客戶的信布局如下：

   (1) 主題：馬上直接跟客戶道歉告知貨要延誤

   We are sorry to inform you that the shipment of O/No.001 will be delayed.

   (2) 說明：接著告知延後的原因是因為這週有颱風

   Because there is a typhoon here this week.

   (3) 結論：請客戶能延長出貨日期

   Please kindly extend the shipping date to the end of next month.

   如此簡單扼要，就將寫信的動機及意思清楚表達。

   一封好的信函或句子，就是不要有多餘的贅字或贅語，能用最少的字數及語句表達清楚意思，達到溝通的目的，即為上乘之作。因為商業上，大家都很忙碌，沒有時間看太冗長繁瑣的句子或文章。特別現在國際貿易交易市場偏重在亞洲或開發中國家，大都為非英語系的國家，商用書信的英文表達盡量使用簡單句型以免造成溝通不良。

4. **文章必須簡潔(brief)、直接(direct)、直入重點(to the point)**

　盡量以最少字數表達清楚意思即可，一些不必要的敘述則可省略

　例：(1)We express our regret at being unable to meet your order.（較囉唆）

　　　　We are sorry we cannot meet your order.（較好）

　　　(2)We contacted your forwarder this morning and they told us there is no vessel available this week.

　　　　（我們今早和你的攬貨公司聯絡，而他們告訴我們這週沒有船）

　　　　此句中的第一句，即屬多餘，可省去，較好的寫法為：

　　　　Your forwarder told us there is no vessel available this week.

5. **當寫一個句子時，盡量不要造成對方的疑惑，每個句子最好給對方答案不要給疑問**

　例：Our delivery time will be more than 90 days due to the hot season.

　　　（由於旺季，我們的交期要 90 天以上）

　　　那麼客人就會問：那麼交期要多久呢？100 天，200 天？所以最好寫：

　　　Our delivery time will be within 120 days due to the hot season.

　　　（交期要 120 天內）

6. **盡量少用不必要的形容詞(Adj.)和副詞(Adv.)**

　因形容詞及副詞是用來追加補充說明句意的修飾詞，如能在基本句型結構中表達清楚，就不要畫蛇添足，加入多餘的修飾字或詞增加混淆。

　例：The question is under active consideration.

　　　（這裡的 active 就顯得多餘了）

　　　The question is being considered.

　　　（以 S＋V 簡單句型簡潔表達較好）

7. **不要使用太多固定式介系詞片語**

　例：in the course of ＝ during（在…期間），如果寫 during the next few days 就比寫 in the course of the next few days 好

8. 不要使用過多添湊語

例：It will be appreciated if you reply soon without hesitation.

（儘快 soon 和不要遲疑 without hesitation 同意思，只要寫其一即可，此句即顯囉唆）

We will appreciate your early reply.（較好）

9. 避免使用易混淆的字眼

例：The writer wishes to acknowledge with thanks receiving your letter.

（此句太冗長，且 writer 語意不清楚）

較好的寫法：We received your letter with thanks. 或 Thank you for your letter.

10. 盡量用主動代替被動（寫出來的句子較簡潔）並使用常見的字

因為國際上不是每個國家都是英語系國家，商業上，盡量以簡單易懂的字句和客人清楚溝通即可

例：The favor of your early reply will be appreciated.（被動）不如

We will appreciate your early reply.（主動）

Will you be good enough	= Please（後者較常用）
at the present time	= now
come to a decision	= decide
in the near future	= soon
terminate	= end
utilize	= use

11. 一句話中，不要使用同一個字，表達不同意思

英文有些單字，有許多不同意思，例如 firm 名詞當「公司」= company，形容詞則是「固定的」；account 名詞是「帳目」，account for 此片語意思是「說明」= explain，所以盡量避免下列寫法：

例：(1)Our firm's sales will explain our firm line to you. 最好改成：

Our company's sales will explain our firm line to you.

（我們公司的業務將為你說明我們固定的項目）

(2)How do you account for the fact that the account is wrong？

最好改成：

How do you <u>explain</u> the fact that the <u>account</u> is wrong？

（你如何說明帳是錯誤的事實）

12.使用具體的字眼

少用似乎(seem)、可能(probably)、好像(likely)、也許(maybe)，這些不確定的字

例：It is more than likely, in fact it is highly probable, that our price will rise.

（我們的價錢很有可能上漲，事實上是非常可能）──不好的寫法

Our price will rise soon.

（我們的價錢很快會上漲）──較好的表達

13.使用完整的句子，少用分詞片語

例：Hoping to hear from you.

此為分詞片語，不是完整的句子，所以最好寫成

We hope to hear from you soon.

14.避免單調

長短句交叉使用或偶爾主被動交互使用，不要從頭到尾都用同一句型或同一個字開頭，以免呆板單調

15.現代商用英文句子新舊寫法對照範例

（■為較囉唆冗長的舊式寫法，▲為較簡潔直接的新式現代寫法）

(1)■In accordance with your request, we are sending you herein two of our latest catalogues.

▲As you requested, we enclose two latest catalogs.

(2)■This is to acknowledge the receipt of your letter of Jan. 15 in which you requested our catalog.

▲Many thanks for your letter of Jan. 15 requesting our catalog.

(3)■Your check in the amount of US$35,000 has been received with thanks.

▲Thank you for your check for US$35,000.

(4)■It will be appreciated if you send Mr. Jones a copy of your latest catalogue.

▲Please send Mr. Jones your latest catalog.

(5)■As per your letter of June 1, we are giving your offer due consideration.

▲We are glad to consider the offer in your letter of June 1.

(6)■If the result you execute this order is satisfactory, we will place further orders with you in the near future.

▲If your performance is satisfactory, we will place you new orders soon.

(7)■We beg to acknowledge the receipt of your letter of August 1.

▲Thank you for your letter of August 1.

(8)■In compliance with your request in your letter of May 1, we are pleased to forward a copy of our latest catalogue.

▲We are glad to send you our latest catalog you reguested on May 1.

(9)■Your e-mail of April 20 was received and contents were noted.

▲Thank you for your e-mail of April 20.

(10)■Although we are accustomed to buying goods with freight prepaid, we agree to try to pay the freight at our end this time. Please deduct this from your CFR quotation and send us your proforma invoice accordingly.

▲Please requote FOB prices and send us your proforma invoice accordingly.

二、文章好壞對照範例

範例 1　通知客戶因最近訂單太多，以後交期要較長

較差的寫法

We have just now received a notice from our factory. As they recently got a rush of orders, so the delivery time must be longer than 90 days.

（缺點：第 1 句多餘，第 2 句會造成疑惑）

較好的寫法

Our factory just informed us that the delivery time would be within 120 days for future shipments due to very tight schedule now.

範例 2　通知客戶下週要出貨，但船公司告知只收兩個 20 呎櫃，不收一個 40 呎櫃，如此會造成運費成本增加，請客戶在以後的出貨，最好事先和攬貨公司查清楚，以免損失。

較差的寫法

We are glad to inform you that we will schedule to ship your goods next week. Today we contact your forwarder "Ideal" to arrange shipping details, but they advised not to accept 40 feet container, only accept two 20 feet containers. We consider it will increase your freight cost. （缺點：第 2 句前半句多餘）

As you are one of our good customers, we do not want to see you get additional cost in future, if you place any orders, please check your forwarder beforehand. （缺點：此段主題不明顯）

較好的寫法

We are glad to inform you that we are going to ship your goods next week. However, your forwarder "Ideal" advised that they have no any 40 feet container but can only offer two 20 feet containers, for which will certainly increase your freight cost. Since you are one of our good customers, we have the obligation to inform you the fact and hope you can check with your forwarder in advance for your future shipments to avoid the loss.

範例 3　告知客戶將趕工於春節前出貨，但是至今尚未收到信用狀，請其告知
　　　　是否已經開出及號碼。

較差的寫法

Regarding your e-mail dated 22nd of Jan, we have <u>advised</u> our factory to work full capacity in order to catch up with your delivery time before Chinese Lunar New Year.

（缺點：advise 只是告知工廠，沒有要求之意思，語氣較弱）

<u>Moreover</u>, we have not received your <u>above mentioned L/C</u> till now. If you have issued, please advise us the L/C no., so that we can earlier get it to arrange shipping details.

（缺點：Moreover 此外，在此應該用 However 然而，語氣才對；above mentioned 的
　　　　用法不當，因為前面都沒提到 L/C，所以要改成 L/C for the above shipment
　　　　才正確）

較好的寫法

Regarding your e-mail of Jan. 22, we have <u>asked</u> our factory to work overtime to catch the delivery time by Chinese Lunar New Year.

<u>However</u>, we have not received <u>your L/C for the above shipment</u> till now. Please confirm if you have issued it and advise us the L/C no. for easily tracing and arranging shipment.

範例 4　國外客戶發來的詢問函

　　　　缺點：文體散漫，廢話太多，主題不清楚

英文原文

RE: Our Inquiry 281-5-12819

Spoke with our Rep yesterday about this customer. By Monday we should receive a sample of customer's existing sample of which they wish to change from membrane technology to conductive rubber.

If we get this business (i.e. Rep has 80% confident factor), it will be to do the complete assembly. Customer wishes to use the same plastic housing since it has been tooled for existing design. Hopefully with small modifications it can be still used for new design.

Customer will install assembly into the plastic case. Once we get the sample, it will be Fed expressed to you for reference. In the meantime, we are enclosing the bill of materials (BOM) for you to begin your quote process.

Customer's anticipated annual volume will be 3,000-5,000 units. They expect the first release to be for 1,000 units.

Customer is requesting a Budgetary quote during the week of 4/18. Thus when you quote, quote 1,000, 3,000, 5,000 pcs plus necessary tooling cost and lead time to produce 1st article sample (see enclosed 9 pages of BOM including Sony SBX1610 infrared remote control receiver and a picture of the existing audio keypad).

Note: We hope to get the specification information on the LCD plus idea of target price for assembly.

P. S.　Customer is currently having various parts of this assembly done by different suppliers, but now wishes to have all parts of the assembly done by one vendor and that's where we come into the picture. Thanks.

以上文章實際重點，只有四點：

(1)附上 9 張材料表(BOM)

(2)請於 4/18 前報出預估樣品費，1-3-5,000 個的單價、模具費及交期

(3)並請提供 LCD 的規格及預估裝配的成本

(4)下週一會用快遞寄出一組客戶現有的樣品供參考

　請注意客戶欲變更現有薄膜技術為導電橡膠，並希望能使用現有的塑膠框，或稍加修改後，仍可用於新設計，因為模具已開

※其他文中所提客戶為何要改變設計的原因，以及業務代表有多大信心爭取此生意等，都是多餘不重要的事，皆可省略，因此較好的寫法如下：

重寫較好表達方式

RE: Our Inquiry 281-5-12819

Enclosed please find 9 pages of BOM including Sony SBX1610 IR Control Receiver and a picture of the existing Audio Keypad.

Please quote by 4/18 your budgetary unit price, tooling cost and lead time for the 1st article sample, 1,000, 3,000, and 5,000 units annually. Also provide us the specification of LCD plus target price for the assembly cost.

We will Fed express you one sample of customer's existing assembly product next Monday for your reference. Please note they would change the design from original membrane to conductive rubber but still use the existing plastic housing, or just slightly modify it for the new design as it has been tooled. Please advise as soon as possible. Thanks.

範例 5　接獲國外客戶來電要求對不良的出貨索賠 HK$30,000.00

老闆指示照下列中文翻譯回覆

中文原文

1. 謝謝您 10/4 的 e-mail

2. 自從我們在 8/15 與貴公司在我們展覽的攤位上見過面之後，便一直沒有任何貴公司的消息，今日收到貴公司傳真要敝公司賠償 HK$30,000 後，實感迷惑，由於我們未曾與貴公司出過任何貨品，故請您們檢查敝公司與貴公司往來文件後，如確是敝公司所犯的過失，敝公司將負責到底，故煩請給我們進一步的消息，我們也將盡快給您滿意的答覆，謝謝！

注意：中文思考邏輯和英文常有很大出入，以上文章中很多句子都是多餘，且過於自謙，例如(1)客人要求賠償，一般覆函第一句，就不要寫「謝謝」；(2)既然確定沒做過生意，就不需對方查看文件，也不需負責到底，因此英文的回覆內容布局應該如下：

(1)主題：很訝異（迷惑）收到貴公司 10/4 的 e-mail 要求我們賠償 HK$30,000

(2)說明：因為我們從未和貴公司做過生意，而僅於 8/15 在展覽會場見過一次面

(3)結論：我們認為您們可能發錯 e-mail，請查看並澄清

重整後正確英文表達方式

Dear Mr. Vicker,

We are confused (surprised) to have received your e-mail of Oct. 4 asking us to return you HK$30,000 for the faulty goods.　Since we have not done any business with your company but just met you one time on August 15 at our booth in Taipei Electronic Show.

So, please check if you mis-sent this e-mail, and clarify.

（或 We thought you might have sent this e-mail to the wrong attention. Please check and clarify.）

Best regards

商用英文信函寫法
Business English Letters

在整個貿易流程中，每個步驟，買賣雙方都有可能隨時和對方作聯絡溝通，從產品推銷／市場開發(Marketing/Promotion)→詢價／規格討論(Inquiry/Specification Discussion)→回覆／報價(Reply/Quotation)→討價還價(Counter Offer)→追蹤／催促(Follow Up)→索樣／送樣(Sampling)→下單／接單(Ordering)→付款／催款(L/C or Other Payment Terms)→信用狀修改(L/C Amendment)→生產(Production)→包裝(Packing)→出貨(Shipment)→保險(Insurance)→押匯(Negotiation with the Bank)→抱怨／索賠(Complaint/Claim)到談代理簽合約(Agency Agreement)，都需要作書信來往討論。

其中以推銷函(Promotion/Sales Letter)、抱怨函(Complaint Letter)和規格討論(Specification Discussion)較為難寫，其他信函就容易多了。因為買賣雙方開始時都不熟識，而且一般大客戶都已有固定合作廠商配合，除非賣方產品很有特色或有特別吸引人之處，否則推銷阻力很大；而抱怨函都牽扯到賠償，要如何將損失降到最低，不僅需要經驗技巧，更需大費周章，大篇幅討論爭辯；而規格澄清討論，有些技術問題，解說較繁瑣，文章寫起來就冗長複雜許多。其他信函只要針對對方問題逐一作答，有的三言兩語即可完成。

在科技進步且工商忙碌的現代，除了送樣和退貨樣品要用郵寄方式，其他往來信函、文件、訂單、目錄、圖檔、圖面規格甚至出貨文件等，為了爭取時效及節省費用，少數用傳真(fax)方式外，大都以電子郵件(e-mail)方式傳送，所以在同一封 e-mail 信函中，亦可常見串掛買賣雙方針對某案例或事件所有書信往來的內容歷史。

以下將各類信函寫法分別說明，並附範例供參考，重點是要熟記前面文法重點、句型結構及中英文對照翻譯時句型結構不同處，加上各類信函專用字彙，以英文文章結構布局，即可寫出一封完好正確的商用英文信函。

一、推銷函(Sales/Promotion Letters)

一個新產品欲打入市場或爭取客戶，其推銷的步驟如下：

1. 選定產品

注意不同地區的人對產品的需求不同，對產品的設計樣式顏色的喜好也不一樣，例如高科技的產品在歐美市場可能賣得不錯，但對低所得收入的開發中國家的人來說，他們迫切需要的可能是日常生活用品；對歐洲人來說，黑白柔和色彩可能顯得高雅，而東南亞熱帶地區的人民則偏好鮮豔亮麗的色彩，所以第一步驟即是先選定一項想推銷的產品（各項產品名稱可參考附件一）。

2. 市場調查

選好所要推銷的產品後，第二個步驟即選定欲推銷的市場，但在推銷之前，最好先做市場調查，查看此產品在這個市場有無下列問題：

(1)有沒有被管制禁止進口。

(2)有沒有被告傾銷(dumping)。

(3)有沒有配額(quota)問題。

(4)有沒有外匯管制（如有，要注意開 L/C 或付款會較慢或甚至有問題，此點可到外貿協會或各國駐台辦事處／大使館／領事館查詢）。

3. 準備資料

了解市場後，確定推銷沒有問題，就開始準備推銷的資料：

(1)公司型錄(Catalog)。

(2)公司簡介(Company Profile)（參見附件二）。

(3)報價單(Quotation/Price List)。

(4)樣品(Sample)。

報價單及樣品視需要而定，或等客戶進一步索取時再提供。

4. 選擇推銷方式

產品推銷的方法很多，基本上可歸納成下列十種：

(1)寄開發信／推銷函(Sales Letter)：最經濟，所以最常用。

(2)刊登廣告：屬被動推銷，等客戶自己找上門，好處是可用當地語言刊登。

(3)參加展覽：直接面對客戶，可當面討論；缺點是成本高，和競爭者同時展出。

(4)拜訪客戶：多半針對老客戶做定期拜訪，或展覽後，順道拜訪，聯絡感情。

(5)貿協介紹／老客戶或朋友引介：此為現成客戶，宜多加把握。

(6)租保稅倉庫(Duty-Bond Depot)：多用於轉口／三角貿易或季節性產品。

(7)郵購(Mail Order)：多用於百貨禮品、裝飾品或小額樣品訂單(sample order)。

(8)設分公司(Branch Office)：用於需提供售後服務(after-sales service)、維修保養(maintenance)的產品，例如機器設備、車輛、電器電腦產品等。

(9)找代理(Agent)：對當地法令不熟或語言溝通有障礙的地區，或銷售上項產品時，可透過代理做銷售、服務及溝通。

(10)寄售(On Consignment)：多為大廠商或零售商採用。

5. 找尋客戶名單方法

如選擇以寄開發信做推銷，要先蒐集客戶名單，才能將寫好的推銷信和目錄一起寄出，找尋客戶名單的方法如下：

(1)到外貿協會的圖書室查閱進出口商名錄／廠商名錄或產品分類書籍，也可經由貿協的電腦查閱（地址：台北市信義路五段 5 號 2 樓世貿中心）。

(2)向商會／工會索取會員名錄。

(3)到各國在台辦事處查詢或請求協助（無邦交的國家）。

(4)透過各大使館／領事館協助（有邦交的國家）。

(5)去函給各國貿協，要求免費刊登需求廣告，或推薦進出口商／廠商名單。

(6)上電腦網路圖書站查詢。

(7)從各相關報章雜誌廣告媒體翻閱。

(8)訂閱外貿協會的《國際商情雙周刊》(*International Trade Biweekly*)，其中的「貿易機會專刊」內，都有不同客戶名單刊出（參見附件三）（地址：台北市基隆路一段 333 號 6 樓；TEL：02-2725-5200 轉 1828）。

(9)可向 Trade Sources 訂閱客戶名單，他們會定期 e-mail 客戶資料給訂閱的公司（參見附件四）。

(10)從國內外各相關展覽中蒐集。

6. 推銷函寫法

拿到客戶名單之後，就開始著手擬稿寫推銷函。寫的時候要有下列的認知：

(1)推銷函是一種廣告宣傳，目的在鼓勵收信者購買特定的產品，或至少要能引起其購買的慾望,所以一封好的推銷函要能先引起收信者注意→停下來閱讀→然後產生購買慾望。

(2)推銷函是大量印刷的宣傳文件,寫給廣泛有興趣的購買者,不是寫給特定一個對象,但是又要寫得讓收信者感覺,像是給其個人的信函般貼切,因此推銷函的寫法就需要比一般回覆函具有更多的技巧。

(3)推銷函布局的方式:

主題(Main Idea):引起興趣的的方法如下,可選其中一、二項,分段敘述
(1)告知讀者賣什麼產品/產品有哪些特色優點/提供什麼特別服務(參範例 2/3/10)
(2)訴諸人性/有什麼附加價值(參範例 9)
(3)此產品在市場上有什麼特別的流行訊息(參範例 11)
(4)如何經由共同朋友/貿協/商會/工會或任何管道引介(參範例 4/5/6/8)

說明(Examples/Explanation):舉例說明支持以上主題的論點
(1)自我介紹公司/工廠的組織規模及經驗(參範例 4/11)(或附公司簡介說明)
(2)產品在如何嚴格品管製程下完成(參範例 11)
(3)舉證有哪些已購買的知名大客戶名單(參範例 4/9)
(4)附上公司產品型錄/報價單供參考(參範例 4/6)

結論(Conclusion)
(1)說服讀者儘快採取行動購買或來信聯絡示知意見,做進一步接觸(參範例 7)
(2)亦可請讀者將此產品也介紹分享給其親朋好友(參範例 8)

MODERN BUSINESS ENGLISH

7. 推銷函常用單字／片語

(1)recommendation (n) / recommend (v) 建議／推薦

(2)Chamber of Commerce 商會

(3)..are interested in importing 有興趣進口

(4)We know (got/learned) your name 我們得知貴公司大名

(5)take this opportunity 利用這個機會

(6)economic depression = economic recession 經濟不景氣

(7)oil crisis 石油危機／ financial storm 金融風暴

(8)high quality products at the lowest prices 物美價廉的產品

(9)experienced exporter/importer/distributor 有經驗的出口商／進口商／經銷商

(10)professional manufacturer 專業製造商

(11)booth = stand（攤位）／ fair = exhibition = show 展覽

(12)features / advantages 特色／ trial order 試銷訂單

(13)monthly capacity 月產能／ sales amount 銷售額

(14)custom-made product 客戶訂製的產品／ know-how 技術

(15)catalog 3-10 頁目錄／ leaflet 單頁目錄／ brochure 小冊子／ circular 傳單／ literature
　　文字說明書／ instruction manual 操作手冊

(16)company profile 公司簡介／ quotation 報價單／ sample 樣品

(17)for your reference 供你們參考／ for evaluation 供評估

(18)relevant = relative = related 相關的

(19)performance 性能／表現

(20)Attached please find our latest catalog.＝ The attached is our current catalog.
　　= We attach our newest catalog.＝ Our newest catalog is attached.
　　如附件是我們最新的目錄／我們在此附上最新的目錄

(21)sample charge and postage 樣品費及郵費

(22)market share 市場佔有／ mass production 大量生產

(23)without hesitation = as soon as possible 不要遲疑＝盡快

(24)set up = establish = build 建立

(25)business relationship = business relations = business connection 商業關係

8. 推銷信函常用的主題句／說明句(Explanation)／結論句(Conclusion)

主題句(Main Idea)

(1) We are glad to know from the xxx magazine that you are interested in our products.

(2) We got your name from the TAITRA（外貿協會）and learned that your are one of the leading importers (exporters) of xxx products in your country.

(3) Through the recommendation of your Commercial Office in USA, we got your name.

(4) Here is an excellent innovation and gift item for you in the coming Christmas season.

說明句(Explanation)

(1) Since 1975, we have been working in this field and have been cooperating with many big companies such as Motorola, HP, Hitachi, etc..

(2) Our company has been engaged in the computer field for more than 20 years.

(3) We possess very professional know-how especially in controlling the quality of logo imprinter.

(4) With the up-dated fully automatic injection machines, we have confidence to meet your every requirement.

(5) With the advantages of durability, precision and efficiency, we believe that you are able to promote our products in your market without worrying about competition.

結論句(Conclusion)

(1) Please take a look on the attached catalog and do not hesitate to keep us informed if you have any comment or inquiry.

(2) The attachments are the images for our latest products for your initial reference.

(3) You are assured of our best quality products and strongest support.

(4) Thank you for your attention and look forward to setting up the business relationship with you soon.

附件一　產品類別名稱表(Product List)

1. **禮品飾品**　Gift items/Ornament
   手環 bracelet，耳環　earring，戒指　finger ring，胸針　broach，領帶 necktie，領帶夾　tie clip/pin，皮帶　belt，髮夾　hair clip/pin，項鍊 necklace，首飾　jewelry

2. **成衣**　Ready-made garmets (clothes)
   運動衫　shirt，毛衣　sweater，夾克　jacket，裙子　skirt，長褲　pants/trousers，短褲　trunks，泳衣　swimsuit，泳褲　swimming trunks，睡衣 pajamas，圍巾　scarf，手帕　handkerchief，手套　glove

3. **鞋襪帽類**　Shoes/Hose/Hat
   鞋襪總稱　footware，雨鞋　rainboot，涼鞋　sandal，高跟鞋　high heel shoes，拖鞋 slippers，球鞋　sports shoes，長短襪總稱　hose，短襪　sock，女用長襪　stocking，帽類　headgears-hat, cap, raincap, eyeshade

4. **手提包**　Handbag，錢包　Purse，皮夾　Wallet，雨傘　Umbrella

5. **日用品**　Daily commodities／家庭用品　Houseware
   梳子　comb，肥皂　soap，刮鬍刀　shaver，牙刷　toothbrush，牙膏 toothpaste，紙巾　paper towel，衛生紙　toilet paper，餐巾　napkin，餐具　dinner set，刀　knife，叉　fork，湯匙　spoon，碗　bowl，碟　dish，盤　plate，杯　cup，茶壺　tea-pot，筷子　chopsticks，指甲刀　nail clipper，指甲挫　nail file，化妝品　cosmetic（乳液 cream，粉餅　powder，眉筆　eyebrow pencil，眼線筆／膏　eyelash pencil/grower，口紅　lipstick，眼影　eye shadow，香水　perfume），瓷器　china ware/pocelain，陶器 pottery/ceramic

6. **家電用品**　Housewares
   洗衣機　washing machine，電熨斗　electric iron，咖啡壺　coffee pot，吹風機 hairdryer，風扇　fan，烘乾機　dryer，冷氣機　air-conditioner，熱水瓶　vacuum flask，烤箱　oven，烤麵包機　toaster，微波爐 micro-wave oven，燈具　lightings，

廚具　kitchenware，鍋子　pan，電鍋　electric pan

7. 家具　Furniture

床　bed，枕頭　pillow，毯子　blanket，蚊帳　mosquito net，窗簾　curtain，壁櫃　closet，櫃子　cabinet，衣架　hanger，桌子　table，椅子 chair，水床　waterbed

8. 車輛／汽車零件　Vehicle/Auto parts

汽車　car，腳踏車　bicycle，腳踏車零件　bicycle parts，車身　body，車架　frame，座椅　seat，頭燈　headlight，尾燈　taillight，鏡子　mirrow，煞車　brake，引擎　engine，輪胎　tire

9. 工具　Tools

手工具　handtools，電動工具　power tools，五金　hardware，小五金　fastener

10. 原料　Raw material

鋼鐵　steel，化學原料／產品　chemical material/products，塑膠原料　plastics material，礦產品　minerals，煤　coal，石油　oil，水泥　cement，木材　wood

11. 通訊產品　Communications-telephones, fax machine, telex machine，電腦及周邊　computers and accessories/peripheral，辦公設備　office equipment，影印機　copy machine，文具　stationery，雜貨　sundry goods

12. 醫藥用品　Medical products

13. 紡織品　Textile

纖維　polyester fabrice，成衣　readymade garment，尼龍　nylon，襯衫　shirt，童裝　children dress，紗　polyester yarn，拉鍊　zippers，蕾絲花邊　lace

14. 運動器材　Sport goods

網球　tennis，網球拍　tennis racket

15. 機器設備　Machine, Equipment，儀器　Instruments

16. 閥　Valve，軸承　Bearing

17. 礦產品　Minerals
石英　quartz，長石　feldspar，磁鐵礦　magnetite，ironoxides，石榴石砂　garnet sand，磁土　china clay

18. 化學原料製品　Chemical Material & Products
塑膠　plastics，橡膠原料　rubber material, foam brushes, polyethylene

19. 皮箱、手提包及袋類　Travel Bags, Handbags

20. 皮革　Leather

21. 包裝材料及容器　Packing PVC
木箱　wooden case，塑膠桶　plastic keg，透明塑膠袋　clear plastic bag，墊板　pallet

22. 食品　Food，烏龍茶　Oolong tea

23. 農畜產品　Agriculture products/Domesticated animal products
烏魚子　mullet rose，動物　animal，冷凍花生　frozen peanuts，冷凍蝦／豆子／雞 frozen shimp/soybean/chicken，洋蔥　onion

24. 消費性產品　Consumer products，電子電器產品　Electronics/Electric products
計算機　calculators，液晶電視機　LCD　TV set，冰箱　refrigerators，電扇　electric fan，收音機　radio，錄影機　video cassettes rewinders(VCR)，電子字典　electronic dictionary，多功能數位相機　multi-function digital video，數位相機　digital camera，迷你口袋型數位相機　mini pocket DV，口袋型電視機　pocket TV，移動式電腦架 magical computer shelf，攜帶型音樂播放機　potable MP3 player，高解析度液晶螢幕 hi-resolution LCD monitor，多媒體伺服器　turbo station，三頻音樂攜帶電話（含MP3

播放機／高畫質相機／動態有聲錄影／藍芽影音傳輸 blue tooth）3G cell phone，多功能無線路由器   wireless multi-function modem router，桌上圖板   graphic tablet／wireless tablet，內建麥克風圖板   build-in microphone tablet，窗型電腦手機   windows mobile for pocket PCs，電玩機   video games machine，電腦空中交談電話機   Sky Talk／Sky Tel／Skype，通用序列匯流排   USB，個人數位助理   PDA(Personal Digital Assistant)，快閃記憶體   flash memory, flash ROM

25.電子零組件   Components
繼電器   relay，供電器   power supply，變壓器   transformer，電阻   resistor，電容器   capacitors，線路板   printed circuit board (PCB)，變頻器   power converter，乾電池   dry battery，多媒體喇叭   multiply speaker

26.燈具   Lighting products／Light fittings

27.廚房用品   Kitchenware: tagging guns, electric torch

28.玩具   Toys

29.嬰兒用品   Baby products

30.建材   Construction material
玻璃   glass

31.衛浴設備   Plumping products

32.禮品   Gift items，裝飾品   Decorated products，人造花   Artificial flowers，能量石手鍊   Magnetic bracelet

33.植物   Plants
花   flowers，草   grass

34. 積體電路產業　IC (Integrated Circuit)

高速靜態隨機存取記憶體　SRAM，動態隨機存取記憶體　DRAM，微控制器　Micro Controller，邏輯性產品　logic product，半導體晶圓　wafer，汽車電視顯示器模組 car TV module，影音系統之半導體零組件　audio and video system IC，光碟系統之半導體零組件　optical signal processing IC，顯示器系統之半導體零組件　scanner and display IC，電源管理系統之半導體零組件　power management IC，通訊系統之半導體零組件　communication IC，多通道語音傳真處理器　multi-channel Voice/FAX Processor

35. 電腦及周邊產業　Computer & Peripherals

電漿顯示器　PDP display，液晶顯示器　LCD display，影像掃描器　image scanner，數位攝影機　digital camcorder，液晶顯示螢幕　LCD monitor，磁碟機　disk drive，印表機　printer，電腦終端機　terminal，全球衛星定位系統　(GPS) global positioning system，地理資訊系統　geographic information system，各種語言自動翻譯系統　multilingual machine translation systems，個人資訊管理系統　personal information management system，數位影像照相機　digital still camera，有／無線影音電話　wire and wireless video phone，多媒體視訊傳輸器　media transiver，指紋監控系統　fingerprint identification control System，網路認證及資訊安全處理設備　internet certification and information security treatment System，電子錢包收銀機　electronic purse terminal，IC 卡讀卡機　smart card reader，磁卡讀卡機　magnetic card reader，電子商務整合系統　e-commerce integration system，全球運籌管理系統　global e-logistics，企業資源規劃系統(ERP) enterprise resources planning，電腦整合製造 (CIM) computer integrated manufacturing，機台連線自動化整合系統　(EAP) equipment qpplication program，伺服器及儲存設備　servers and storage system

36. 通訊產業　Telecommunication

衛星電視接收設備　satellite TV Receiving equipment，數位微波傳輸設備　digital microwave RF equipment，無線個人行動通訊設備　wireless communication equipment，無線區域網路設備　wireless LAN，GPS 接收機　GPS receiver，GPS 導航設備 GPS navigator，PHS/WiFi 雙網手機　PHS/WiFi double-network cellphone，寬頻無線通訊設備　broadband wireless communication equipment，光纖通訊系統　fiber optical communication System，數位數據機　multi digital data unit，路由器　router，

寬頻視訊電話　video phone，智慧型手機　smart phone, iPhone, iPad

37.光電產品　Electro-Optical

光纖網路系統　fiber-optic data collection network (FIBNET)，太陽能電池　solar cell，二極體　LED，晶粒　chip，液晶面板　LCD panel，數位影像系統　digital imaging system，平面顯示器　flat panel display，光濾波器　optical filter，光放大器　optical amplifier，可充電式鎳氫電池　rechargeable NI-MH batteries，雜訊濾波器　noise suppression filter，電感　inductor，高密度扼流線圈　choke，溫度感測器　temperature sensor，高密度晶片排組　high density resistor array

38.精密機械產業　Automation

純水系統　pure water system，廢水回收處理　waste water treatment，血液血糖分析儀　blood glucose self-monitoring system，物流配送自動化整合系統　warehouse, distribution center material handling automation system，晶圓篩選機　wafer sorter，運送晶圓機器人　wafer transfer robot

39.生物技術產業　Biotechnology

生物晶片　BIOCHIP，人工水晶體　INTROCULAR lens，關節潤滑液　intra-articular injection，防組織沾黏產品　anti-adhesion products，基因體晶片　DNA microarray，奈米電子技術　NANO electronics tech，生物性農藥　microbial BIOPESTICIDE，生物肥料　microbial fertilizer，飼料添加劑　animal feed additive

注意：一般買賣產品的基本類別大致分為：

(1)原料(Raw Material)：例如金屬、原木、礦產、化學原料等。

(2)零件(Components)：例如電子／電腦零件、各種車輛零件等。

(3)半成品(Semi-Products)：例如電腦周邊設備、機器零組件、加工品等。

(4)成品(Finished Products)：例如吃／穿／用的所有物品。

(5)整廠輸出及技術轉移(Turn-key plant project & Know-how)。

附件二 公司簡介樣本(Company Profile)

Company Profile

Samwell International Inc. established in 1975

Address: 317-2, Sec. 2, An Kang Rd., Hsinten Taipei, Taiwan

E-mail address: info@samwellg.com.tw

Capital: US$10,000,000

Chairman: Sam S. Lee

Factories: Samtech Corporation established in 1984

　　　　　Sambest Corporation established in 1988

Factory space: 68,000 Square Feet

Main Products: CMS (Components Manufacturing & Supply)

　　　　　　　EMS (Electronics Manufacturing & Service)

　　　　　　　PSS (Product & System Integration Service)

Quality System: ISO9002, QS9000, ISO9001, ISO/TS 16949:2002

Awards: 1996: 25 years outstanding service

　　　　1999: In recognition of many years of consistently outstanding service

　　　　2004: In appreciation of the outstanding contribution of Samwell

　　　　　　　International Inc. to HP Taiwan Ltd.

Revenues: 2009 : 25 million US$　　2010: 28 million US$

　　　　　2011: 32 million US$　　2012: 35 million US$ (Target)

Principal Customers: MSI, Intecom, T.I., Motorola, HP, etc.

Organization:

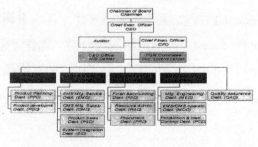

For more information, please visit our website: www.Samwellg.com

附件三 《貿協國際商情雙周刊》「貿易機會專刊」樣本
(International Trade Biweekly)

 TAITRA Database

本欄貿易機會收錄了外商來函、經濟部及貿協100餘個駐外單位提供之貿機，
依產品別，羅列了擬採購及推銷之產品項目，範圍涵蓋43個產業，
為廠商拓展國際市場的捷徑。

外商擬向我採購貨品

本欄貿易機會如有錯誤訊息，
請傳真 Fax：02-27576045 或 E-Mail：inform@taitra.org.tw
經確認後，本刊將贈七折貿協書廊購書券一張。

 農產品及食品

001 Dried Ume (Plum)
Mr. Mon
Morita Shokuzai Kaihatsu
Kenkyusho
1-2-3, Imafukuminami, Joto-Ku,
Osaka, 536-0006, Japan
Tel:81-6-69331450
Fax:81-6-69311941
E-Mail:mon@ajizukuri.co.jp
東京台貿中心提供

002 Peaches, Vegetable Oils And Fats, Soya Bean Oil Cake And Meal, Tea, Vinegar, Trees, Plants, Stocks And Cuttings
Mr. Moe Irshad
Metro International Trading Company
P.O. Box 71 1395 Lawrence Ave.
W. North York, Ontario M6L 1A7,
Canada
Tel:1-416-6521117
Fax:1-416-6564910
E-Mail:mohdirshad@rogers.com
駐加拿大經濟組提供

003 Frozen Seafood (Any Type), Commercial Fishing Articles And Related Products
Mr. Fermin Hernandez
Ind. Pesquera Y De Servicios
Marlin Azul
Lote No.7 Cll. Central,
Buena Vista,
Puerto Iztapa,
Escuintla,
Guatemala
Tel:502-59775836
Fax:502-59775836
E-Mail:
pesqueramarlin@yahoo.com.mx
汕埠台貿中心提供

004 Mineral Water
Odinachi Kosim
12 Rue D' Thompson B.P.4654
Lome, Togo
Tel:228-9152032
Fax:228-9152032
E-Mail:kosmac2@yahoo.com
行銷處專案服務組提供

 化學原料及製品

005 Dimethyl Form Amide, Purity With 99.85%
Mr. Srinivas K.S.
Pioneer Chemical Inds. Pvt. Ltd.,
119, B-Wing, Gokul Arcade
Premises, Garware Chowk,
Vileparle (East), India
Tel:91-22-55518729
Fax:91-22-28203907
E-Mail:pioneersri@gmail.com
網路中心採購組提供

006 Diminezene 2.36 Sachets
Ms. Sk Mukerji
Skm Pharma Pvt Ltd.
Tf-10 City Point, Infantry Road,
Bangalore 560001, India
Tel:91-80-22863448
Fax:91-80-22860476
E-Mail:info@skmpharma.com
網路中心採購組提供

007 Linear Alkyl Benzene, Looking For The Proper Supplier Who Is Able To Offer For 1000mt Of Lab On The Basis Of CFR Korea Or FOB Available Port
Mr. Henry Yeo
Lanto K Co., Ltd.
13th Floor Samchang Plaza, 173
Dowha-Dong,
Mapo-Ku, Seoul,
Korea
Tel:82-2-7039180
Fax:82-2-7039184
E-Mail:hiyeo@lantok.co.kr
網路中心採購組提供

▼智利商 MEDOVIC Y CIA LTDA 經本刊再次聯繫後擬採購下列產品：

產品說明		
General Description: (including product application & specifications)	1.PLASTIC RESINS (HDPE, LLPE, PP, PVC EMULSION, OTHERS) This Raw Material is for use in producing films, tubes, others. 2.BOPP FILM for flexible packaging, clear. metaiized, pearlescent, etc. 3.HDPE FILM for Inside Liner in Paper Sacks	◎ 公司基本資料： Company Name:MEDOVIC Y CIA LTDA Year of Establishment : 1935 Address: Bombero Nunez 181, CP 6622954, Santiago, Chile Business Type: SALES AGENT Website : www.medovic.cl Tel: 56-2-7771550 Fax: 56-2-7378948 Email: gerencia@medovic.cl Contact Person/Position: Zarko Medovic
Product Features/ Functions:	Resins = Raw materials for producing plastic products BOPP Film=for flexible packaging laminations	
Additional Info:	All supplied in Containers of 20 or 40Ft.	
Target Cost:	International competitive levels	

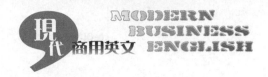

附件四　貿易來源樣本(Trade Sources)

From: ienquiry@tradesources.com

To: srcosmos@ms67.hinet.net

Sent: Friday Feb. 24, 2012 5:10PM

Subject: TRADE SOURCES PRODUCT ENQUIRY SERVICE

Dear Sir/Madam,

The following buyers are looking for the respective products, which are compatible with your business. Please kindly contact them for details:

(If you cannot view the information, please try to change it to Chinese.)

Product Enquired: Toy animal, plastic/rubber

Company Name: Divertitoys Mexico

Address: local d-11, Plaza Patria

City: Zapopan, Jal

Country: Mexico

Contact: Hector Echegollen

Title: Manager

Tel: 52-53-3641-5932

Fax: 52-53-3610-0395

E-mail: echegollen77@hotmail.com

Wish you a successful business!

Best Regards,

Trade Sources Group Ltd.

http://www.tradesources.com

範例 1　美國進口商透過美國進口商協會台北辦事處尋找供應商

Dear Sirs,

North American Importers Association, 1987, New York (NAIA) is a member importer association. With more than 50,000+ importers as members from North America and worldwide, our mandates are to source international quality suppliers to serve their import needs and offer multi-language exposure marketing, to promote the cooperation of both members and suppliers.

 If you are willing to be our preferred supplier in Taiwan , and be referred to our importers, please kindly reply this e-mail with following information :

1. Company Name (English Name) :

2. Website:

3. CEO (Name Ms. Or Mr.) :

4. Phone No :

5. E-mail :

6. Where is your production base?

7. Do you accept OEM / ODM business?

8. Any certificates? (such as ISO、CE、UL...) :

9. How many years in the business?

10. What's your major line?

11. Do you have agent in USA or anywhere in the world?

12. Please provide 1-3 pictures of your main products for display in our website.

Please get back to us as soon as possible.

Best Regards,

North American Importers Association

www.usacan.org

文章結構

主旨：北美進口商協會(NAIA)是一個由北美及全球超過 50,000 位進口商會員於 1987
年在美國紐約成立之組織。協會最主要任務是找尋全球優良供應商及產品，
提供給全球進口商的會員們參考及採購或合作。並提供多國語系市場曝光，
促成進口商與供應商之合作關係。

說明：如果您對於成為本協會之產品供應商有興趣，以及期望與本會所擁有之進口
商會員建立合作聯繫關係，請回覆下列相關資料：

1. 公司英文名稱：

2 公司網站：

3. 負責人：

4. 聯絡電話：

5. 電子信箱：

6. 產品製造地／國？

7. 是否接受貼牌代工？

8. 是否有相關認證？（例如 ISO、CE、UL...）：

9. 有多少年的生產經驗？

10. 主要產品種類？

11. 在美國或其他國家是否有代理商？

12. 請提供 1～3 張主要產品圖片以便在我方官網展示。

結論：請盡快和我方聯絡。

生字／片語

1. mandates　使命／任務
2. to source international quality suppliers　找尋全球優良供應商
3. offer multi-language exposure marketing　提供多國語系市場曝光
4. OEM　原廠委託製造；ODM　原廠委託設計（＝貼牌代工廠）
5. certificates　認證：ISO　國際標準組織；CE　歐規；UL　美規
6. major line　主要項目／主要產品
7. agent　代理

範例 2　液晶顯示器(LCD)製造商在其網站上刊登的行銷廣告文

LCD & Touch Panel Integration　ISO9000(EN29000) & ISO 13485:2003since 2009

Our LCD products are widely applied in a broad range of applications and markets.

With advancing technologies in digital, dot matrix and graphic LCD panels and modules, we are always committed to provide LCD products with the highest level of contrast, wide viewing angles, long life, and high reliability.

With 35 years' engineering design experience, we can help to avoid potential issue during integrating LCD and touch screen assembly. Our LCD and touch screen are assembled in our in house clean room to ensure perfect performance.

For more information, please contact us.　(E-mail: sales@samwellg.com)

文章結構

主旨：具國際標準認證的液晶顯示器和觸控面板整合。

說明：本公司的液晶顯示器產品被廣泛應用於各種產品及市場上。具有最先進的數字、點陣及繪圖顯示板和模組技術，本公司保證提供具有高對比、寬視野角度、長壽及信賴度高的 LCD 產品。

具有 35 年工程設計經驗，我們可幫助避免整合 LCD 和觸控面板組裝時可能產生的潛在問題。在本廠自己無菌廠房內組裝，可擔保完美性能。

結論：請和我方聯絡索取詳細資料。

生字／片語

1. LCD & Touch Panel Integration　液晶顯示器和觸控面板整合
2. advancing technologies　先進技術，digital　數字，dot matrix　點陣，graphic　繪圖，modules　模組，the highest level of contrast　高對比，wide viewing angles　寬視野角度，long life　長壽，high reliability　高信賴度
3. potential issue　潛在問題，in house　廠內，perfect performance　完美性能

範例 3 推銷多媒體喇叭／數位影音等科技產品 (Multiply speaker/MP3/DV)

From: "hiteck" <hitech@msa.hinet.net>
To: "computer" <computer.imp@earthlink.net>
Subject: High-Tech Products
Attn: Import Manager
Dear Sirs,

We are a professional manufacturer in producing multiply speakers, digital audio & video items, MP3, MP4, DV(multi-function digital video), 3G Cell Phone, Hi-Resolution LCD Monitor, Skype and more high-tech items.

This year we specially developed many new stylish items and sincerely invite you to visit our booth no. 123, Hall 1 at Computex Taipei. We welcome you to have a seat and taste our special coffee at the same time and expect your valuable comments to our items. For more information, please visit our website: www.hitech. com.

Best regards
Cindy Lee/Hi-Tech

文章結構

主旨：本公司為一專業高科技產品製造工廠生產多媒體喇叭、數位音響及影視產品、音樂播放機、多功能數位相機、三頻音樂攜帶電話、高解析度液晶螢幕，電腦空中交談電話機及許多高科技產品。

說明：今年本公司特別發展許多新型產品，竭誠邀請您到我們在台北電腦展第一館的攤位 123 來參觀。

結論：歡迎您對本公司的產品提出寶貴意見，並請到我們攤位坐坐，喝杯特製咖啡。如需進一步資料，請上我們網站:www.hitech.com 參觀。

生字／片語

1. professional manufacturer　專業的製造工廠
2. multiply speakers　多媒體喇叭，digital audio & video items　數位音響及影視產品，MP3/MP4　音樂播放機，DV(multi-function digital video)多功能數位相機，3G Cell Phone　三頻音樂攜帶電話，Hi-Resolution LCD Monitor　高解析度液晶螢幕，Skype　電腦空中交談電話機
3. Computex Taipei　台北電腦展；booth　攤位；valuable comments　寶貴的意見

範例 4　推銷「導電矽膠片」（Silicon Rubber Pad）

From: Mary<merrybest.mary@msa.hinet.net>

To: ASE Turkey

Subject: Silicon Rubber Pads

Dear Sirs,

We are glad to know from CPU magazine that you are interested in our Silicon Rubber Pads and attach herewith our relevant catalogs for your initial reference.

Since 1975, we have been working in this field and have been cooperating with many big OEMs such as Texas Instruments, Motorola, Hitachi, etc.

Moreover, it was our great honor to have won a "Supplier Excellence Award" from Texas Instruments and an "Appreciating Award" from Bontempi, Italy recognizing our company as a trustworthy supplier. All of these prove that we not only provide the most competitive prices, best quality and performance, but also have the ability to help the customers solve all of their problems happened during the design stage to the finished products.

Since we have expanded our factory space and installed more up-dated fully automatic injection machines, we have confidence to meet your every requirement under our advantageous position.

Please do take a look on the enclosed catalog and do not hesitate to let us know of your comments or any inquiries for either your existing products or new products.

Many thanks for your attention to the above and look forward to having the pleasure to serve you very soon.

Very truly yours

Merrybest Int'l Co.

Mary Huang/V. President

文章結構

主題：在第一段告知從廣告媒體引介，得知貴方有興趣本公司導電矽膠片的產品，並附上相關目錄供初步參考。

說明：(1)在第二段自我介紹，介紹本公司在此業界已有多年經驗並舉例和許多大廠一直有良好的合作關係。

(2)在第三段舉例本公司曾獲得國外大買主選為「優良供應商獎」及「感謝獎」，以此證明本公司可提供最佳價錢、品質、服務及各項表現，並協助客人解決其從設計到成品過程中遇到的所有問題。

(3)在第四段告知正在擴大產量及使用最新的機器設備，有信心滿足客人需求。

結論：最後兩段請客戶參照所附目錄，表示意見或告知其任何現有產品或新產品的詢價，最後感謝其注意此信函，並盼望有榮幸為其服務。

生字／片語

1. relevant ＝ relative/related　相關的
2. for your initial reference　供貴方初步參考
3. OEM (Original Equipment Manufacturer)　原始設備廠／原始委託製造
4. Supplier Excellence Award　優良供應商獎
5. Appreciating Award　感謝獎
6. recognizing (recognize)　認可
7. trustworthy supplier　值得信賴的供應商
8. the most competitive prices ＝ the best prices　最有競爭性的價錢
9. performance　表現／性能
10. design stage　設計階段
11. finished products　成品
12. expanding　擴大
13. factory space ＝ production capacity　工廠產能
14. up-dated fully automatic injection machines　最新型全自動射出機
15. have confidence ＝ are confident　有信心
16. requirements　（規格／條件）要求
17. advantageous position　有利的立場
18. take a look　看看
19. do not hesitate　不要遲疑
20. comments　意見
21. inquiries (inquiry)　詢價
22. existing products　現有產品
23. look forward to　盼望

範例 5　主動寫信給已知進口商推銷閥類(Valves)產品

To: Union Engineering Co., Ltd

Dear sirs,

This is King-Tech Valve Precision Industry Co.

From a friend, we are glad to know that you are looking for a reliable valve supplier to cooperate with your company.

Since 1999, we have been working in this field and specialize in manufacturing and supplying various kinds of valves as the image attached. Please take a look on our website: www.kingtech-valve.com to see if any item interests you.

We are looking forward to your inquiry or any comment to the above and hope to have the opportunity to serve you soon.

Best regards

Mary Huang/ VP of Export Dept.

文章結構

主旨：這是寶閥精密工業公司，從朋友處得知貴公司正在尋找一家可靠的閥類供應
　　　商來和你們合作。

說明：自從 1999 年以來，本公司一直從事於此行業並專門於生產及供應如所附圖檔
　　　的各種閥類產品。請上我方網站 www.kingtech-valve.com 看看是否有任何項目
　　　令貴公司感興趣。

結論：我們期盼貴公司的詢價或對以上的任何意見，並希望有機會為你們服務。

生字 / 片語

1. reliable valve supplier　可靠的閥類產品供應商
2. to cooperate with your company　和貴公司合作
3. have been working in this field　一直從事於此行業
4. specialize in manufacturing and supplying　專門於生產及供應
5. various kinds of　各種；image　圖檔
6. inquiry　詢價；comments　意見
7. have the opportunity to serve you　有機會為你們服務

範例6　推銷「小五金──螺絲釘／螺帽／螺栓」(Fasterner-Screws/Nuts/Bolts)

WAN FASTENER MFG
E-mail address: wanfastener@evl.net

Long Wood Trade Co. Ltd.   Our Ref.: FS-212
369 Cherry St. London
England

Attn.: Import Manager

Subject: Screws, Bolts and Steel

Dear Sirs,

From Traders' Express, we got your name and learned that you are one of the leading importers of the subject products, which are the main export items of our company during the past fifteen years.

The attachment is our latest catalog showing you the items and detailed specifications we are available to supply. Please just indicate the items which interest you, we will immediately send you our best offer and free samples for your further evaluation.

Thank you for your attention to the above and look forward to setting up the business relationship with you very soon.

Yours faithfully
Wan Fastener Mfg.
Simon Wu
Export Manager

文章結構

主題：第一段告知由 Traders' Express 雜誌得知客戶大名及其為主要小五金產品的主要進口商之一，而本工廠外銷此類產品已有 15 年經驗。

說明：第二段中告知附上最新目錄，其標示可供應的項目及詳細規格，請客戶指出感興趣的規格，將會馬上寄出最好的報價及免費樣品供進一步評估（因小五金規格尺寸瑣碎繁多，要有客人特定規格的詢價，才能提供報價及正確樣品）。

結論：感謝客戶注意以上資料，並期盼早日和其建立商業關係。

生字／片語

1. Screws/Nuts/Bolts/Steel　螺絲釘／螺帽／螺栓／鋼鐵

2. one of the leading importers　主要的進口商之一
   （注意：one of ＋複數名詞）

3. Attached please find our latest catalog.　如附請見本公司最新的目錄
   ＝ Please find our latest catalog as attached.
   ＝(The) attached is our current catalog.　如附是本公司的最新目錄
   ＝ We enclose/attach our newest catalog.　我們附上本公司最新目錄
   ＝ Our newest catalog is enclosed/attached.　附上本公司最新目錄
   （以上四種寫法都可變換應用）

4. showing you the items and detailed specifications（當 Adj. Ph.）
   ＝ in which we showed you the items and detailed specifications.（當 Adj. Cl.）
   在目錄中顯示所有項目及詳細規格

5. we are available to supply（當 Adj. Cl.）我方可供應的

6. indicate ＝ specify ＝ point out　指出

7. further evaluation　進一步的評估

8. look forward to ＝ expect　期盼（注意：look forward to 的 to 是介系詞，後面要跟 N 或 V-ing）

9. set up (setting up)＝ establish ＝ build　建立

10. business relationship ＝ business relations ＝ business connection　生意關係

範例 7　英國經銷商推銷電子零件 L.E.D.

From: alex@fortune.com

To: minray@ms33.hinet.net

Subject: The supply of L.E.D.

F.A.O.: Mr. James Lin/Min-Ray Enterprises Ltd.

Dear James,

　　Our company is based in the U.K. and is a distributor of electronic components throughout the world. About two weeks ago, we were advised by your Chamber of Commerce that you were looking for some L.E.D's Part No. HLMP6500 on which we quoted an alternative to, and were found to be very competitive.

　　We were also advised that the purchase order for a quantity of 45 Kpcs of L.E.D.'s would be placed by your company. It is with this transaction in mind that we would like to contact you.

　　I would very much like for you to contact us on the above matter to discuss further.

Best regards

Alex English

文章結構

> 主題：第一段自我介紹並告知寫此信的目的
> 　　　本公司基地在英國，經銷各種電子零件到世界各地，兩週前，由貴國商會告知貴公司正在尋找電子零件 L.E.D. P/No. HLMP6500，本公司可提供同等替代品，價錢頗具競爭性。
> 説明：本公司亦被告知貴公司欲下訂單 45,000 個，本公司即因此筆交易希望和貴公司聯絡。
> 結論：本公司極希望貴公司來函和我方進一步討論上述零件事件。

生字／片語

1. F.A.O.＝ for the attention of　收件人
2. Our company is based in U.K.　本公司基地（大本營）在英國
3. distributor　經銷商
4. Electronic Components　電子零件
5. throughout the world ＝ in the world ＝ worldwide　遍布全世界
6. alternative　同等替代品
7. purchase order　訂購單
8. 45K pcs ＝ 45M pcs ＝ 45,000 pcs（注意：商業上，K 或 M ＝ 1000）
9. place (the order)＝ issue (the order)　下訂單的下可用 place 或 issue
10. transaction　交易
11. to discuss further　作進一步討論
12. Best regards　致上最大敬意（多用在 e-mail 的結尾語）
13. Chamber of Commerce　商會

範例 8　SPA 設備工廠寫信給貿易中心，請求推介當地進口商名單

From:　"Sunlight"　<sunlight.spa @ msa.hinet.net>

To: Trade Center USA

Subject: Recommendation for the Importers of Ultrasonic Bubble Massage Bath

Dear Sirs,

At the Foreign Trade Development Conference, we got to know that your Trade Center helps the commercial and industrial circles get connections with one another to start their business deals. So, we would highly appreciate to get the same help from you, too.

As the leading manufacturer of Ultrasonic Bubble Massage Bath, we just developed a new product with high-tech, low-noise processing unit with trouble-free assembly. We have got CSA, UL and CE approval. It is a kind of Home SPA for total body massage-relaxes and relieves stress/fatigue. After taking a SPA in the bathtub at home with various kinds of herbal bath oil like lavender, chamomile, rosemary, juniper, pine needle, eucalyptus and hayseed, it is easy to charge the energy of the body and improve the health.

Would you be kind enough to introduce us the importers or the companies who are interested in importing this equipment and send us the name list accordingly, or pass our name and e-mail address to those companies who can contact us directly.

Thank you very much for your great assistance in advance and may we hear from you soon.

Best regards

Mary Huang/Export Manager

MODERN
BUSINESS
ENGLISH

Part 4 （一、推銷函）
商用英文信函寫法

153

文章結構

主旨：本公司由外貿協會得知，貴貿易中心協助各工商界人士相互聯絡，做成生意，因此本公司將非常感激能獲得相同的協助。

說明：本公司為氣泡超音波按摩浴設備的主要製造廠，最近剛發展出一項高科技、低噪音及簡易安裝的新產品。我們已獲得美加品檢認證及歐規電檢認證。這是一種家庭式按摩浴設備，提供全身按摩——可鬆弛並解除壓力及疲勞。當在家中浴缸泡澡時加入各種不同精油像是薰衣草、甘菊、迷迭香、杜松、赤松針、尤加利及茂草花等，洗過按摩浴後，將很容易恢復體能並改進健康。

結論：請介紹有興趣進口此設備的進口商或公司，並將名單寄給我們。或將本公司名字及電子郵址轉給那些可直接和我方聯絡的公司。在此先感謝大力協助並盼盡快獲得回音。

生字 / 片語

1. Recommendation　推薦
2. Foreign Trade Development Conference　外貿協會
3. Trade Center　貿易中心
4. Ultrasonic Bubble Massage Bath　氣泡超音波按摩浴設備
5. commercial and industrial circles　工商業界
6. get connections with one another　相互聯絡
7. business deal　商業交易；importers 進口商；name list 名單
8. high-tech, low noise processing unit　高科技、低噪音產品
9. trouble-free assembly　簡易安裝
10. CSA, UL and CE approval　美加品檢認證及歐規電檢認證
11. body massage　身體按摩
12. relaxes and relieves stress/fatigue　鬆弛並解除壓力及疲勞
13. herbal bath oil　精油：lavender　薰衣草、chamomile　甘菊、rosemary　迷迭香、juniper　杜松、pine needle　赤松針、eucalyptus　尤加利、hayseed　茂草花
14. charge　充電　energy of the body　體能
15. great assistance (＝ help)　大力協助；in advance　事先

範例 9　推銷高級手錶和珠寶類產品(Watches and Jewelry)

From: Geneva <Harvey.Phen@aur-hiph.de>

To: eybvzk.iegqhc@msa.hinet.net

Subject: High Quality Watches and Jewelry

High quality products indulge your dignity with an elegant time piece or jewelry that is meticulous in design, exquisite in style and rich in beauty. All famous and most popular brands such as Tiffany, Rolex, Prada, Cartier, Omega, Chanel, plus luxurious items are available. The excellent workmanship and performance are assured. Custom-made designs are also welcome. Click here and grab them.

文章結構

主題：最精緻設計、優美款式及華麗美感的高貴鐘錶或珠寶等高級產品可滿足您的身分。

說明：所有著名及流行品牌，例如 Tiffany, Rolex（勞力士）, Prada, Cartier, Omega, Chanel（香奈兒）加上奢華產品都可供應。保證最佳手工及性能。

結論：歡迎客製定做設計品。請按此處並抓取您想要的。

生字／片語

1. High quality products　高級產品
2. indulge your dignity　滿足您的身分
3. elegant time piece or jewelry　高貴的鐘錶或珠寶
4. meticulous in design　精緻設計；exquisite in style　優美款式
5. rich in beauty　華麗美感；luxurious items　奢華產品
6. Tiffany, Rolex（勞力士）, Prada, Cartier, Omega, Chanel（香奈兒）（品牌名稱）
7. excellent workmanship and performance　最佳手工及性能
8. Custom-made designs　客製定做設計品
9. Click here and grab them　請按此處並抓取您想要的

範例 10　推銷新款防振防水手機(Anti-Shock and Water Proof Cell Phone)

From: top.phone@aur-amter.de

To: allmembers@msa.hinet.net

Subect: Anti-Shock and Water Proof Cell Phone VG-01

We recently developed an anti-shock and water proof cell phone VG-01. Users will be able to control Skype just by using the keys on the phone. It also has echo eliminating technology to improve the voice quality. It also supports MSN Messenger and Yahoo Messenger. Other features include the followings: MP3, Bluetooth link technology, VGA Camera with video capability, speakerphone and stereo speakers, GPS software, high resolution and multimedia, etc. The VG-01 is set for release in the near future. www.topphone.com

文章結構

主題：本公司最近發展一隻防震防水的手機 VG-01。

說明：用戶可使用電話按鍵控制 Skype（空中談話器），此手機具有回音消除技術可改進聲音品質，它亦可支援 MSN Messenger 及 Yahoo! Messenger（網路即時通訊軟體），其他特色包括 MP3（音樂播放）、藍芽連接技術、視頻圖像陳列含錄影功能照相機、喇叭擴音器及立體音響喇叭、衛星導航軟體、高解析畫面及多媒體等。

結論：VG-01 即將上市。網站：www.topphone.com

生字／片語

1. anti-shock and water proof cell phone　防振防水的手機
2. Skype　空中談話器；echo eliminating technology　回音消除技術
3. improve the voice quality　改進聲音品質
4. MSN (Microsoft Network) Messenger　網路即時通訊軟體
5. features　特色；MP3 (music player)　音樂播放
6. Bluetooth link technology　藍芽連接技術；VGA (Video Graphics Array) Camera with video capability　視頻圖像陳列含錄影功能照相機
7. speakerphone　喇叭擴音器；stereo speakers　音響喇叭
8. GPS (Global Positioning System) software　衛星導航軟體
9. high resolution and Multimedia　高解析畫面及多媒體

範例 11　推銷「汽車零件」(Auto Parts)

Subject: Auto Parts

Attn: Import Manager

Dear Sirs,

As you know, due to the worldwide economic depression, most of the buyers are looking for the qualified Auto Parts at the lowest possible prices.

Since we have just upgraded our machinery, our fully automatic production process can certainly reduce the labor cost and avoid the man-made mistake for the mass production. So, we can assure you the best quality Auto Parts at the most competitive prices, for which will make you can compete very well in the market when you sell our Parts in The States. Please refer our catalog and price list attached for all details.

We are also glad to inform you that we will be exhibiting at WESCON in San Francisco, Oct. 20-25. Our booth numbers are 546/548. You are welcome to come over to our booth to take a look, so that we can prove what we said to you. Moreover, we may be able to have the further discussion for our mutual profitable business there, too.

Thank you very much for your attention to the above and look forward to seeing you soon.

Very truly yours,
Mobile Auto Mfg.
Miki Yamaha/Sales Manager

文章結構

主題：第一段提供客戶市場資訊

　　　如您所知，由於世界性經濟不景氣，大部分的買主都在找尋物美價廉的汽車零件。

說明：(1)第二段自我介紹能力

　　　由於我們剛升級本廠的機器設備，全自動化生產製程可於大量生產時，自然降低人工成本，避免人為疏忽，所以我們可保證提供貴公司物美價廉的汽車零件，並使貴公司在美銷售我方零件時，極具競爭能力，細節請參照如附的目錄及價目表。

　　　(2)第三段歡迎客人親臨展覽會場面洽

　　　我們很高興通知您本公司將參加 10/20-25 在舊金山舉行的 WESCON 展覽，攤位號碼是 546/548，歡迎蒞臨參觀，以證明我們上述所說，此外亦可進一步當面討論雙方共同有利的生意合作。

結論：感謝客戶對以上的注意，並期盼很快和其見面。

生字／片語

1. the worldwide economic depression　世界性經濟不景氣

2. the qualified Auto Parts at the lowest possible prices　物美價廉的汽車零件

3. we have just upgraded our machinery　我們剛升級我方機器設備

4. fully automatic production process　全自動化生產製程

5. reduce the labor cost　降低人工成本

6. avoid the man-made mistake　避免人為疏忽

7. mass production　大量生產

8. compete (V)　競爭；competitive (Adj.)　有競爭的

9. exhibiting (exhibit)(V)　展出；exhibition/show/fair (N)　展覽

10. San Francisco　舊金山

11. booth ＝ stand　攤位

12. moreover ＝ in addition ＝ besides ＝ furthermore　此外

13. further discussion　進一步的討論

14. mutual profitable business　雙方有利的生意

練習 12　推銷函練習

1. 將下列的主題句／説明句／結論句翻譯成英文

主題句(Main Idea)

(1)經由貴國商會(Chamber of Commerce)的推薦，本公司想知道貴公司是否有興趣進口汽車零件(Auto Parts)。

(2)本公司經由 CPU 雜誌得知貴公司大名，並希望藉此機會和貴公司建立商業關係(business relationship)。

(3)如您所知，由於世界經濟不景氣(worldwide economic depression)及石油危機(oil crisis)，大部分買主都正在尋找物美價廉的產品。

説明句(Exploration)

(4)自從 1975 年以來，我們一直從事於電腦產品業界。

(5)經由多年研究發展(R&D)，本公司已成為消費性產品(consumer products)主要的出口商及製造商。

(6)具有耐用(durability)、精準(precision)及成本效益(cost effectiveness)等特色，我們相信貴公司可藉由在貴市場推銷我方產品而獲利。

(7)由於價錢合理、交貨準時及優良品質，本公司已在此行業贏得領先地位。

(8)本公司附上最新的目錄(catalogue)、報價單(quotation)及公司簡介(company profile)供貴公司參考。

結論句(Conclusion)

(9)我們盼望盡快收到／聽到貴公司回音。

(10)感謝您對以上的注意，並希望很快和貴公司建立生意關係。

(11)請指出貴公司有興趣的產品，我們將馬上提供相關報價及樣品供評估。

(12)我們期盼盡快和貴公司開始合作。

2. 請依中文劃線部分填入適當英文

主題：我們（是）很高興從漢諾威展覽上得知貴公司是德國主要的電腦產品的進口
商，並希望利用這個機會介紹給貴公司，我們今年剛發展出來的一個高科技
的產品。

說明：具有多年研究發展的經驗及許多專業的工程師，我們能夠不斷地提供客戶最
新的產品，並為客戶創造最好的商機。
如附是我們的相關的目錄及報價單供參考。

結論：感謝您的興趣並希望很快和你們建立商業關係。

a. computer products	g. Hannover Fair
b. high technical product	h. take this opportunity
c. developed	i. establish business relationship
d. constantly	j. professional engineers
e. create	k. R&D
f. business opportunity	l. relevant

We are glad to know from the _____(1)_____ that you are the leading importer of
_____(2)_____ in Germany, and would like to _____(3)_____ to introduce
you a _____(4)_____ which was just _____(5)_____ by us this year.
With many years' experience in _____(6)_____ and many _____(7)_____ ,
we can _____(8)_____ offer the newest products and _____(9)_____ the
best _____(10)_____ for the customers. Enclosed are our _____(11)_____
catalog and quotation for reference.

Thank you for your interest and hope to _____(12)_____ with you soon.

填入答案：

(1)	(2)	(3)	(4)	(5)	(6)
(7)	(8)	(9)	(10)	(11)	(12)

3. 請自由發揮寫一封完整的鞋子（shoes）的推銷函
註明工廠在大陸，可提供物美價廉的產品。

4. 將下列的句子翻譯成英文

(1)作為一家經驗豐富的出口商，本公司自 1990 年以來一直從事電腦產品行業，並和本地廠商關係良好。

(2)本公司為最大的玩具供應商之一，我們有信心令客戶滿意。

(3)本公司是聲譽良好的主要電子產品廠商，由於品質優良，本公司產品廣受歡迎。

(4)我們每月產能約 100 萬個，平均每月銷售額約美金 1,000 萬元。

(5)請指出有興趣的規格或編號，以便於我們報最好的價錢給貴公司。

(6)本公司產品占有 60%世界市場。

(7)這是客戶定製的產品。

(8)這是一張試銷訂單。

(9)這是我們最有競爭性的價錢。

二、詢問函(Inquiry Letters)

1. 寫詢問函的目的

(1)進口商主動詢問某種特定產品的報價／索取產品的資料（目錄／規格／樣品）。

(2)接到賣方推銷函後，感到興趣而去函詢價或索取資料。

(3)某特定產品的特殊規格探詢廠商是否可供應。

(4)詢問維修零件的價錢、交期。

(5)詢問特定產品報價及條件同時，亦諮詢當代理的可能性。

2. 尋找國外廠商的方法

(1)到外貿協會（世貿中心 2 樓）查閱各國廠商名錄或出口商名錄。

(2)向各國在當地的辦事處查詢或請其推薦（無邦交國家）。

(3)向各國駐當地的大使館／領事館請求協助或推薦（有邦交國家）。

(4)向工會／商會索取廠商名錄。

(5)去函各國貿協，要求免費刊登需求廣告或推薦廠商。

(6)上電腦網路圖書站查閱。

(7)從各相關產品的廣告雜誌查閱。

(8)訂閱貿協的「貿易快訊」。

(9)從國內外各相關產品的展覽上蒐集廠商資料。

3. 詢問函寫法

　　寫詢問函者的身分都是買主(buyer)，可能是進口商、代理商或國內製造商，如買方是個知名度高的公司或大買主，一般詢問函就很簡單扼要，直入重點，只用兩三行文句，直接告知對方想要詢問索取什麼資料；但是如果賣方(seller)是個大廠或知名度高的廠商，或特殊產品的供應商，那麼詢問函中最好加上自我介紹及需求量，以獲取較好的報價及支持。

詢問函布局的方式如下：

主題(Main Idea)

(1)直接告訴對方正在尋找什麼產品或要索取什麼資料（目錄／報價／規格／樣品等）
（參範例 3/5/6/9）

(2)自我介紹公司的規模、銷售能力或需求量（參範例 1/7）

(3)經由何種管道（貿協／商會／工會／雜誌／朋友介紹等）得知其大名（參範例
2/10/11）

說明(Examples/Explanation)

(1)告知索取上述資料的原因／目的或用途（參範例 8）

(2)告知要購買的數量／規格／條件（例如價錢條件／包裝要求／付款方式等）（參
範例 6/8/9）

結論(Conclusion)

(1)請儘快提供或寄來上述資料，如有問題，儘快告知或轉介紹可供應者（範例
3/6/7/11）

(2)謝謝其合作，並等候回音

---

4. 詢問函常用單字 / 片語

(1)inquiry 詢問 / urgent inquiry 緊急的詢價

(2)best price 最好的價錢 / price list 價目表 /quotation 報價單

(3)target price / idea price (ideal price)目標價

(4)delivery time 交期 / shipment ＝ shipping time 出貨時間

(5)lead time ＝ manufacturing time ＝ production time 生產時間

(6)minimum order quantity ＝ min. quantity ＝ min. order 最低訂購量

(7)sample 樣品 / quantity 數量 / quality 品質

(8)RFQ ＝ Request for quotation 詢問報價

(9)market share 市場占有 / sales channel 銷售管道

(10)exclusive distributor 獨家經銷商 / sole agent 獨家代理

(11)terms and conditions 條件 / payment terms 付款條件

(12)as soon as you can ＝ as soon as possible ＝ by return ＝ right away 盡快

(13)can supply from stock 可從庫存供貨

(14)Proforma Invoice (P/I)預約發票

(15)initial order ＝ trial order ＝ sample order 試銷訂單/樣品訂單

(16)associated company 關係企業

(17)from the advertisement 從廣告上

(18)experienced sales representative 有經驗的業務代表

(19)arrange a meeting 安排一個會議

(20)before placing the order / before we place the order 下訂單以前

(21)factory capability 工廠產能

(22)Q.C. (quality control) 品管 / Q.A. (quality assurance) 品保

(23)revised drawing 修改的圖面 / instruction manual 操作手冊

(24)for our evaluation 供評估/for our reference 供參考

(25)Thank you for your attention to the above. 謝謝您對以上的注意

(26)We look forward to starting the business cooperation with you soon. 我們期盼和你們開始生意關係

5. 詢問函常用的主題句 / 説明句 / 結論句

主題句(Main Idea)

(1) From your trade office here, we know you are a manufacturer of notebook computers and are interested in distributing your products in our market.

(2) We are interested in your electronic products and hope to receive your latest catalogue and price list soon.

(3) From your website, we know that you just developed a latest sneaker with a computerized motor in the sole and feel interested in importing this product.

(4) As a leading manufacturer of Valves, we are looking for the water hammer arresters.

(5) We are deeply interested in your new developed hi-tech product.

説明句(Explanation)

(1) Please kindly advise us your terms and conditions for the sole agent in Asia.

(2) We have confidence in promoting this new product well through our strongest sales channels in Asia.

(3) Please provide us a quotation for quantity of 1K, 2K and 3K pieces.

(4) Please kindly send us the catalogues and price list for the machine you produce.

(5) Please offer us your best prices and delivery time for the commercial grade ingots.

結論句(Conclusion)

(1) We are waiting for your P/I by return.

(2) Can you bid it and supply them from stock?

(3) I will visit your factory next week and would like to receive your offer soon.

(4) Please kindly advise if you are available to provide us the above products or if you can recommend us the manufacturers who are able to provide them.

(5) Thank you for your attention to the above and look forward to starting the cooperation with you soon.

範例 1　進口商主動詢問有關多功能抽水閥及壓力平衡閥(Multi-function
　　　　Valve and Balanced Direct Acting Pressure Valve)

From: richard@armlink.com

To: sb.wow@msa.hinet.net

Subject: Valves

Dear Sirs,

We are an international pump manufacturer with our headquarters in Toronto. Someone has passed your product catalogue to us. We are very interested in your products like multi-function valve and balanced direct acting pressure valve. May we ask if you are able to send us some brochures on those products along with the delivery time and prices in US dollars?

Best regards

Richard Armstrong

Technical sales Representative- Far East

www.armstrongpumps.com

文章結構

主題：我們是一家國際幫浦工廠，總部在多倫多。某人將貴公司目錄遞交給我們。

說明：我們對你們的產品非常有興趣，像是多功能抽水閥及壓力平衡閥。

結論：我們想詢問貴公司是否可以寄給我們這些產品的宣傳小冊、交期和美金報價
　　　嗎？

生字 / 片語

1. international pump manufacturer　國際幫浦工廠

2. headquarters　總部；in Toronto　在（加拿大）多倫多

3. catalogue　目錄；brochures　宣傳小冊

4. products　產品；multi-function valve　多功能抽水閥

5. balanced direct acting pressure valve　壓力平衡閥

6. delivery time　交期；prices　價錢；prices in US dollars　用美金報價

範例 2　進口商看到網站，洽詢有關運動鞋(Sneaker)產品

From: scosmos.mary@msa.hinet.net
To: sales@adidas.com
Subject: Inquiry for Smartest Sneaker

Dear Sirs,

From your website, we got to know that you just developed a latest Smartest Sneaker with a computerized motor in the sole and feel interested in importing this product. As you mentioned, it is made to "sense" an athlete's cushioning and support needs based on weight, pace and surface and then adjust itself accordingly. Besides, it's available as running or basketball shoes. We know your retail price is US$250/pair, but please kindly quote us your best delivery time and the best export prices based on both FOB and CIF Keelung by sea for the quantity of one 20 feet container. We are looking forward to your quotation.

Best regards

文章結構

主題：從貴公司網站上，我們得知貴公司剛開發出一項在鞋底具有電腦馬達的最新
　　　「最敏捷運動鞋」，並對進口此產品有興趣。

說明：貴公司提及此運動鞋可根據體重、速度和表面檢測運動員的軟墊和支撐的需
　　　要做自動調整，此外亦可用來當跑步鞋或籃球鞋。我們知道零售價是 250 美
　　　元一雙，但是請報最好交期及 20 呎整櫃量的最好外銷價錢。
　　　根據 FOB（離港價）和 CIF 海運到基隆（含運費保費在內價）的報價。

結論：我們期待貴公司的報價單。

生字／片語

1. the latest Smartest Sneaker with a computerized motor in the sole　在鞋底具有電腦馬達的最新「最敏捷運動鞋」

2. sense　檢測／感應；pace　速度；retail price　零售價

3. athlete's cushioning and support needs　運動員的軟墊和支撐的需要

4. delivery time　交期；FOB = Free on Board　離港價／船上交貨價

5. CIF = Cost, Insurance and Freight　含運費保費在內價

6. 20 feet container　20 呎貨櫃；look forward to ＋ N/V-ing　期待／期盼

範例 3　進口商向德國廠商洽詢「水鎚捕捉器」(Water Hammer Arresters)

MIIN-RUEY ENTERPRISE INC.

No. 805, Bai-Men Rd., Yu-Shan Town, Kuen-Shan, China

E-mail address: sb.wow@msa.hinet.net

GMA WEGAND GMBH

From: sb.wow@msa.hinet.net

To: GMA WEGAND GMBH

Dear Sirs,

Re: Inquiry for Water Hammer Arresters

As a leading manufacturer of Valves & piping assembly, we are looking for the Water Hammer Arresters, which are to be used in our products.

Please kindly advise if you are available to provide us the above products or if you can recommend us the manufacturers who are available to provide them. If yes, would you be kind enough to send your catalog or brochure to our above address for our evaluation. After selection, we will advise you the model and quantity we are interested in for making the quotation.

Thank you for your attention to the above and look forward to starting the cooperation with you soon.

Best regads

Mary Huang / Import Manager

Miin-Ruehy Enterprice Inc.

www.sb.wow.com

文章結構

> 主題：自我介紹，並直接告知正在找尋的產品
>
> 　　　作為閥類及管線裝配的主要製造商，本廠正在尋找用於我方產品內的「水鎚捕捉器」。
>
> 說明：請告知貴公司是否可供應上述產品，或可推薦有能力供應的工廠。
>
> 　　　如能供應，請將貴公司型錄寄至本公司上址供評估參考，經挑選後，我們會告知有興趣的規格及數量，以利貴方提供報價。
>
> 結論：謝謝您對以上的注意，並期盼很快能和貴公司開始合作關係。

生字／片語

1. inquiry　詢問

2. Water Hammer Arresters　水鎚捕捉器（產品名稱）

3. a leading manufacturer ＝ a main manufacturer　一個主要的製造商

4. provide ＝ supply ＝ offer　提供

5. you are available to provide ＝ you can provide　貴方能提供

6. catalog　圖文並茂的數頁目錄

7. brochure　整本有詳細規格的小冊子型錄

8. for our evaluation ＝ for our reference ＝ for our study　供我方評估／參考／研讀

9. after selection　挑選後

10. we are interested in ＝ Adj. Cl.　形容前面的 N(the model and quantity)

11. for making the quotation ＝ for quoting ＝ so that you can make the offer　以便於報價

12. to the above　對以上之事

13. look forward to　期盼（注意：to 是介系詞，後面跟 N 或 V-ing）

14. starting the cooperation with you ＝ doing the business with you ＝ setting up the business relationship with you　開始和貴方的合作關係

MODERN
BUSINESS
ENGLISH

Part 4 (二、詢問函)
商用英文信函寫法

169

範例 4　進口商詢問感應器(Sensors)

Dear Peter,

Please kindly advise us if you can supply the sensors from your stock or if you still produce the sensors for your FPA instrument in your factory at present. If yes, please advise us your best price and delivery time by return.

As said, the sensors of this instrument are easily broken and we need about 4 pcs of the sensors to make the instrument work.

We are looking forward to your positive reply soon.

Best regards

文章結構

主題：請告知貴公司是否可從庫存提供感應器或是否貴廠目前仍有生產 FPA 儀器的感應器。如是，請盡快告知最好的價錢及交期。

說明：如所告知，此儀器的感應器很容易損壞，為使此儀器能順利操作，我們約需要四顆感應器。

結論：我們期盼貴公司盡快給予肯定的答覆。

生字 / 片語

1. sensors　感應器；supply the sensors from your stock　從庫存提供感應器的貨
2. instrument　儀器
3. best price and delivery time　最好的價錢及交期
4. by return ＝ soon　盡快
5. are easily broken　容易損壞
6. make the instrument work　使儀器能操作
7. positive reply　肯定的答覆

範例 5　進口商詢問手機充電收音機(Dynamo Radio)的進口代理條件

From: Box Wu

To: sales@dynamoradio.com

Subject: To be the Sole Agent for your Dynamo Radio

Dear Sirs,

We are deeply interested in your new developed hi-tech product, Dynamo Radio, which is an environmental protection radio for charging cell phone and for using Siren (SOS) in emergency shown on your website. Please kindly advise us your terms and conditions for the sole agent in Asia.

We have confidence in promoting this new product well through our strongest sales channels in Asia and in getting a good market share for you. Please let us know the details of your Agent Agreement soon.

Best regards

Box Wu / Electronic Products Importer

文章結構

主題：我們非常有興趣貴公司網站上所刊登的新開發高科技產品——手機充電收音機——此環保型收音機可做為手機充電及緊急求救的警報器用。

說明：請告知欲當貴公司亞洲地區獨家代理的條件為何？
　　　我們有信心經由我方在亞洲極堅強的銷售管道來推銷此新產品，並可為貴公司爭取很好的市場占有率。

結論：請盡快告知代理合約的細節。

生字／片語

1. Dynamo Radio　手機充電收音機；for charging cell phone　手機充電用
2. new developed hi-tech product　新開發的高科技產品
3. environmental protection radio　環保型收音機
4. for using Siren (SOS) in emergency　緊急求救的警報器用
5. sole agent in Asia　亞洲地區獨家代理；sales channels　銷售管道
6. market share　市場占有；Agent Agreement　代理合約

範例 6 進口商詢問測試儀器(Test Instrument)

To: sales@onion.com

Dear Sirs,

From Mr. Thirty of Logic, we know your company produces many kinds of instruments. Since we are looking for the test instrument for testing our balancing valves, please kindly advise us if you are available to supply the testing instrument similar to Logic's FlowPlus as per the picture attached. If so, please send us your catalog or detailed information for reference. If you are unable to assist us with the above inquiry, kindly advise us of a supplier who may be able to.

We are looking forward to your reply to the above soon.

Best regards
Mary Huang/King-tech Valve

文章結構

主題：從 Logic 公司的 Thirty 先生處，我們得知貴公司生產各種儀器。本公司正在尋找測試我們平衡閥的測試儀器，請告知貴公司是否可提供類似附圖的 Logic 的 FlowPlus 的儀器。

說明：如是，請將貴公司的目錄或詳細資料寄給我方參考。如貴公司對以上詢價無法幫忙，可否告知可供應的供應商。

結論：我們期盼貴公司盡快回覆以上。

生字 / 片語

1. produce 生產；test instrument 測試儀器
2. balancing valve 平衡閥；similar to 類似於
3. as per the picture attached 參照附圖
4. catalog or detailed information 目錄或詳細的資料
5. for reference 供參考；look forward to 期盼（後面＋ N or Ving）
6. assist ＝ help 幫忙；to the above 對以上（之事）

範例 7　美商向台灣工廠詢問「機械設備」(machines)型錄及報價

From: neol@bnbmachine.com
To: taiwan@saanshin.com
Subject: Electro-mechanics Equipments

Dear Sirs,

　　Our company is dedicated to the trading of electromechanics equipments, some of them made by ourselves, some imported or made by third parties.

　　Due to the high variety of our trade, we have importation and exportation departments, Hardware and Software departments, and we have also got a metallurgical section, which can make us able to supply the technical maintenance for the equipments we trade.

　　Please kindly send us the catalogues and price list for the machines you produce, so we can check if it is possible for both of us to establish the business relations and profitable deals.

　　We are waiting for the information from you soon.

Best regards
Noel White/Manager
B&B Machinery Trade Co.
www.bnbmachinery.com

文章結構

主題：利用第一及第二段作自我介紹及產品能力介紹

　　(1)本公司專注於電子機械設備的貿易，某些設備由本廠所製造，有些進口或由第三者製造。

　　(2)由於我們貿易的多元化，我們有進出口部門，硬體和軟體部門，也有冶金部門，使本公司對所貿易的設備能提供技術保養。

說明：請寄來貴方所生產的設備的型錄及價目表，以便我方查看是否貴我雙方有建立商業關係及有利交易的可能性。

結論：等候貴方資料。

生字／片語

1. dedicate　專注於

2. electromechanics equipments　電子機械設備

3. third parties　第三者

4. due to ＝ owing to ＝ because of　由於／因為（後接名詞或片語）

5. high variety of our trade　我方貿易的多元化

6. importation and exportation departments　進出口部門

7. Hardware and Software departments　硬體及軟體部門

8. metallurgical section　冶金部門

9. technical maintenance　技術保養

10. establish the business relations ＝ set up the business relationship ＝ build the business connections　建立商業關係

11. profitable deals　有利的交易

範例 8　台商向澳洲工廠詢問「鋁錠及鋁廢料」的報價
(Aluminum ingots and Aluminum scrap)

From: michael@taiwanmetal.com
To: export@fnsmetal.com
Subject: Aluminum Ingots

Dear Sirs,

Re: Aluminum Ingots

Through the introduction of Australian Chamber of Commerce in Hong Kong, we were advised of your company and your ability to export aluminum ingots.

We are at present acting on behalf of an associated company that is engaged in the manufacture of aluminum doors and window frames, to locate alternative sources for this commodity. Therefore, would you please be kind enough to advise if your company is capable of offering 99.7% commercial grade ingots in either 16 kg or 22 kg bars. Our associate at present consumes 500 M/T per month, and have had over 10 years experience in the manufacture of the above mentioned products.

We would also be interested to know if your company has the ability to offer tense 95 aluminum scrap.

Please let us know the above information as soon as you can.

Best regards
Michael Kao

文章結構

主題：第一段告知從何處得知供應商大名，經由香港的澳洲商會介紹，我們得知貴 公司大名，及貴公司外銷鋁錠的能力。

說明：(1)本公司為一專業鋁門窗製造廠的關係企業代表，目前正協助其尋找此產品 的替代供應商。因此，請告知貴公司是否有能力供應99.7%純度的商業級鋁 錠，每條重16公斤或22公斤皆可，我們的關係企業目前每月消耗500噸， 並有10年製造上述產品的經驗。

　　　　(2)我方也有興趣知道是否貴公司亦能提供95張力的鋁廢料。

結論：請儘快告知以上資訊。

生字 / 片語

1. Aluminum Ingots　鋁錠

2. Aluminum scrap　鋁廢料

3. Through the introduction of Australian Chamber of Commerce in H.K.
　經由香港的澳洲商會的介紹

4. ability to export aluminum ingots　外銷鋁錠的能力

5. at present ＝ now ＝ presently　目前

6. acting　扮演

7. on behalf of　代表

8. an associated company　關係企業公司

9. is engaged in the manufacture of aluminum doors and window frames
　從事於鋁門窗的製造

10. to locate alternative sources　尋找替代的供應商（＝貨源）

11. commodity　商品

12. commercial grade ingots　商業級的鋁錠

13. 16kg or 22 kg bars　每一條重 16 或 22 公斤

14. consume ＝ use　消耗

15. 500 M/T per month　每月 500 噸　(M/T ＝ Metric Tons)

16. the above mentioned products　上述的產品

17. as soon as you can ＝ as soon as possible　儘快

範例 9　老客戶詢問「抑制閥」(Check Valve)的標價

From:"dr" <dr@bvcusa.com>

To:<mary0105@ms33.hinet.net>

Subject: Request for Quote

Dear Mary,

Dan Ruppert with Butterfly Valves (the USA Valve Rep) here.

This is an urgent inquiry and due end of next week, can you bid it and supply them from stock?

SPECS OF VALVES	QTY
1. CHECK VALVE, SILENT, WAFER, 2"	342 sets
2. CHECK VALVE, SILENT, WAFER, 4"	227 sets

Best regards

Dan

文章結構

主題：這是 Ruppert 美國蝶閥代表。這是一個緊急的詢價。

說明及結論：下列詢價須於下週底前報出，貴公司能標價並從庫存出貨嗎？

閥規格	數量
1. 靜音薄型抑制閥 2"	342 組
2. 靜音薄型抑制閥 4"	227 組

生字／片語

1. butterfly valve　蝶閥
2. urgent inquiry　緊急詢價
3. due　到期
4. bid　投標／標價
5. QTY = quantity　數量
6. can supply from stock　可從庫存供貨

範例 10　進口商看到廣告來洽詢 PDA（個人數位助理）

Fm: sb.wow@msa.hinet.net

To: export@de.com

Dear Sirs,

We have obtained details of your esteemed company from the Asian Electronic Products Buyers Guide. It will be appreciated if you can kindly send us your Proforma Invoice indicating your best FOB price and delivery time for an initial order of one set of your PDA. I will visit your factory within the next weeks and hope you can arrange a meeting during my stay. We are waiting for your P/I by return.

Best regards

Leo Chen

文章結構

主旨：我們從亞洲電子產品買主指南上獲得貴公司資料。

說明：將感謝貴公司能將預約發票傳給我們，上面註明一台 PDA 樣品訂單的 FOB 最好價錢及交期。

結論：我將於下幾週拜訪貴廠，在我停留時間希望貴方能安排一個會談。我們等候貴公司盡快送來預約發票。

生字／片語

1. esteemed company = respected company　令人尊敬的公司

2. Asian Electronic Products Buyers Guide　亞洲電子產品買主指南

3. It will be appreciated = We will appreciate it　我們將感謝

4. P/I (Proforma Invoice)　預約發票／預付發票；indicating　註明

5. FOB price (Free on Board)　船上交貨價；delivery time　交期

6. PDA (Personal Digital Assistant)　個人數位助理

7. arrange a meeting　安排一個會談；initial order = sample order　樣品訂單

8. by return = soon = at once = right away　馬上／盡快

MODERN BUSINESS ENGLISH
現代 商用英文

範例 11 客戶看到廣告,詢問筆記型電腦產品(Notebook Computers)

From: Stone Sofka <Ssofka@intertest.net>
To: Gram@<moi5555@moi.com.tw>
Subject: Inquiry for Notebook Computers

Dear Sirs,

We read your advertisement on Asian Magazine, January issue, and are interested in your Notebook computers.

Please inform us the full technical features of all your Notebook computers and send us the quotation based on FOB Taiwan and CIF Athens, as well as the rest of the terms of cooperation with your company.

We have a great experience in the promotion of computers in Greek market. Our Director and Sales Manager have worked for the exclusive distributor for many big computer companies for 10 years with great success.

We are interested in distributing your company in Greece for all kinds of your computers.

Please send above information by return e-mail.

Thanks & Best regards
Stone Sofka

文章結構

主題：從一月號《亞洲雜誌》上看到貴公司廣告並對你們的筆記型電腦有興趣。

說明：索取報價條款並自我經驗介紹

(1)請告知我方貴公司所有筆記型電腦的技術特色，並根據 FOB 台灣及 CIF 雅典報價給我們，並告知其他與貴公司合作的條款。

(2)在推銷電腦產品到希臘市場上我們具備很好的經驗，我們的董事及業務經理在擔任許多大型電腦公司的獨家經銷商方面非常成功。

結論：(1)我們有興趣在希臘經銷貴公司所有電腦機型。

(2)請用 e-mail 儘快傳送以上資料。

生字 / 片語

1. advertisement　廣告

2. Asian Magazine　亞洲雜誌

3. January issue　一月號

4. notebook computers　筆記型電腦

5. technical features　技術特色

6. based on FOB Taiwan and CIF Athens　根據 FOB 台灣及 CIF 雅典價

7. the rest of the terms　其他條款

8. experience　經驗

9. promotion　推銷(N)

10. Greek market　希臘市場

11. director　董事

12. sales manager　業務經理

13. exclusive distributor　獨家經銷商

14. great success　極大成功

15. distributing　經銷

16. by return ＝ at once ＝ right away　馬上

範例 12 美國客戶詢問「鍵盤」(Keyboard)報價

From: Alex Johnson <johnson@alexelectronics.com>
To: sales@maximkeyboard.com
Subject: Inquiry for Keyboard

Dear Sirs,

We are in a prototyping phase of a new project and are looking for a keyboard for this project. Because of speed, we are looking for an available (existing) keyboard which can meet our needs.

Basically the keyboard has to be a QWERTY type about half the size of a standard IBM PC type keyboard. The maximum number of keytops is 64. Since we will build inside the case also some electronics which need a board space of about 120 cm^2 (approx. 20×6cm) and a component height of about 16 mm, the case has to have the enough space to hold this.

We would like to receive your offer at FOB Hong Kong for :
1. Immediate need: 100 pcs
2. Large quantity: 50,000 pcs
Please also advise us your soonest delivery time, packing method and payment terms.

I will visit your factory within the next weeks and would like to receive your offer as soon as possible, so that we can arrange a meeting during my stay.

Best regards
Alex Johnson

文章結構

主題：第一段告知詢問的目的

我們正在進行一個新產品的原型階段，並正在尋找合用的鍵盤，由於時效，我們想找一個現成合用的鍵盤。

說明：第二段告知所需規格；第三段要求報價條件

(1)基本上此鍵盤必須是QWERTY（打字機排列法），尺寸為標準IBM個人電腦型鍵盤的一半大小，最多64個字鍵，由於我們將在殼內嵌裝一些電子零件，所需板子尺寸為120平方公分＝20×6 cm，零件高度為16 mm，此外殼需有足夠空間來支撐。

(2)我們希望收到貴公司報 FOB 香港價，按下列數量：

①馬上需求 100 pcs

②大量需求 50,000 pcs

請同時告知最快交期、包裝方式及付款條件。

結論：我將於下幾週拜訪貴廠，希望儘快收到貴方報價，以便在我停留時間可安排會談。

生字／片語

1. prototyping phase of a new project　新產品（計畫案）的原型階段
2. because of speed　由於時效問題
3. existing keyboard　現有鍵盤
4. basically　基本上
5. QWERTY type　打字機排列法
6. standard IBM PC type keyboard　標準 IBM 個人電腦型鍵盤
7. Maximum number of keytops　最大字鍵數目
8. build inside the case　內嵌於殼內
9. electronics　電子零件
10. board space　板子尺寸
11. component　零件
12. delivery time ＝ shipping date　交期
13. packing method　包裝方式
14. payment terms　付款條件
15. arrange a meeting　安排會談

範例 13　進口代理索取「裝飾品」(Ornament)工廠製程及品管錄影帶

To: Mr. Jack Lee/Fashion Ornament Mfg. Ltd.

The biggest customer we are promoting now is very interested in your new design of the fashion Ornament. However, they need us to provide the information about your factory before placing the order.

(1)They would like a Video Tape of the factory showing the manufacturing facility and overall factory.

(2)They want to know your factory quality control/quality assurance plan.

(3)Any other information/documentation which will provide additional input concerning the factories capability

Please advise by return when the above can be submitted.

Best regards
William Jones

文章結構

主題：我們正在推銷的最大客戶對貴公司新設計的流行飾品感到興趣，但下訂單以前，他們要求我們提供工廠資料。

說明：(1)他們希望有一卷顯示生產設備及整個工廠的錄影帶。
　　　(2)他們要看工廠的品管及品保計畫。
　　　(3)任何可以提出有關工廠能力的其他附加資料或文件。

結論：請即告知何時可提供以上資料。

生字／片語

1. Fashion Ornament　流行飾品
2. before placing the order　下訂單以前（before ＝介系詞＋ N or V-ing）
3. Video Tape　錄影帶
4. manufacturing facility and overall factory　生產設備及整個工廠
5. quality control (Q.C.)　品質管制；quality assurance(Q.A.)　品質保證
6. information/documentation　資料／文件
7. factories capability　工廠的能力

練習 13　詢問函練習

1. 將下列翻譯成英文

(1)這是一個緊急的詢問，請　用傳真　回覆（by fax ＝方法，放句尾）

(2)我們　從廣告上　得知貴公司的大名（from the advertisement ＝地點）

(3)請提供目錄及報價單以供評估

(4)請報最好的船上交貨價錢和交期

(5)請告知　試銷訂單　的　最低訂購量　（N1 的 N2→N2 for N1）

(6)我們盼望儘快和貴公司建立商業關係

(7)我們是　這家工廠　的　關係企業公司　（N1 的 N2→N2 of N1）

(8)請告知是否你們可以　從庫存　供貨（from stock ＝地點，放句尾）

(9)我們將下給你們一張樣品訂單，請寄預付發票給我們

(10)我們希望當你們的獨家經銷商，請告知貴公司的條件

(11)我下週將　拜訪貴廠，請安排一個會談（next week ＝時間，放第一句尾）

(12)下訂單以前，我們希望收到你們的樣品

(13)因為客戶修改圖面，請重新報價

(14)我們的業務代表告知客戶的目標價是 US$10/pc.

(15)請寄操作手冊給我們

(16)Q.C./Q.A./ASAP/P.I./RFQ 的英文全名及中文意思

2. 請在空格內填入適當的英文生字或片語：

(1) 我們是日本最大的 3C 產品進口商。

We are the biggest _____ of 3C _____ in Japan.

(2) 本公司有 100 個有經驗的業務，也有很好的銷售管道。

We have 100 _____ sales representatives and also have very good _____.

(3) 我們有興趣進口你們的產品。

We are _____ in _____ your products.

(4) 請告知獨家代理的條款。

Please advise us your _____ for a sole agent.

(5) 請報給我們貴公司最好的價錢及交期。

Please offer us your best _____ and _____.

(6) 謝謝你們對以上的注意並期盼盡快和貴公司開始生意合作。

Thank you for your _____ to the above and look forward to _____ business cooperation with you soon.

3. 請自由發揮寫一封完整詢問函（產品自定）

三、報價及回覆函(Quotations & Replies)

1. 寫回覆函或報價，要注意下列幾個重點

   (1)收到客戶詢問函後，最好當天或隔天馬上回覆，以示效率及誠意，因為一般客戶大都會同時洽詢數家供應商，除了價錢，也會考慮服務效率及合作誠意。

   (2)回覆函是所有商用信函中最簡單的一種，只要針對客戶的詢問一一作答，如果其中有一兩項問題，無法馬上回覆，記得在回函中也要告知對方大約可回答的預計時間，以免客戶引頸長盼。

   (3)回覆函雖是最簡單的信函，卻也是最重要的信函，因為這是客戶決定訂單的關鍵，因此在回覆同時，也要利用一點推銷技巧，強調自己的價錢、品質及服務都比其他同業好，也要讓客戶覺得我方是個值得信賴，可以合作的最佳人選，也可在回函中，追加提供一些相關資料。

2. 回覆函的內容有下列幾點

   (1)根據客戶要求，提供報價單／樣品／規格圖面或相關資料。

   (2)商會或貿協回函提供進出口商名單。

   (3)回覆產品技術規格問題。

3. 客人索取樣品(Sample)，有四種回覆

   (1)樣品費和郵費全部免費

   (Both the sample and postage are free.)

   (2)樣品免費，但要客人支付郵費，因為樣品便宜，但是很重而使郵費高

   (The sample is free, but you have to pay for the postage because it is very heavy.)

   (3)請先支付樣品和郵費，因為材料成本高

   (Please pay sample charge and postage in advance because the material cost is high.)

   (4)請客戶先付樣品費和郵費，正式下訂單時，再退還或從訂單中扣除

   (Please pay the sample charge and postage in advance. However, we will return them to you when ordering, or deduct them from the order.)

4. 回覆函寫法

主題：第一段先感謝確認收到詢問函（參範例 2/3/6/7）。

說明：(1)按照客戶要求，寄上或附上所要的報價單／目錄／樣品或規格資料（參範例 3/7/9）。

　　　(2)也可順便再自我推銷，介紹強調自己或產品的優點特色（參範例 8）。

　　　(3)附帶提供一些額外的相關資料（參範例 6）。

結論：(1)希望儘快收到對方意見或訂單情形（參範例 3/6/7/8）。

　　　(2)如還需要其他進一步資料也請儘快來函（參範例 6）

5. 報價單(Quotation/Offer Sheet)或價目表(Price List)的基本內容

注意：客戶詢價大都很簡單，有時只要求儘快報價，並沒有提到任何報價條件，但是報價時，一定要將所有報價條件按公司的標準一一列出，因為每個條件都與成本有關，而報價一但被客戶接受，這些條件就變成訂單上所有的條件，不易變更（參附件二／三／四）。

(1) 編號(Model No./Style No./Part No./Item No.)：按型錄上的編號打出

(2) 品名規格(Specification/Description)：簡單敘述產品名稱及規格

(3) 單價(Unit Price)：報價時，一定要註明清楚下列六點：

　①幣別(Currency)：US$, DM, HK$, GBP, EUR（注意 NT$不要對外報價）

　②單價(Unit Price)：例如：US$10/PC CFR New York by sea

　③單位(Unit)：pc, set, pair, dozen, kg

　④價錢條件(Price Terms) / 國貿條規(INCOTERMS 2010)

　　Rule I：任何或多種運送方式的規則（7 種）

　　EXW　　Ex-Works / Ex-Factory 工廠交貨條件規則

　　FCA　　Free Carrier 貨交運送人條件規則

　　CPT　　Carriage Paid To 運費付訖條件規則

　　CIP　　Carriage, Insurance Paid To 運保費付訖條件規則

　　DAT　　Delivered At Terminal 終點站交貨條件規則

　　DAP　　Delivered At Place 目的地交貨條件規則

DDP　　Delivered Duty Paid 稅訖交貨條件規則

Rule II：海運及內陸水路運送的規則（4種）

FAS　　Free Alongside Ship 船邊交貨條件規則

FOB　　Free on Board 船上交貨價條件規則

CFR　　Cost and Freight 運費在內條件規則

CIF　　Cost, Insurance and Freight 運保費在內條件規則

注意：(1)FAS/FOB/FCA：價錢成本相同但風險轉移點不同；FOB/CFR/CIF這三個條件和FCA/CPT/CIP三個條件則是價錢成本不同但風險轉移點相同。

(2)離廠價＋搬運費(carrying fee)＋內陸運費(inland freight)＋報關費(broker charge)＋海關通關費(customs charge)＋銀行費用(banking charge)＝FOB離港價＋Freight運費＝CFR含運費在內價＋Insurance保險＝CIF含運保費在內價。

⑤目的地港(Destinations)：FOB Keelung（FOB/FCA 後為出口港或出口機場）CFR/CIF/CPT/CIP後為目的地港或機場（不要只寫國名，因為國外不同港口，運費不同），也要註明裝運方式（空運／海運／郵包／快捷）

例：CFR New York by sea freight

⑥裝運方法(Shipping Method)：by air/by sea/by air parcel/by courier 快捷（可用 DHL, Fed-Express, UPS 或 Express Mail）

(4)交期(Lead Time/Shipping Date/Shipment/Delivery Time)：報價時大都寫天數，等正式訂購才註明出貨日期

例：Delivery Time: Within 30 days after receiving the L/C.

(5)最低訂購量(Minimum Order Quantity)：報價時要註明數量

(6)有效期(Validity)：註明天數 30 days 或 subject to our final confirmation when ordering（下訂單時，再讓我們做最後的確認）

(7)包裝方式(Packing Method)：如知包裝明細就列出，如不清楚就寫標準外銷包裝

例：Packing: Standard Export Packing.

(8)付款條件 (Payment Terms/Terms of Payment)：常用的有下列幾種：

①不可撤銷保兌及可轉讓即期信用狀

By irrevocable, <u>confirmed and transferable</u> Letter of Credit at sight.

注意：保兌(confirmed)及可轉讓(transferable)為選擇性，視情況而定，confirmed L/C 開狀行一定要是大銀行或是有大銀行作擔保，多用於第一次交易的新客戶，可確保 100%付款。

transferable L/C 用於三角貿易或轉口貿易，可將部分金額轉開到海外的工廠，例如：美國客戶開 US$10,000 的 L/C 到台灣公司，只要 L/C 上註明 transferable，就可將部分金額，例如 US$5,000 轉開到大陸工廠，餘額留在台灣的銀行帳上。

②遠期信用狀(Usance L/C)

By irrevocable Letter of Credit at 30 days sight in our favor.

通常在 at sight 中間插入票據期限 30 days, 60 days, 90 days…

遠期信用狀上要註明利息費用是：

買方付(The interest is for buyer's account)或賣方付(The interest is for seller's account)，或各付一部分(30 days interest is for applicant's account and 60 days interest is for beneficiary's account) (applicant = buyer; beneficiary = seller)

注意：開 L/C 的方法有三種：

　　(1)全電信用狀(Full cable L/C)＝正本。

　　(2)短電信用狀(Short cable L/C)＝通知＋正本另外郵寄過來。

　　(3)郵寄信用狀(Mailed L/C)＝正本。

③付款交單 Document against Payment(＝D/P)

先出貨，然後將出貨文件直接／或透過銀行寄給客戶，客戶收到文件後馬上付款，此為託收(Bill for Collection)的一種，用於大客戶或老客戶。(Note: CAD ＝ Cash against Documents ＝ D/P)

④承兌交單 Document against Acceptance(＝D/A)

D/A 付款後面要註明天數，30 days, 60 days, 90 days…

例如：Payment：by D/A 30 days（＝出貨後 30 天付款，從提單 B/L 上的出貨日起算）此為託收的另一種，也多用於大客戶或信用好的客戶。

注意：報價時，要將利息費用加入成本內（如以銀行年利率10%計算，每30天，大約占成本1%）

⑤記帳／月結 Open Account (= O/A)

此付款方式多用於長期採購，或每月經常有多批出貨的客戶。

⑥分期付款 Installment

此付款方式多用於機器設備、模具、工程案或土地／房屋買賣等，此付款方式可使買賣雙方互相牽制，買方就不用擔心貨到無法使用，賣方也不用擔心買方交貨不付款，特別因 L/C 付款方式對買賣雙方所需支付給銀行的費用太高（含開狀費、通知費、看單費、手續費、郵電費與利息等），現在大部分交易都改採用此分期付款方式（百分比或分期次數可由買賣雙方自行約定）代替 L/C，例如：

Payment terms: 30% down payment(deposit)（定金三成）

40% before shipment（出貨前付四成）

30% upon receiving the goods（收到貨馬上付清尾款）

Payment method: by T/T (Telegraphic Transfer)（付款方式用電匯）

⑦寄售 On Consignment

注意：寄售是賣掉才付款，或賣多少給多少，沒賣掉就不付款，所以此付款方式多用於大廠（資本較雄厚）或零售商。

⑧交貨同時收現金 Cash on Delivery (= COD)

此為一般國內交易的付款方式，即貨送到同時就收現金票

⑨下訂單同時即付現金 Cash with Order (= CWO)

此方式多用於市場缺貨或搶貨，買方只好拿現金去排隊

(9)付款方法(Payment Methods)

除了信用狀以外，其他付款條件：D/P, D/A, O/A, CWO, COD 或 On Consignment 的付款方法，可採下列其一：

①電匯 Telegram Transfer/Cable transfer(T/T) / Bank Transfer/Wire

②信匯 Mailed Transfer(M/T) / Letter Transfer (L/T)

③銀行本票 Bank Check

④銀行匯票 Bank Draft

⑤郵政匯票 Money Order

注意：T/T 是付款方法，不是付款條件，所以在 Payment 後，不可以只打 T/T (Payment by T/T)，因為如此常會造成付款糾紛。如要強調以 T/T 付款，

一定要加註付款的時間，例如：Payment: by T/T before shipment 或
Payment: by T/T within 30 days after shipment。

(10)保險(Insurance)：CIF/CIP 報價，賣方要將保險費計入成本內。客戶如沒特別要求，一般都投保全險(All risk)。保險基本上分三種：

ICC(A)＝ All Risk 全險

ICC(B)＝ ICC (C)FPA 平安險＋ WPA 水漬險

ICC(C)＝ FPA 平安險

兵險(Ware Risk)、罷工險(Strike)、罷工暴動內亂險(SRCC)、竊盜未能交貨險(TPND)等則依客人要求需另外加保。

投保金額＝出貨發票金額＋ 10%

保費＝投保金額×保率（約在 0.3%～1%間；視保險公司及產品而定）

例如：出貨發票金額為 US$10,000；

投保金額＝ US$10,000×1.1 ＝ US$11,000

保費＝ US$11,000×0.5%＝ US$55.－

附件一 INCOTERMS 2010 定義及運用

1. EXW-Ex Works(insert named place of delivery)工廠交貨條件規則（加填指定交貨地）

- 適用任何運送方式或超過一種以上（複合運送）的運送方式。
- 適合國內貿易（於國際貿易場合：使用 FCA 較妥）。
- 定義：本規則意指賣方在其營業處所或其他指定地（工廠或倉庫）貨物交付買方處置時，即屬賣方交貨。
- 賣方無須將貨物裝上任何收貨的運送工具，無須辦理輸出通關手續。
- 賣方最好敘明指定交貨的約定地點。
- 買方負擔自指定交貨的約定地點接管貨物起所產生的一切費用及風險。
- 本條件為賣方負擔最小的義務。
- A2/B2：通關手續、費用及風險由買方負擔。
- A5/B5：風險轉移：買方自賣方交貨起，負擔貨物滅失或毀損的一切風險。
- A6/B6：費用的劃分：賣方支付貨物的一切費用直至交貨為止。
  買方須支付交貨後一切費用包括一切關稅稅捐及辦理通關的手續費等。

注意：EXW 價＋搬運費＋內陸運費＋報關費 (broker charge)＋海關通關費(customs charge)＋銀行費用(banking charge)＝FOB/FCA + Freight = CFR/CPT + Insurance = CIF/CIP

2. FCA -Free Carrier(insert named place of delivery)貨交運送人條件規則（加填指定交貨地）

- 適用任何運送方式或超過一種以上（複合運送）的運送方式。
- 定義：本規則意指賣方於賣方營業處所或其他指定地，將貨物交付買方所指定的運送人或其他人，並將貨裝載上買方所提供的運送工具，即為賣方交貨及風險轉移。
- 如於賣方營業處所交貨，敘明指定交貨地方內的地址：
  FCA No.38 Courts Albert, 1er, Paris, France, Incoterms 2010
- A6/B6 費用劃分：賣方需辦妥輸出通關手續及負擔輸出應付的一切通關手續費、關稅、稅捐及其他費用。（但不負責輸入通關手續及進口稅）

3. CPT-Carriage Paid To(insert named place of destination)運費付訖條件規則（加填指定目的地）

- 適用任何運送方式或超過一種以上（複合運送）的運送方式。
- 定義：本規則意指賣方於一約定地方，把貨物交付其指定的運送人或其他人，且支付將貨物運送至指定目的地所需運費。
- 風險移轉：風險於貨物交付第一運送人時移轉。除非契約中另有敘明。
- A2/B2 賣方負責通過任何國家運送時所須的一切通關手續並自負風險及費用（買方負責辦理貨物輸入，及其通過任何國家運送的一切通關手續並自負風險及費用）。
- A3/B3 賣方需負擔至指定交貨的目的地的運費（買方負擔保險及費用）
- A6/B6 費用劃分：賣方需負責貨物裝載費用及至目的地的運費及需辦妥輸出通關手續及負擔輸出應付的一切通關手續費、關稅、稅捐及其他費用（但不負責輸入通關手續及進口稅）。

4. CIP-Carriage and Insurance Paid To (insert named plceof destination)運保費付訖條件規則（加填指定目的地）

- 適用任何運送方式或超過一種以上（複合運送）的運送方式。
- 定義：本規則意指賣方於一約定地方，將貨物交付賣方所指定的運送人或其他人，賣方須支付將貨物運至目的地所需的運送費用及保險費用。
- A3/B3 賣方僅需投保最低承保範圍的保險（ICC C），保險金額為出貨金額的110%，買方如擬獲得較大承保範圍的保障（例如 ICC A、ICC B、WAR RISK、STRIKE），需清楚地與賣方做此約定，或自行安排額外的保險。
- 在 CPT、CIP、CFR 或 CIF 條件下，當賣方將貨物交給運送人時，賣方即履行交貨義務而不是貨物到達目的地時。
- 若利用相繼運送人送至約定的目的地，則風險於貨物交付第一運送人時移轉。
- A6/B6 費用劃分：賣方需負責貨物裝載費用及至目的地的運保費及需辦妥輸出通關手續及負擔輸出應付的一切通關手續費、關稅、稅捐及其他費用（但不負責輸入通關手續及進口稅）。

5. DAT-Delivered At Terminal（insert named terminal at port or place of destination)終點站交貨條件規則（加填目的港或目的地的指定終點站）

●適用任何運送方式或超過一種以上（複合運送）的運送方式。

●定義：本規則意指賣方於指定目的港或目的地的指定終點站（包括任何地方，例如碼頭、倉庫、貨櫃場或公路、鐵路或航空貨物站），從到達運送工具卸貨，交由買方處置時，即屬賣方交貨。

●如買方要求賣方將貨物由終點站運至另一地點，則使用 DAP 或 DDP
如需要辦理出口通關，賣方需辦理輸出通關。（但不需負責輸入通關手續及進口稅）

●A5/B5 風險轉移：賣方負擔風險至交貨為止。

●A6/B6 費用劃分：賣方需支付一切費用至交貨為止。

6. DAP-Delivered At Place(insert named place of destination)目的地交貨條件規則（加填指定目的地）

●適用任何運送方式或超過一種以上（複合運送）的運送方式。

●定義：本規則意指在指定目的地，將準備卸載的貨物交由買方處置時，即屬賣方交貨。賣方負擔將貨運至指定地的一切風險（如需賣方負擔卸貨費用，合約上要另註明）。

●如需要辦理出口通關，賣方需辦理輸出通關（但不需負責輸入通關手續及進口稅）。

●A5/B5：風險轉移：賣方負擔風險至交貨為止。

●A6/B6：費用劃分：賣方需支付一切費用至交貨為止（包括輸出時應付的一切關稅、稅捐及其他費用，以及貨物通過任何國家運送時的費用）。

7. DDP-Delivered Duty Paid(insert named place of destination)稅訖交貨條件規則（加填指定目的地）

●適用任何運送方式或超過一種以上（複合運送）的運送方式。

●定義：本規則指賣方在指定目的地，將已辦妥輸入通關手續而尚未從運送工具卸下的貨物交由買方處置時，即屬賣方交貨。賣方須負擔將貨物運至該地為止的一切風險及費用，包含在目的地國家的任何輸入「稅負」（包括辦理通關手續的義務及風險，及支付通關手續費用、關稅、稅捐、加值稅及其他費用）。

● A5/B5 風險轉移：賣方負擔風險至交貨為止。

● A6/B6 費用劃分：賣方需支付一切費用至交貨為止（包括輸出輸入時應付的一切關稅、稅捐及其他費用，以及貨物通過任何國家運送時的費用）。

注意：EXW 表示賣方負擔最小的義務。

　　　而 DDP 則表示賣方負擔最大的義務。

8. FAS- Free Alongside Ship(insert named port of shipment)船邊交貨條件規則（加填指定裝運港）

● 本規則僅使用於海運或內陸水路運送。

● 定義：本規則意指賣方將貨放置於由買方所指定的裝船港船舶邊（例如：在碼頭上或駁船上）即屬賣方交貨。即自該時點起買方須負擔貨物滅失或毀損的一切費用及風險。

● 賣方須辦理貨物輸出通關手續（The FAS term requires the seller to clear the goods for export.）。

9. FOB-Free On Board(insert named port of shipment)船上交貨條件規則（加填指定裝船港）

● 本規則僅使用於海運或內陸水路運送。

● 定義：本規則意指賣方在指定裝船港將貨物裝載於買方所指定的船舶上時，即屬賣方交貨（購貨時亦如此）。

● 風險及費用：當貨物裝載於船舶上時，貨物滅失或毀損的風險即行移轉，買方自該時點起須負擔一切費用及風險。

● 賣方須辦理貨物輸出通關手續及負擔相關費用。

● 若為貨物於裝載於船舶前交貨，例如貨櫃場交貨，則應使用FCA（貨交運送人條件）。

10. CFR-Cost And Freight(insert named port of destination)運費在內條件規則（加填指定目的港）

● 本規則僅使用於海運或內陸水路運送。

● 定義：本規則意指賣方將貨物裝載於船舶上時，即屬賣方交貨。

● 風險及費用：當貨物裝載於船舶上時，貨物滅失或毀損的風險即行移轉，賣方必須訂約並支付將貨物運至指定目的地港所需的費用及運費。

- 因費用和風險轉移點不同，如買方認為必要，可於契約中註明裝船港。
- 賣方須辦理貨物輸出通關手續。
- 若為貨物於裝載於船舶前交貨，例如貨櫃場交貨，則應使用 CPT（運費付訖條件）。

11. CIF-Cost Insurance And Freight(insert named port of destination)運保費在內條件規則（加填指定目的港）

- 本規則僅使用於海運或內陸水路運送。
- 定義：本規則意指賣方將貨物裝載於船舶上時，即屬賣方交貨。
- 風險及費用：當貨物裝載於船舶上時，貨物滅失或毀損的風險即行移轉，賣方必須訂約並支付將貨物運至指定目的地港所需的費用及運保費。
- 因費用和風險轉移點不同，如買方認為必要，可於契約中註明裝船港。
- 賣方須辦理貨物輸出通關手續。
- 若為貨物於裝載於船舶前交貨，例如貨櫃場交貨，則應使用 CIP（運保費付訖條件）。
- 賣方須訂立保險契約，並支付保險費。賣方僅須投保最低承保範圍的保險。買方如擬獲得較大承保範圍的保障，需清楚地與賣方做此約定，或自行安排額外的保險（CIF 出貨金額＋ 10%＝保額）。
- 賣方須辦理貨物輸出通關手續。
- 若當事人不願以越過船舷交付貨物，則應使用 CIP（運保費付訖條件）。

INCOTERMS 2010 國貿條規之運用：

1. FOB/FCA 後面為出口港/機場 （例如：FOB Keelung / FCA New York Airport）
2. CFR/CIF; CPT/CIP 後面為目的地港口/機場 （例如：CIF New York）
3. FAS/FOB/FCA：成本相同，但風險轉移點不同。
4. FOB/CFR/CIF：成本不同，但風險轉移點相同。
5. FCA/CPT/CIP：成本不同，但風險轉移點相同。
6. DAP/DAT 都是送達客人指定目的地或終點站，包括運輸途中的運保費，跨國界或不跨國界（國內交貨）都可使用。賣方需負責出口通關報關費用，但不需負責進口通關報關費用及進口稅。DAP 賣方不負責卸貨；但 DAT 賣方則須負責卸貨。
7. EXW 是賣方唯一不需負責出口通關報關費用。離開工廠以外的費用皆由買方負擔。
8. DDP 是賣方唯一需負責進口通關報關及進口稅等費用。

附件二 報價單樣本(Quotation)

From: Cindy Lee
To: Peterson

QUOTATION

Dear Sirs,

Many thanks for your inquiry for our Silicone Rubber Pads and we are glad to quote you our best price and terms as below:

Item	Description	Quantity	Unit Price	Amount
P/No. SW-01	Silicone Rubber Pad 12 Keys	10,000 pcs	FOB Keelung US$1.20/pc	US$12,000

Packing method: Standard Export Packing.

N.W.: 10kgs   G.W.: 12kgs/carton

Shipment: Within 45 days after order confirmed.

Payment: 30% down payment and the rest 70% by T/T before shipment.

Minimum Quantity: 10,000 pcs.

Validity: 30 days or subject to our final confirmation when ordering.

Very truly yours,

Samwell International Inc.

_____

Cindy Lee

Export Manager

附件三　價目表樣本(Price List)

PFM Price-list

To: King-Tech Precision

Dear Sirs,

We are glad to offer you our best prices and terms as follows:

Testing Instruments

Quantity	PFM-1	PFM-2	PFM-3
1-20	@12,000 kr.	@ 15,000 kr.	@ 18,000 kr.
21-50	11,000	13,000	17,000
51-80	10,000	12,000	16,000
81-100	9,500	11,000	15,000
100-up	9,000	10,500	14,500

Basic terms:

1. All prices in Swedish Krona (SEK), are based on EX Works (INCOTERMS 2010) our factory in Sweden.

2. Shipment: in 6 weeks after order confirmed and receipt of T/T payment.

3. Payment: by T/T before shipment.

4. Validity: within 3 months or until our further notice.

Very truly yours,

PFM Instrument Mfg.

Mats Storm / Export Manager

附件四　預約發票樣本(Proforma Invoice)

KING-TECH VALVE PRECISION INDSTRY INC.

No.47, Lan 127, Sec. 2, Sinsheng N. Rd., Taipei, Taiwan

E-mail address: sb.wow@msa.hinet.net

PROFORMA INVOICE

To: Bull Mart Sdn. Bhd.

P.O.Box 4560 Petaling, Malaysia

Payment: By T/T before shipment.

Shipment: Within 6 weeks after receipt of payment

Via sea freight from Shanghai to Malaysia.

Bank Swift Code: TPBKTWTP555

P/O Number. BM-105

Date: 2012/02/02

Shipping Mark:

BMB

PORT KLANG

C/No.1-UP

MADE IN CHINA

Item	Description	Quantity	Unit Price	Total Amount
	Balancing Valve Model 41-05		USD	CIF Port KLANG USD
1	1"	100 PCS.	27.00	2,700.00
2	1-1/2"	50 PCS.	40.00	2,000.00
3	2"	50 PCS.	70.00	3,500.00

Total Amount:: US$ 8,200.00

Say Total Amount: U.S. Dollars Eight Thousand and Two Hundred only.

Seller:

KING-TECH Valve Precision Ind. Inc.

Signed by: *Kingsmang Joong*

Buyer:

Bull Mart Sdn. Bhd.

Signed by: *Bull Mart*

6. 報價及回覆函常用單字／片語

(1)offer sheet／quotation／price list 報價單/價目表

(2)as requested ＝ as you requested ＝ as per your request 如所要求

(3)as follows／as below 如下

(4)free sample／sample free of charge 免費的樣品

(5)sample charge 樣品費／postage 郵費

(6)for approval（樣品的）確認／in advance 事先（放句尾）

(7)order position 訂單情況／unit price 單價

(8)packing method 包裝方式：standard export packing 標準外銷包裝

(9)bulk packing 散裝／individual packing 個別包裝

(10)inner box 內盒／export carton 外箱／container 貨櫃

(11)net price 淨價／without commission 不含佣金

(12)payment method 付款方法：by T/T 電匯／Bank Check 銀行本票

(13)payment terms 付款條件：by irrevocable L/C at sight in our favor. 用我方為受益人的 不可撤銷即期信用狀

(14)down payment ＝ deposit 訂金／the rest ＝ the balance amount 餘額/尾款

(15)after sales service 售後服務／stock 庫存

(16)ship the goods ＝ make the shipment 出貨

(17)out of production 不生產／sample order 樣品訂單

(18)shipping method 出貨方法：by air freight 用空運／by sea freight 用海運 by air parcel post 用郵包/ by courier 用快捷／by express 用快遞

(19)the country of origin 原產地／It is made in Japan 產地是日本

(20)minimum order quantity 最低訂購量

(21)price 價錢／delivery time 交期／validity 有效期

(22)regarding ＝ as to ＝ as for 有關

(23)further information 進一步資訊

(24)Please advise us without hesitation. ＝ Please don't hesitate to advise us. 請不要遲疑告 知我們

7. 報價及回覆函常用主題句／說明句／結論句

主題句(Main Idea)

(1)Thanks for your e-mail of 1/16 with interest in our products and would like to quote you our best price and terms as follows.

(2)Thank you for your e-mail inquiry, but sorry to inform you that we are not available to offer you the item listed in your e-mail.

(3)Further to our telephone conversation this morning, attached please find the relevant quotation for the A1 project for your evaluation.

(4)In response to your e-mail inquiry dated 6/18/2007, we enclose our P/I showing you all relevant terms and conditions for the cooling fans.

(5)Many thanks for your inquiry for our Mother Board.

說明句(Explanation)

(1)As requested, we are glad to quote you as below.

(2)Pricing is estimated in US$ and subject to fluctuation.

(3)Packing: Standard export packing.

(4)Payment: 30% down payment by T/T when ordering, and the rest 70% by L/C before shipment.

(5)Delivery time: within 30 days after order confirmed.

(6)Minimum order quantity: 1000 sets

(7)Validity: 30 days or subject to our final confirmation when ordering

結論句(Conclusion)

(1)Please advise your order position soon.

(2)Thanks for your attention and look forward to hearing from your soon.

(3)Please do not hesitate to contact us if you need any further information or if you have any comment.

(4)Upon receipt of your payment by T/T, we will forward the sample to you right away.

範例 1　展覽回來依客戶要求提供禮品報價（Gift Item）

Dear Andy,

Thank you very much for visiting our booth at the Chicago Gift Item Show last month and showing your interest in our gift item PK-01. In reply to your inquiry, we are glad to offer you our best price and terms as follows:

Model no.　　　　Unit Price　FOB Macao

PK-01　　　　　US$35.00/set

Minimum order quantity: 1000 sets.

Payment terms: by irrevocable L/C at sight in our favor.

Shipment: within 4 weeks after order confirmed and receipt of the L/C.

Validity: within 20 days or subject to our final confirmation when ordering.

We are looking forward to your order position soon.

Best regards,

Lucy Lin / Innovation Gift Item Mfg..

文章結構

主旨：非常感謝您上個月參觀本公司在芝加哥禮品展的攤位，並表現出對我們的禮品 PK-01 有興趣。如所要求，我們很高興報給您最好的價錢及條件如下：

說明：型號: PK-01；　單價：每組 35 美元 FOB 澳門

　　　最低訂購量：1000 組。

　　　付款條件：不可撤銷我方為受益人的即期信用狀。

　　　出貨：訂單確認並收到信用狀後 4 週內。

　　　有效期：20 天內或訂購時再讓我方做最後確認。

結論：我們期盼盡快收到貴公司訂單。

生字／片語

1. booth = stand　攤位；gift item　禮品
2. as follows = as below　如下；in reply to = in response to　回覆
3. minimum order quantity　最低訂購量；payment terms　付款條件
4. shipment　出貨；validity　有效期；order position　訂單情況
5. subject to our final confirmation　再讓我方做最後確認

範例 2　進口商回覆範例 1 並提出要求

Dear Lucy,

Thanks for your offer for your model PK-01. However, in order to be competitive in the market, please try your best to make the cost down and offer us a further discount. Besides, we hope to know:

1. Do you accept your products with our logo?

2. Do you accept the same minimum order quantity for the products with our logo.

3. Can you accept the payment by D/A 30 days instead of L/C?

4. Please advise us your terms and conditions for a sole agent.

Please confirm the above soon as we really want to start business with you.

Best regards,

Andy Jobs

King-Mart Importer

文章結構

主旨：謝謝您對型號 PK-01 的報價，但是為了能在市場上競爭，請盡力降價並再報給我方多一些折扣。 此外，我們希望知道下列幾點：

說明：1. 貴方是否接受在產品上打上我方的商標？
　　　2. 針對我方商標的產品，是否接受一樣的最低訂購量？
　　　3. 可否接受付款條件為 30 天承兌交單代替信用狀？
　　　4. 請告知獨家代理的最好條件。

結論：請盡快確認以上因為我們真的想和貴公司開始做生意。

生字 / 片語

1. in order to be competitive in the market　為了能在市場上競爭
2. cost down　降價；further discount　進一步折扣
3. logo　商標
4. D/A 30 days (Document against Acceptance)　30 天承兌交單
5. terms and conditions　條件；sole agent ＝ exclusive agent　獨家代理

範例 3　回覆客戶詢問「化學原料」(Chemical Product)的報價

From: excellent@msa.hinet.net

To: brown@evatechnology.com

Subject: PA9000-5AA Chemical Material

Dear Mr. Brown,

Thanks for your e-mail of January 9.

As requested, we will arrange to send you 1 kg of PA9000-5AA chemical material by air parcel post tomorrow for quality approval.

As to the best price and delivery time, we would like to quote as below:

Price: US$1.14/kg at CIF London by sea freight

Quantity: 20 tons monthly

Payment: By irrevocable L/C at sight in our favor.

Packing: Bulk packing in plastic keg.

Shipment: in two weeks after receipt of your payment.

Validity: 30 days

Please advise your order position soon.

Best regards

Lucy Wang

Sales Manager

MODERN
BUSINESS
ENGLISH

Part 4

（三、報價及回覆函）
商用英文信函寫法

203

文章結構

主題：謝謝貴公司 1/9 的 e-mail。

說明：(1)如所要求，我們將於明日用郵包寄 1 公斤 PA9000-5AA 的化學原料給貴公司
　　　　做品質確認。

　　　(2)有關最好的價錢及交期，報價如下：

　　　　價錢：US$1.14/kg CIF 倫敦海運價

　　　　數量：每月 20 噸

　　　　付款：受益人為我方的不可撤銷即期信用狀

　　　　包裝：散裝於塑膠桶內

　　　　出貨：收到付款後兩週內

　　　　有效期：30 天

結論：請儘快告知訂單情況。

生字／片語

1. Chemical Material　化學原料

2. As requested　如所要求（寫法很多種）

　　＝ As you requested ＝ As per your request ＝ With regard to your request

　　＝ According to your request ＝ With reference to your request

　　＝ In accordance with your request

3. by air parcel post (＝ by APP)　用郵包

4. for quality approval　供品質確認

5. As to ＝ As for ＝ Regarding ＝ With reference to ＝ With regard to　有關

6. in our favor　受益人為我方

7. bulk packing　散裝；individual packing　個別小包裝

8. plastic keg　塑膠桶

9. order position　訂單情況

範例 4　回覆客戶詢問放大鏡(Magnifier)的報價

From: scosmos.mary<scosmos.mary@msa.hinet.net>

To: optician<optician@aol.com>

Subject: Reply to your inquiry for Magnifier

Dear Mr. Schoenjohn,

As requested, we enclose herewith an image of our standard magnifier and quote as below:

Item No. CT-212B Magnifier

Standard size: 188 mm×65 mm×0.5 mm (L×W×T)

FOB price: US$0.18/pc (without logo)

Logo imprint charge: US$0.02/pc for each color

Min. quantity: 5,000 pcs.

Shipment: within 30 days after order confirmed

Payment: by T/T before shipment

Packing: 50pcs/inner box, 100 pcs/export carton

(NW: 14 kgs; GW: 15 kgs; Measurement: 0.8 cuft for each carton)

Validity: 25 days from the date quoted.

Please check if the above meet your requirement; if not, please send us your image or picture by e-mail for quoting.

Thanks for your kind attention and look forward to hearing from you soon.

Best regards

Mary

V.P./Soar Cosmos International Inc.

文章結構

> 主題：如所要求，我們在此附上一張我們標準的放大鏡圖樣並報價如下：
>
> 說明：項目編號：CT-212B 放大鏡
>
> 　　　標準尺寸：188 mm×65 mm×0.5 mm（長×寬×厚度）
>
> 　　　FOB 價錢：每個 0.18 美元（不含商標印刷）
>
> 　　　商標印刷價：每一個每種顏色 0.02 美元
>
> 　　　最低數量：5,000 個
>
> 　　　交期：訂單確認後 30 天內
>
> 　　　付款：出貨前以電匯付款
>
> 　　　包裝：50 個裝一個內盒；100 個裝一個外銷紙箱
>
> 　　　　　（每箱淨重 14 公斤；毛重：15 公斤；材數：0.8 立方呎）
>
> 　　　有效期：報價日算起 25 天
>
> 結論：請查看以上是否符合貴公司需求，如不符合；請以電子郵件傳來貴公司的圖樣或照片以利報價。謝謝您的注意並盼望很快收到您的回音。

生字 / 片語

1. as requested ＝ as you requested ＝ as per your request　如所要求

2. herewith ＝ herein　在此

3. image　圖樣

4. standard magnifier　標準的放大鏡

5. as below ＝ as follows　如下

6. L×W×T ＝ Length×Width×Thickness　長×寬×厚度

7. logo imprint charge　商標印刷價格

8. min. quantity ＝ minimum quantity　最低數量

9. after order confirmed ＝ after you confirm the order　訂單被確認後

10. T/T: Telegram Transfer　電匯

11. inner box　內盒；export carton　外銷紙箱

12. NW ＝ Net Weight　淨重；GW ＝ Gross Weight　毛重；Measurement　材數／材積

13. cuft.＝ cubic feet　立方英呎；CBM ＝ cubic meter　立方公尺

14. meet your requirement　符合貴公司的需求（requirement ＝條件的要求）

範例 5 商會回函介紹尼龍纖維廠商

From: bureau@trade.com

To: matthew@samwell.com

Subject: Nylon 66 filament

Dear Matthew,

Regarding your request for nylon 66 filament, I have sent your letter to Monsanto Chemical Company and you should be hearing from them shortly. I spoke with one of their export managers and was told that they don't ordinarily export this item. However, they are interested in pursuing the idea with you. Their e-mail address is monsanto@earth.net and contact person is July Highfill.

If they are unable to help, please let me know, Dupont Chemical is another large company headquartered in Wilmington, Deleware. They may be of some assistance. The contact person is Dr. Linseman and his e-mail address is linseman@dupont.com.

We hope the above can be of the help to you.

Best regards

Sharon Lulu

International Representative

Bureau of International Trade and Development

Florida Department of Commerce

331 Collins building, 107 West Gaines Street

Tallahassee, Florida 32399-2000

www.bureauofintltrade.com

文章結構

主題：有關你們詢問尼龍纖維 66 一事，我已將你們的信轉傳給 Monsanto 化學公司，你們應該可以很快收到他們的回音。

說明：我與他們的外銷經理之一談過，他們說沒有定期外銷此產品，但是他們有興趣與你們進行此構想，他們的電子郵址是 monsanto@earth.net，聯絡人是 July Highfill。

如果他們無法幫忙，請讓我知道，Dupont 化學公司是另一家大廠，總部在 Wilmington, Deleware，他們可能會協助，聯絡人是 Dr. Linseman，電子郵址是 linseman@dupont.com。

結論：希望以上對貴公司有所幫助。

生字 / 片語

1. Nylon 66 Filament　尼龍纖維 66
2. Monsanto Chemical Company　孟山多化學公司
3. export manager　外銷經理
4. ordinarily　通常
5. pursuing (pursue)　進行
6. Dupont Chemical　杜邦化學公司
7. headquartered (V), headquarter (N)　總部
8. Wilmington, Deleware　地名
9. assistance　協助
10. contact person　聯絡人

範例 6　電腦產品報價

To: Korean Computer Import Co.

Attn: Mr. Pa

Re: Quotation for Computers

Dear Mr. Pa,

Thanks for your interest in our computer products.

Attached please find our revised quotation with following modifications:

(1)Please correct the specification of HDD to 1×40 MB.

(2)We add the price for a printer as you requested.

Please note the prices we quoted you for monitors and printers are the original factory prices without any profit, for which we just render you our service.

Also, please don't worry about the quality as all peripherals such as mother board, case, keyboard, etc. we chose for you are with the best quality and updated styles.

Regarding the payment you mentioned, under the mutual trust, we don't request any down payment or deposit money from you. The payment term shown on our enclosed quotation is Cash on Delivery. So, you just have to pay us the cash upon receipt of the goods. However, if you decide to place us the order, we need to have your formal order or to sign back our sales contract.

We hope everything is clear to you now. If you still have any question, please let us know at any time.

Best regards

Mary Huang

MODERN
BUSINESS
ENGLISH

Part 4 （三、報價及回
覆函）
商用英文信函寫法

209

文章結構

主題：謝謝您對本公司電腦產品的興趣

說明：(1)如附請見我們修正下列幾點的更新報價單：

　　　①請更正硬碟規格為 1×40 MB。

　　　②如所要求，我們增加 1 台印表機的價錢。

　　(2)請注意我們所報價的螢幕及印表機是原廠價，不含任何利潤，這是我們提
　　　供給貴公司的額外服務。

　　(3)同時，請不要擔心品質，因為所有我們為貴公司選擇的周邊產品如主機板、
　　　外殼、鍵盤等，都是最佳品質及最新版本。

　　(4)有關貴方所提付款，基於雙方共同信任，我們不要求任何定金，在我們如
　　　附報價單上的付款條件是交貨付現(COD)，因此當貴公司收到貨品的同時，
　　　再行付款即可。然而，當你們決定訂貨時，我們需要你們的正式訂單，或
　　　簽回我們的銷售合約書。

結論：希望一切都已清楚，如尚有任何問題，請隨時告知。

生字／片語

1. computer products　電腦產品
2. revised quotation　修正的報價
3. modification (N)，modify (V)　修改
4. correct the specification　更正規格
5. HDD　硬碟；printer　印表機；monitor　螢幕（顯示器）
6. original factory prices　原廠價
7. without any profit　毫無利潤
8. just render you our service　僅提供服務
9. peripherals　電腦周邊；mother board　主機板；case　外殼；keyboard　鍵盤
10. updated styles　最新版本；at any time　隨時
11. under the mutual trust　基於雙方共同的信任
12. down payment/deposit money　定金
13. Cash on Delivery (COD)　交貨付現
14. formal order　正式訂單；sales contract　銷售合約書
15. amend (V)　修改（信用狀），amendment (N)
　　revise (V)　修正（圖面／價錢），revision (N)
　　correct(N/V)　改正（錯誤），adjust (V)　調整（價錢）

範例 7 導電橡膠片報價

From: zamac@msa.hinet.net
To: tamoni@itwindustria.com
Subject: 104 keys condusctive rubber pad

Dear Mr. Tamoni,

We thank you and confirm the receipt of your e-mail dated Jan. 10, with the drawing 132 for the 104 keys conductive rubber pad.

After we carefully studied your specifications, we would like to quote you our best prices based on following two alternatives：

    (1)104 keys conductive rubber pad

        made by LSR(Liquid Injection Method)

        FOB Taiwan at US$2.60/pc

        Tooling Cost: US$11,000 for a 3 cavities tool

    (2)104 key conductive rubber pad

        made by molded method

        FOB Taiwan at US$3.00/pc

        Tooling Cost: US$6,000 for a 2 cavities tool

        Tooling time: 45-60 days

        Payment for tooling cost: 1/2 to be paid in advance when ordering and the rest 1/2 to be paid upon sample approval

        Payment for shipment: by irrevocable L/C at sight.

Thanks for your consideration and don't hesitate to let us know of your comments to the above soon.

Best regards
Betty Yang/VP

文章結構

主題：謝謝您並確認收到貴公司 1/10 的 e-mail 與 104 鍵導電橡膠片的圖面 132。

說明：仔細研讀你們的規格後，我們報下列兩種選擇的最好價錢：

(1)104 鍵導電橡膠片——液體射出成型法

　　FOB 台灣價：每片 US$2.60

　　模具費：US$11,000（3 穴）

(2)104 鍵導電橡膠片——壓模成型法

　　FOB 台灣價：每片 US$3.00

　　模具費：US$6,000（2 穴）

　　開模時間：45-60 天

　　模具付款：1/2 定金（下訂單時付一半），樣品確認後付尾款 1/2

　　出貨付款方式：不可撤銷即期信用狀

結論：謝謝您的考慮，請不要遲疑讓我們知道您對我們以上報價的意見。

生字 / 片語

1. conductive rubber pad　導電橡膠片

2. confirm　確認

3. alternatives　二者擇一；replacement　賠貨／換貨；substitute　替代品

4. LSR(Liquid Injection Method)　液體射出成型法

5. molded method　壓模成型法

6. tooling cost　模具費

7. tooling time　開模時間

8. in advance　事先

9. the rest　餘款（＝ the balance）

10. upon sample approval　樣品確認同時

11. Thanks for your consideration　謝謝您的考慮

12. comments　意見

範例 8　電腦周邊──主機板(Mother Board)的報價

SPOT M/B SUPPLIER

To: Mr. Scott Aloi/SANTO Tech Co.

Re: Mother Board (M/B)

Many thanks for your inquiry about our M/B.

The attachment is the catalog of our updated M/B for your study. Since our products all are made under very strict quality control and all are of high compatibility, we would suggest you to order a sample for test, then you will certainly like its performance. Our Proforma Invoice for the sample charge and postage is attached. Upon receipt of your payment by T/T, we will forward the sample to you right away.

Moreover, we also provide very good after sales service, if you have any technical problems, our engineers would be glad to serve you at any time.

We are awaiting your sample order soon.

Best regards

文章結構

主題：感謝您對我們主機板的詢價。

說明：附件是我們最新的主機板目錄供研讀，由於我們的產品都是在嚴格品管之下完成，且具有高度相容性，我們建議您訂購一個樣品測試，您將會喜歡我們的產品性能，附上樣品費及郵費的預付發票，當一收到貴公司電匯付款，樣品將馬上寄出。此外，我們也提供最好的售後服務，如果您有任何技術問題，我們的工程師將很樂意隨時為您服務。

結論：等候貴公司的樣品訂單。

生字／片語

1. made under strict quality control　在嚴格的品管之下做成

2. of high compatibility　具有高度相容性；engineer　工程師

3. after sales service　售後服務；technical problem　技術問題

範例 9 銀質原料(Silver Material)的替代品報價

Dear Neil,

Thanks for your e-mail inquiry, but sorry to inform you that .999 silver material is not available here. The finest one here is .925 silver, and the relevant quote is as below:

Material: .925 silver; Diameter: 40 mm; Thickness: about 2.0 mm;

Min. quantity: 1000 pcs.

FOB Price: US$13.50/pc

Packing: One piece per poly bag and 20 pcs in a standard export carton

Shipment: within 25 days after sample approval

Payment: 1/2 by T/T upon order confirmed, and 1/2 by T/T before shipment

Please check if the above meet your request, or let us know your comments.

Best regards

Mary/VP

文章結構

主題：感謝你們的 e-mail 詢價，但是抱歉告知這裡沒有.999 銀質原料。

　　　這裡最好的銀質原料為.925，其相關報價如下：

說明：材質：.925 銀；直徑：40 mm；厚度：約 20 mm；最低量：1,000 片；

　　　離港價：每片 US$13.50；包裝：每片裝一個塑膠袋，20 片裝一個標準外銷紙箱；出貨：樣品確認後 25 天內；付款：訂單確認時電匯 1/2 貨款，出貨前再電匯 1/2 尾款。

結論：請查看以上是否符合貴公司需求，或請示意見。

生字 / 片語

1. the finest one　最精細的；relevant quote　相關報價
2. diameter　直徑；thickness　厚度；Min. quantity　最低量
3. standard export carton　標準外銷紙箱；comments = opinion　意見
4. by T/T (Telegram Transfer)　用電匯；by bank check　用銀行本票
5. upon order confirmed　當訂單一被確認時；before shipment　出貨前

範例 10　澳洲廠商回覆奶粉（Milk Powder）報價

From: bobfitz@milkaustralia.com

To: mikepau@milkpowder.com

Subject: Milk Powder

Dear Mike,

We have just received an offer for the product of Milk Powder.

The origin is Eastern Europe, and the quality is "Extra Grade", which specs is as per the attachment.

Packing: 4 layers paper bags with approx. 16/17 MT per 20' container

Price: US$1,500/MT FOB Rotterdam

Offer is as received without our commission.

Please take above into consideration.

Best regards

Bob Fitz

文章結構

主題：我們剛收到奶粉產品的報價。

說明：產地是東歐，品質是「特級品」，規格參附件。

　　　包裝：4 層紙袋裝，裝一個 20 呎整櫃，大約 16～17 噸。

　　　價錢：每噸美金 1,500 元 FOB 鹿特丹。

結論：以上報價是原廠價，不含我們的佣金，請考慮。

生字 / 片語

1. Milk Powder　奶粉
2. A.M. = above mentioned　以上所提
3. The origin is Eastern Europe　產地是東歐
4. Extra Grade　特級品
5. 4 layers paper bag　四層紙袋
6. container　貨櫃（通常一個 20 呎貨櫃最重不得超過 18 噸）
7. as received　如所收到（原廠價）
8. without our commission　不含我方佣金

練習 14　報價及回覆函練習

1. 將下列翻譯成英文

（注意：劃底線部分的 Adj.及 Adv.的位置要調整

N1 的 N2→N2 of /for/in N1；時間／地點／方法的副詞要放句尾）

(1)謝謝您 <u>10/28</u>　的　詢價，我們很高興報價如下：（＝詢價 of 10/28）

(2)如所要求，我們用郵包寄上一個免費樣品以供確認品質（by APP 放句尾）

(3)付款條件是　<u>受益人為我方</u>　的不可撤銷　<u>即期</u>　信用狀

(4)請　<u>儘快</u>　告知訂單情況

(5)包裝是標準外銷包裝

(6)以上報價是淨價不含我方的佣金

(7)我們要求 30%定金，<u>出貨後 30 天內</u>　付清尾款

(8)我們提供很好的售後服務

(9)我們有庫存，當一收到貴方　<u>用電匯</u>　付款，我們可以　<u>馬上</u>　出貨

(10)如附是我們的預付發票，請　<u>用銀行本票</u>　付樣品費及郵費

(11)如需要進一步的資料，請　<u>儘快</u>　告知不要遲疑

(12)很抱歉，此產品已不生產了

(13)我們希望　<u>很快</u>　收到你們的樣品訂單

(14)產地是日本

(15)最低訂購量是一個 20 呎的貨櫃

2. 請依下列大意寫一封英文報價回覆函

主題：謝謝您 1/15 的 e-mail，表示對於我們的毛衣(Sweaters)有興趣。

說明：如所要求，我們很高興報價如下：

價錢：每打美金 35 元 FOB 台灣淨價

交期：收到付款後 30 天內

包裝：標準外銷包裝

付款：不可撤銷保兌即期信用狀

最低訂購量：1000 打

有效期：30 天

有關樣品，收到樣品費後，我們將　<u>馬上</u>　寄出。

結論：如尚需進一步資料，請不要遲疑來信告知，我們等候貴公司的訂單。

MODERN BUSINESS ENGLISH
現代 商用英文

3. 依照中文大意填入正確英文單字 / 片語：

(1)謝謝你們對本公司產品感興趣。

Thank you for your _____ in our _____ .

(2)如所要求，我們很高興報價如下：

As _____ , we are glad to quote you _____ _____ :

(3)我們的最低訂購量是 10,000 個。

Our _____ _____ _____ is 10,000 pcs.

(4)如尚需進一步資訊，請不要遲疑來信告知。

If you still need any _____ _____ , please advise us without

_____ .

(5)我們期盼聽到貴公司訂單情況。

We are looking forward to _____ your _____ _____ .

4. 請填入適當的語詞以完成下列書信

A. consideration	B. conclusion	C. gross weight	D. surprise
E. Measurement	F. confirmed	G. interest	H. standard
I. as below	J. quotation	K. following	L. confirm

Dear Tom,

RE: Quotation for Rubber Shoes

Thanks for your e-mail dated 6/18/07 showing your ____(1)____ in our rubber shoes.

As requested, we are glad to quote you our best price and terms ____(2)____ :

Item No. CT-110

FOB price: US$10/pair

Min. quantity: 5,000 pairs

Shipment: within 30 days after order ____(3)____ .

Payment: by T/T before shipment

Packing: Standard export packing. (GW: 12kgs; ____(4)____ : 0.5cuft/carton)

Validity: 25 days from the date quoted.

Please take the above into ____(5)____ and confirm your order soon.

Best regards

填入答案

| 題號 | (1) | (2) | (3) | (4) | (5) |
| 答案 | | | | | |

四、討價還價函(Bargain/Negotiation Letters)

討價還價的商議原則如下：

1. 賣方(the seller)不論利潤高低，總是要堅稱自己的報價利潤很低，沒利潤，甚至低於成本(under the cost)。

2. 買方(the buyer)的原則剛好相反，不論賣方價錢再低，總是還要嫌貴，或假稱其他競爭者報價更低，或預算有限，加減再砍殺一點價錢。

   注意：買方既來殺價，其實已有意向賣方購買，只是想多留一點利潤給自己；賣方即使利潤再低，最好意思意思降一點，給買方面子。但不要一次降價太多，否則買方會認為賣方利潤很高，每次都會故意殺價。

3. 除了議價以外，買方也可能要求賣方在付款條件方面做讓步，例如改用 D/A、D/P 方式，不要用 L/C，或要求縮短交期，或贈送免費備品等。

4. 賣方在回覆時有四個選擇狀況

   (1)不接受

   如不接受要告知原因，例如品質較好，材料成本高(good quality, costly material)；原料漲價(raw material cost has been raised)；人工上漲(labor cost has been increased)；新台幣升值(NTD appreciation/NTD has been appreciated)；匯率浮動劇烈(fluctuation of the exchange rate)；買方的還價低於成本太多，無法接受……等。

   (2)接受

   如接受買方的還價(counter offer)，不論價錢是否仍有利潤，不要答應的太爽快，大都會聲稱自己已無利潤或低於成本，接受的原因是想與買方建立生意關係（新客戶）；或因其為老客戶／長期大客戶，特別犧牲接受，特別幫忙。

   (3)條件式接受

   表示願意接受買方的還價，但是有附帶條件，例如請買方增加訂單數量；改變付款條件（先用 T/T 事先付款，或將遠期改為即期信用狀，或先給部分定金等）；或將多批出貨改為一次出貨，以節省報關手續及銀行費用；或建議其買較便宜的替代品(alternative)或別的機型。

   (4)折衷還價

   賣方有時為了表示誠意(to show our good faith)，雖無法照買方開價完全接受，但仍折衷多少降一點價錢，例如買方要求給 10%折扣，賣方折衷還價，只願意接受 5% 或 3%折扣。

5.買方還價函寫法

> 主題：先感謝賣方的報價及寄來的相關資料，但很抱歉其價錢偏高，或比同行的報價高許多。（參範例 1/2/7）
>
> 說明：舉出嫌貴的原因，例如景氣不好、市場競爭激烈、欲購買的數量很大、後續還有許多大訂單、預算有限。或直接告知目標價……等。（參範例 1/3/5）
>
> 結論：請賣方再考慮是否接受買方開價，或重新報較好的價錢條件。（參範例 3/6/7/10）

6.賣方回覆還價函的寫法

> 主題：先表示很遺憾或很抱歉聽到買方不滿意自己的價錢，或認為自己價錢偏高。（參範例 2/4）
>
> 說明：(1)先表示自己已報最好的價錢，此報價幾乎已無利潤。（參範例 4）
>
> (2)再表示自己經仔細考慮後，願意特別接受／無法接受／條件式接受／或折衷還價，原因及條件參照以上第 4 點寫法。（參範例 9）
>
> 結論：(1)如接受，請買方儘快確認訂單。（參範例 9）
>
> (2)如不接受，向其道歉，並表示希望下次有機會能再合作。（參範例 4/11）
>
> (3)如條件式接受或折衷還價，請其考慮，儘快告知意見。（參範例 2）

7. 討價還價函常用單字／片語

(1)re-quote/re-offer　重報／a better price　較好的價錢

(2) a 20 feet container/a 20-foot container　一個 20 呎櫃

(3)target price／idea price/ideal price　目標價

(4)competitor　同行／競爭者

(5)raise the price a lot ＝ increase the price a lot　漲價漲很多

(6)local customers　國內的客戶／good relationship　關係良好

(7)strong position　很強的地位 in the market　在市場上

(8) annual sales amount ＝ yearly turnover　年營業額

(9)state-run company　國營公司

(10)sole supplier ＝ only supplier　唯一的供應商

(11)high value product　高價值的產品

(12)sample charge and postage　樣品費及郵費

(13)wage　薪資／overhead cost ＝ overheads　管銷費用

(14)profit　利潤／discount　折扣

(15)one sample for each model　每種各一個樣品

(16)sole agent ＝ exclusive agent　獨家代理

(17)terms and conditions　條件

(18)to show our faith ＝ to express our sincerity　為了表示誠意

(19)NTD to USA has been appreciated for 10%　台幣對美元升值 10%

(20)appreciation　升值／devaluation　貶值

(21)under the cost　低於成本

(22)freight cost　運費／broker charge　報關費

(23)customs charge　海關費用／duty　關稅／tax　發票稅

(24)life cycle ＝ life time　壽命期／forecast quantity　預估量

(25)free sample ＝ sample free of charge　免費的樣品

(26)for quality approval　供品質確認

(27)to get a good market share　得到一個好的市場佔有

8. 討價還價函常用的主題句／說明句／結論句

主題句(Main Idea)

(1)Thanks for your quotation, but we found you raised the price a lot.

(2)We received your quotation but we found that your price is 10% higher.

(3)We regret to inform you that we can't accept your offer.

(4)We are sorry to hear that you think our offer too high.

(5)Could you re-offer us a lower price as our quantity is 20,000 pcs.

說明句(Explanation)

(1)Please re-quote us a better price for a twenty feet container.

(2)Our target price is US$10/pc.

(3)Your price is 10% higher than that of the other competitor (supplier).

(4)As you know that NTD to USD has been appreciated for more than 10%, the price we offered you is already our bottom price.

(5)Since the local wages and overhead cost have been much increased, we can't accept the order under the cost.

(6)To show our faith and to support you, we would specially accept 5% discount for the order quantity of 20,000 pcs.

結論句(Conclusion)

(1)Please take the above into consideration and decide your order soon.

(2)We are sorry we couldn't help you in this case. However, we hope we can cooperate with you in the future deals.

(3)Please kindly confirm your acceptance to the above soon.

(4)Please try your best to make the cost down and let us know of your revised quotation soon.

(5)We hope our above support will help you get a good market share.

範例 1　買方來函抱怨字鍵(Keytops)報價貴 10%，請賣方重報較好價格

From: richard@keyway.com

To: anneyang@goodwell.com

Subject: Your quotation for keytops

Dear Anne,

Thanks for your quotation dated Jan. 15 for the keytops with USA version.

We find your price is 10% higher. Could you re-quote us a lower price as we are interested in buying 40,000 sets of keytops.

We are awaiting your answer by return.

Best regards

Richard Gill

文章結構

主題：謝謝您在 1/15 提供美國版字鍵的報價。

說明：我們發現你們的價錢高了 10%，能否重報較低的價錢給我們，因為我們有興趣購買 40,000 組的字鍵。

結論：等候你們儘快回音。

生字／片語

1. Keytops　（鍵盤的）字鍵

2. USA version　美國字體版本

3. re-quote a lower price ＝ re-offer a better price　重報較好價錢

4. set　組／台（單位）

5. We are awaiting your answer by return.

　＝ We are waiting for your answer by return.　我們等候你們盡快回音

範例 2 出口商回覆範例 1，對買方的還價採折衷方式

From: anneyang@goodwell.com

To: richard@keyway.com

Subject: Your mail of January 18 for Keytops

Dear Richard,

We are sorry to hear that you think our last offer for the above too high.

After careful re-checking, we would specially reduce 5% for the order quantity of 40,000 sets to show our sincerity.

As you know, NTD to USD has been appreciated for more than 10%, the price we re-offered above is already our cost.

Please take the above into consideration and decide your order soon.

文章結構

主題：很抱歉聽到貴方認為我們對以上產品的報價太高。

說明：我們仔細重算後，為了表示誠意，針對 40,000 組的訂單量願意特別降價 5%，如您所知，新台幣兌美金已升值超過 10%，以上重報的價錢已是我方的成本價。

結論：請考慮以上，並儘快決定訂單。

生字 / 片語

1. to show our sincerity = to express our faith　為了表示誠意

2. NTD to USD has been appreciated for more than 10%　新台幣兌美金已升值超過 10%

3. re-offered　重報；consideration(N)　考慮；decide　決定

MODERN
BUSINESS
ENGLISH

Part 4 （四、討價還價函）

商用英文信函寫法

223

範例 3　買方還價並請查看是否有便宜的替代品──電腦外殼
　　　　（Computer Case）

From: jefflauwe@unioncomputer.com

To: Jerrod

Subject: Your Offer for Computer Case

Your FOB price of US$18/set is not good enough. We must have a price of US$15/set to get the business. We suggest you re-examine the ESD shield price, also study the alternative of using a wire for ESD protection or perhaps paint the underside of the frame to reduce the cost.

Please reply the above before this Friday.

Best regards

Jeff Lauwe

文章結構

主題：有關你們電腦外殼 FOB 報價每組美金 18 元太貴，為了爭取訂單，我們必需要求每組美金 15 元。

說明：我們建議你們重新檢查防幅射板的價錢，也請查看是否可以用替代的鐵線當作防幅射保護，或者也許在面板內部塗漆以降低成本。

結論：請於週五前回覆以上。

生字／片語

1. computer case　電腦外殼
2. to get the business ＝ to obtain the order ＝ to secure the order　爭取（取得）訂單
3. re-examine ＝ re-check　重新檢查
4. ESD Shield　防幅射板；ESD protection　防幅射保護
5. alternative　另一種選擇方式
6. wire　鐵線；paint　油漆；underside of the frame　面板內部
7. to reduce the cost　以降低成本

範例 4　賣方回覆範例 3，不接受買方的還價，因為低於成本

From: Jerrod

To: jefflauwe@unioncomputer.com

Subject: Computer Case

We regret to hear that you can not accept the price we offered.

After we carefully checked your target price, we are really sorry to inform you that US$15/set is under our cost. To avoid the quality problem happened after shipment, we will not consider adopting any other alternative.

Sorry we couldn't help you in this case, however, we hope we can cooperate with you in the future deals.

文章結構

主題：有關電腦外殼，很遺憾聽到貴方無法接受我方報價。

說明：仔細檢查貴方期望的目標價後，抱歉告知我方無法接受，因每組美金 15 元實在低於成本，而為了避免出貨後發生品質問題，我方將不考慮採用任何其他替代方式。

結論：抱歉無法協助此案；然而，希望在未來生意上能再與貴公司合作。

生字 / 片語

1. We regret ＝ We are sorry ＝ We are regretful　我們很抱歉
2. target price ＝ idea price ＝ ideal price　目標價／理想價
3. under our cost　低於我方成本
4. To avoid the quality problem　為了避免品質問題
5. happened after shipment　在出貨後發生
6. adopting any other alternative　採用任何其他替代方式
7. in this case　在此案
8. in the future deals　在未來的生意
9. cooperate(V)，cooperation(N)　合作

範例 5　進口商告知單價可接受，但是模具費太高──線路板組裝 （PCB Assembly）

From: pinhorse@tonton.com

To: johnmore@taiwanpcb.com

Subject: PCB Assembly/Your Quotation TP-0325

Thanks for your above quotation. The price for PCB assembly seems OK, but our customer said that he had been quoted tooling cost of under US$6,000.- by the other competitor.

We suggest you re-quote an inexpensive aluminum tool capable of only 40,000 pcs maximum life time.　Then we will re-quote to the customer on the same basis.

Please reply by return.

文章結構

主題：有關貴公司報價單 TP-0325 線路板裝配。
　　　謝謝貴公司以上報價，線路板裝配費似乎可接受，但是客戶說他被別家同業報予低於 6,000 美金的模具費。
說明：我們建議你們重報一個便宜的鋁模，只要能生產最多 40,000 個的壽命期即可，然後我們將根據同樣條件報給客戶。
結論：請儘快回覆。

生字／片語

1. PCB(Printed Circuit Board)　線路板；Assembly　裝配
2. competitor　同業／同行／競爭者
3. tooling cost　模具費
4. an inexpensive aluminum tool　一個便宜的鋁模
5. capable of only 40,000 pcs maximum　能生產最多 40,000 個
6. life time ＝ life cycle　壽命期
7. on the same basis　根據同樣條件

MODERN BUSINESS ENGLISH 現代商用英文

範例 6　進口商還價要求 15%折扣

From: davidwu@camel.com

To: rainchild@leecmedical.com

Subject: Cooled Incubator/Your offer of 1220

We received your above quotation, but we found that you raised the price a lot. However, we still tried very hard to negotiate with our customer with your new offer.

According to our previous deals, you always granted us 15% discount from your formal quotation. So, we offered the same discount to the buyer this time and got the order finally.

Please kindly confirm your acceptance of this order at GBP586.50 by return. Thanks for your help in advance.

文章結構

主題：我們收到你們 12/20 冷凍細菌培養器報價，但是發現你們價錢上漲許多。然
　　　而我們仍盡力以你們新的報價與客戶交涉。

說明：按照我們過去的交易，你們總是從正式報價單中，再給予我方 15%折扣，因
　　　此這次我們也報相同的折扣給客戶，最後拿到了訂單。

結論：請即確認願意以 586.50 英鎊接受此訂單，在此先感謝您的協助。

生字 / 片語

1. You raised the price a lot = You increased the price a lot　你們將價錢上漲許多
2. tried very hard = did our very best　盡最大努力
3. negotiate　交涉；GBP = Great British Pound　英鎊(= Sterling)
4. According to our previous deals　根據我們以前的交易
5. granted(grant)　給予；formal quotation　正式報價單
6. We offered the same discount　我們報予相同的折扣
7. confirm your acceptance　確認貴方接受

範例 7　進口商告知價錢偏高太多，除非大幅降價，否則將接不到訂單

　　　　產品：主機板(Mother Board)

SANTO TECH CO.

Houston, USA

To: Mr. Lois Lin/Spot M/B Supplier　　　　　　　From: Scott Aloi

Re: Your mail of 2/15 for M/B

Dear Mr. Lin,

As we previously mentioned, your price quoted for M/B is much higher than that of the other competitor. We have found out that the actual difference is US$1.50/set.

The customer would like to give us the opportunity to get the business because of our very good relationship with them, but your price must be "within reason" or somewhat in line with the competition.

Since the customer estimates the usage to be 15,000 sets, the difference US$1.50×15,000 sets ＝ US$22,500.- , this is asking a lot!

We suggest that you review your cost very carefully because with a difference of US$22,500.-, we will definitely not get the business.

We are awaiting your explanation and reply soon.

Best regards

Scott Aloi

General Manager

文章結構

主題：如前所述，你們主機板報價比另一家競爭者高太多，我們已經查出每片貴 1.50
美金。

說明：由於我方的極良好關係，客戶願意給我們生意機會，但是你們的報價必須「合
理」或在某種競爭水平之內。由於客戶預估用量為 15,000 組，差額美金 1.50
元×15,000 組＝美金 22,500 元，那就太獅子大開口了！

結論：我們建議你們仔細重審成本，因為差額美金 22,500 元是鐵定拿不到訂單的，
等候你們儘快回覆說明。

生字／片語

1. As we previously mentioned　如我們之前所提

2. your price quoted for M/B is much higher than that of the other competitor　貴方主
機板報價比另一競爭者高太多

   注意：(1)quoted for M/B(Adj. Ph.)＝which was quoted for M/B(Adj. Cl.)

   　　　　(2)much higher 高太多，用 much 表示

   　　　　(3)than that of the other competitor

   　　　　　＝ than the price of the other competitor

   　　　　　　（that of 不可省略，否則變成競爭者、人與價錢比較）

   　　　　　＝ than the other competitor's price（另一種寫法）

   　　　　　＝ than the other competitor's（price 可省略）

3. actual difference　實際差額

4. opportunity　機會

5. very good relationship　極良好關係

6. within reason　在合理範圍內

7. somewhat in line with the competition　在某種競爭水平之內

8. estimates the usage　預估用量

9. This is asking a lot!　這要求太多＝獅子大開口

10. review　重新審閱

11. We will definitely not get the business.　我們鐵定拿不到訂單

範例 8 客戶欲增購數量，但要求較低報價

To: National Precision Industry Ltd.

Dear Philips,

We are going to buy three more units of your Printing Machine PM-3. Since we have bought one unit of the same machine early this year, if we add three units more into this order to make the total order quantity reach four units, please kindly confirm if you can specially accept the unit price US$50,500/unit which you offered for the quantity of 5-10 units.

We will confirm our order upon receiving your confirmation to the above.

Best regards,
Hunter Lee / SB Importer

文章結構

主旨：本公司欲增購 3 台貴公司 PM-3 印刷機。

說明：因我們於今年年初已購買過 1 台，現想增購 3 台一樣的機器，使訂單量達到共 4 台。請告知是否可以惠予接受 5～10 台的單價 50,500 美元。

結論：當我們接到貴公司對以上的確認，會馬上確認訂單。

生字 / 片語

1. Printing Machine　印刷機
2. add　增加
3. make the total order quantity reach four unit　使訂單量達到 4 台
4. confirm our order　確認我方訂單
5. upon receiving your confirmation to the above　收到貴方對以上的確認

範例 9　回覆範例 8 因匯率浮動，賣方無法接受較低報價

To: SB Importer

Dear Hunter,

We are glad to know that you are going to buy three units more of our Printing Machine PM-3, but we are sorry to inform you that the best price we can offer you is to keep US$52,800/unit for the quantity of 1-4 units according to our revised price list effective from this January due to the fluctuation of the exchange rate.

However, if you really hope to get the unit price US$50,500/unit, we would suggest that you increase the order quantity to 5 units to get the better offer.

Please consider the above and confirm your order soon.

Best regards

Philip Dollars/ Export Manager

National Precision Industry Ltd.

文章結構

主旨：很高興得知貴公司欲再增購 3 台本公司 PM-3 印刷機，但是很抱歉告知因匯率浮動關係，我方最好價錢是仍維持每台 52,800 美元，參照今年 1 月生效對於 1～4 台的最新修正價目表。

說明：然而，如貴公司真得想取得每台 50,500 美元較好的價格，我們建議你們將數量增加到 5 台。

結論：請考慮以上，並盡快確認你們的訂單。

生字 / 片語

1. according to　按照；revised price list　最新修正的價目表
2. effective from this January　從今年 1 月開始生效
3. increase the order quantity　增加訂單量
4. to get the better offer　以便取得較好的價錢
5. please consider the above　請考慮以上

範例 10　進口商要求重報小量價錢，並告知預購目標價

　　　　產品：無線電話(Cordless Telephone)

 ENGLISH ELECTRONICS LTD.

London, U. K.

To: Nelly Hu/Phone Trade Co.

Re: Cordless Telephone Model 286

Dear Nelly,

Thank you very much for your speedy reply. Our customer has come back to us saying that they can not take 10,000 sets of Cordless Telephone Model 286. Please re-quote the price based on 2,000 sets.

Also, please note your price of US$11.70/set is not competitive enough, because once we add freight, broker charge, customs charge and duty, etc., our landed price is US$14.65/set. Besides, we and our rep. want to make profit of US$2/set. Based on this, we need a price of US$10/set.

Please start negotiating with the factory and advise the best price ASAP.

Best regards

Melanie Youth

Purchaser

文章結構

主題：謝謝您快速回覆，有關無線電話型號 286，客戶告訴我們他們無法購買 10,000 台，請根據 2,000 台的數量，重新報價。

說明：同時，請注意你們的價錢每台US$11.70 沒有競爭性，因為當我們加上運費、報關費、海關費用及關稅等，登陸價就已高達每台US$14.65，此外，我們與我們的業務代表每台要賺US$2，基於此，我們需要的價錢是每台美金 10 元。

結論：請進行與工廠交涉並儘快告知最好的價錢。

生字／片語

1. Cordless Telephone　無線電話
2. speedy reply ＝ quick reply　快速回覆
3. based on ＝ for ＝ according to　根據
4. not competitive enough　不夠有競爭
5. add　加上 (V)
6. freight ＝ freight charge　運費
7. broker charge　報關費
8. customs charge　海關費用
9. duty　關稅
10. landed price　登陸價
11. rep. ＝ representative ＝ sales rep.　業務代表
12. make profit　賺利潤
13. negotiating (negotiate)　交涉
14. ASAP ＝ as soon as possible　儘快

範例 11　出口貿易商表示數量太小，無法報價，請進口商放棄此詢價

　　　　產品：熱軋鐵板（Hot Rolled Steel Sheet）

From: paulpu@koreametal.com

To: howardli@samwellg.com

Subject: Hot Rolled Steel Sheet for Phillipine

Dear Howard,

We regret to inform you that we are not in a position to ask for the quotation from the Sole Manufacturer POSCO with such a small quantity of 69 M/T, which is estimated by us approximately US$25,000 / 69 tons.

Please be advised that POSCO is a State-Run Company and the Sole Manufacturer of this item in Korea. This item is politically on long term basis and they don't respond to spot inquiries from trading companies.

So, we would suggest you not to promote this item for following reasons:

(1)The customer already has stable source of supply. This fact leads us into hard competition and also makes us difficult to get long term contract.

(2)This item is not a high-value product, E.G. US$25,000/69 tons. If we don't make contract for thousands of tons, we can not make money.

However, the most interesting business will be definitely Stainless Steel Sheet. Please concentrate in promoting this item which is much more profitable for both of us.

Sincerely Yours,

Paul Pu

文章結構

> 主題：很抱歉我們無法以 69 噸如此少的量，從唯一的工廠 POSCO 取得報價，此 69 噸，預估大約美金 25,000 元。
>
> 說明：請瞭解 POSCO 是韓國國營公司，也是此產品唯一的工廠，此項產品是根據長期政略性的，他們並不提供貿易公司個別詢問的報價（因鋼鐵生產有限，採配額供應，一般只報價給固定採購的長期客戶）。
>
> 結論：因此，我們建議你們不要推銷此項產品，原因如下：
>
> (1)此客戶已有固定貨源，這個事實會導致我們很難競爭，也很難拿到長期合約。
>
> (2)此項目並非高價值的產品，例如 69 噸才美金 25,000 元，如果我們沒有拿到上千噸的合約，是不可能賺錢的。
>
> 　　然而，最令人感興趣的生意當然是不銹鋼板，請專注推銷對你我雙方都有利的這項產品。

生字 / 片語

1. Hot Rolled Steel Sheet　熱軋鐵板
2. We are not in a position to ask for the quotation　沒有立場要求報價
3. approximately ＝ approx.　大約
4. State-Run Company　國營公司
5. Sole Manufacturer　唯一的工廠
6. politically　政略性地
7. on long term basis　根據長期訂購原則
8. spot inquiries　個別詢價（單一詢價）
9. promote this item　推銷此項產品
10. stable source of supply　穩定（固定）；source ＝ supplier　供料貨源（廠商）
11. This fact leads us into hard competition　此事實將導致我們難以競爭
12. long term contract　長期合約
13. high-value product　高價值產品
14. thousands of tons　數千噸；E.G. ＝ for example　例如
15. definitely　肯定地；Stainless Steel Sheet　不銹鋼板
16. concentrate　專注於；much more profitable　更有利潤的

範例 12　進口商請工廠不要調漲太多，因市場競爭激烈
　　　　　產品：毛毯(Blankets)

TAIWAN MUTE TRADE CO.

To: England Blanket Mfg.

Attn: Mr. George Black

Re: Your Offer dated Jan. 5 for Blankets

Dear Mr. Black,

Please note it is already very difficult to promote your blankets under your original offered prices. If you raise again the prices, it will be more difficult for us to compete in Taiwan market.

As you know, the devaluation of NT$ to EUR was about 8-10% in the past three months. Besides, the local wages and overhead cost have been much increased, so, in fact, we almost have no profit in selling your products here.

Please try your best to make the cost down in every respect and let us know of your revised quotation soon.

Truly yours,
Linda Ho/Assistant to President

文章結構

> 主題：請注意，用你們原來所報的價錢，推銷你們的毛毯已經是很困難，如果你們再漲價，我們在台灣市場將更難競爭。
>
> 說明：如你們所知，新台幣對歐元在過去 3 個月已貶值了大約 8～10%，此外，國內薪資及管銷費用也大幅上漲，因此，事實上，我們在此銷售貴公司的產品已無利潤。
>
> 結論：請盡力於各方面降低成本，並儘快讓我們知道貴方修正的新價。

生字／片語

1. Blankets　毛毯
2. under your original offered prices　用原來所報的價錢
3. raise the price　漲價
4. compete (V)，competitive (Adj.)，competition (N)　競爭
5. in Taiwan market　在台灣市場
6. as you know　如你所知
7. devaluation　貶值（注意：新台幣貶值＝外幣升值，對出口商較有利）
8. appreciation　升值（注意：新台幣升值＝外幣貶值，對進口商較有利）
9. local wages　國內薪資
10. overhead cost　管銷費用
11. almost have no profit　幾乎沒有利潤
12. try your best ＝ do your best　盡力
13. cost down　降價
14. in every respect ＝ in all respects　各方面
15. revised quotation　修正的報價
16. EUR　歐元
17. NTD to EUR　新台幣對歐元

練習 15　討價還價函練習

1. 將下列翻譯成英文

（注意：劃底線的部分，位置要調整）

(1)我們認為你們的價錢太貴

(2)請重報一個 20 呎整櫃　的　較好的價錢 (N2 for N1)

(3)我們的目標價是每個美金 10 元，請告知是否貴方可接受

(4)你們的價錢（是）比同行（的價錢）貴 8%

(5)我們發現你們　將價錢　上漲　許多

(6)我們　與國內的客戶　有很好的關係

(7)我們　在市場上　有很強的地位

(8)我們 1 年的營業額是美金 6,000 萬元

(9)我們是國營公司，也是　此產品　的　唯一的供應商(N2 of N1)

(10)這是高價值的產品，請付樣品費及郵費

(11)由於薪資及管銷費用（被）大幅上漲，我們幾乎沒有利潤

(12)請寄　每種　各一個的樣品　給我方（＝一個樣品 for 每種）

(13)請告知　獨家代理　的　條件（N2 for N1）

(14)如所要求，我們在此附上預約發票

(15)我們很抱歉聽到貴方無法接受我方價錢

(16)為了表示誠意，我們願意給你們 3%的折扣

(17)由於新台幣兌美金已經升值了 10%，我們不能 低於成本 接受訂單
　　（低於成本＝ under the cost 放句尾）

(18)運費／報關費／海關費用／關稅

(19)ASAP／E.G./ GBP/ REP.英文全名及中文意思

(20)同業／壽命期／機會／降價／貶值／升值／預估量／傳單

2. 在空格內填入適當的英文生字或片語：

(1)謝謝你們寄來成衣的報價單，但是我們發現你們的價錢太貴。

Thanks for your ＿＿＿＿＿＿ for the ready-made garments, but We found your

price is too ＿＿＿＿＿＿ .

(2)因為我方訂購量為 10,000 打，請重報較好的價錢給我們。

Since our _____ _____ is 10,000 dozens, please _____ a

better _____ to us.

(3)我方的理想價是每打 30 美元。如能接受，請寄預約發票給我們並提供一個免
費樣品供確認品質。

Our _____ _____ is US$30/dozen. If it is _____, please

send P/I to us and offer a _____ sample for quality _____.

3. Write a reply letter on behalf of the seller to the above counter offer of US$30/dozen.
You can make your own decision to accept or not to accept it.

4. 將下列句子翻譯成中文

(1)The customer would like to give us the opportunity to get the business because
of our very good relationship with them, but your price must be "within reason" or
somewhat in line with the competition.

(2)We have already had the regular suppliers at present. Unless you can offer us much
cheaper price with acceptable quality, we will not consider buying from you.

(3)As you know, NTD has been appreciated for 5% and the labor cost has been increased for 10% this year. If you couldn't confirm your order before the end of this
month, we could not but revise the price.

(4)Since the freight cost will be adjusted from early August, we will only accept the
shipment to be made by the end of July. Or, the price has to be adjusted according to
the new freight cost.

(5)Since this is our first deal, it is our principle to accept the payment only by L/C.

(6)We regret to inform you that we could not make a quotation or take any order
at this moment because our production capacity has been fully occupied until
the end of July.

五、索送樣品或規格確認函
(Sampling or Specification Approval Letters)

1. 索取樣品及規格確認通常在價錢談妥後，才索取及提供，狀況有下列幾種：

   (1)下正式訂單前，先送樣品給客戶確認品質規格（工廠標準產品）。

   (2)下正式訂單後，再送樣品規格給客戶確認。

   （此為客戶特定規格的產品，要特別做樣品，通常會要求客戶先決定訂單，才下去做樣品，待樣品確認後再正式生產）

   (3)接到客戶模具訂單，開模具後，先送樣品供確認。

   (4)下樣品訂單後，才送樣品供確認。

   （有些高單價的產品，像電視／冰箱／電腦／錄音機……等，一般都是請客戶下樣品訂單 sample order，付款可用電匯 T/T、信匯 M/T 或銀行本票(Bank Check)，收到錢後，用郵包(by air parcel post)、快遞(by courier)或空運(by air)方式寄去，確認後，客戶再下正式大量訂單）

   (5)下訂單前或後，先送規格供確認。

   （如果買賣的是機器設備時，就只需送規格圖面供確認，稱為 document approval，確認後再下去正式生產）

2. 有關樣品費(Sample Charge)及郵費(Postage)是否要求，請參看「三、報價及回覆函」的說明。

3. 送樣品或規格供確認時，要特別注意：

   (1)「樣品品質＝出貨品質」(Sample quality ＝ Shipment quality)

   如於生產或出貨時，發現與確認樣品有所不同時，一定要事先或出貨前，告知客戶，取得同意，或重送樣品供再確認，否則就會有遭客戶退貨的可能。

   (2)樣品沒確認前，不要先行出貨。

   (3)有時剛好沒有客戶要的規格尺寸的樣品時，就先送類似的樣品，供品質參考，等正式訂購時，再送正確樣品供正式確認。

4. 索取樣品或規格函寫法

主題：買方告知接受報價，希望儘快送樣品規格供確認(We confirm the acceptance of your prices quoted on xxx. Please send us 3 free samples for approval.)

說明：樣品確認後，將馬上決定正式訂單(We will confirm our order upon sample approval.)

結論：等候貴方樣品(We hope to receive your sample soon.)

5. 送樣品或規格函寫法

主題：收到貴方來函索取樣品或規格。（參範例 2）

說明：(1)按照要求，正在準備樣品／規格圖面中，將於×××時間寄去。（參範例 2）

(2)抱歉目前沒有現樣，要等有別的客戶訂購生產時，才能順便做樣品寄去，先寄類似樣品供品質參考。（參範例 8）

(3)樣品成本高，收到樣品費後，馬上寄去。（參範例 4/7）

(4)請客戶先付樣品費，正式下訂單時，再退還此費，或從訂單中扣除。（參範例 5）

(5)正在開模具，預計×××時可送樣品供確認。（參範例 6）

結論：(1)請儘快確認樣品，並確認正式訂單。（參範例 2）

(2)如有問題，請儘快通知。（參範例 4/5）

6. 客戶收到樣品後，如果沒有問題，大都會直接下訂單；如有問題，也會再來信要求修改或重送樣品。

7. 送樣品時，在包裹內，要付上發票(Invoice)，如為高單價的樣品，在發票上的金額最好不要打太高，或送樣前先問客戶發票金額要打多少，依客人要求打出後並在發票上及包裹地址的下方註明：無商業價值的樣品(Sample of no commercial value.)（參附件一）。

MODERN
BUSINESS
ENGLISH

Part 4 (五、索送樣品或
規格確認函)
商用英文信函寫法

241

附件一　樣品發票樣本(Invoice)

INVOICE

Date: _____

Invoice No. 120225

From: King-Tech Valve Precision Industry Inc.

No.47, Lane 127, Sec. 2, ShinSheng N. Rd.

Taipei, Taiwan

To: LONGPLUS

Aeropole de Charleroi, 1, rue Ader

B-2510 Gosselies, Belgium

Item	Description	Quantity	Unit Price	Total Amount
1	Spare Parts for Valve Testing Instrument	1 set	EUR 30.-	EUR 30.-

Total Amount: EUR 30.-

SAY TOTAL AMOUNT EUROS THIRTY ONLY.

" SAMPLE OF NO COMMERCIAL VALUE."

Very Truly Yours,

King-Tech Valve Precision Industry Inc.

_____

8. 索送樣品或規格確認函常用單字／片語

(1)free sample ＝ sample free of charge　免費的樣品；handmade sample　手製的樣品

(2)We have no sample available.　我們沒有樣品可提供

(3)for quality reference　供品質參考；for sample approval　供確認樣品

(4)sample charge and postage　樣品費及郵費

(5)sample order　樣品訂單；confirm the order　確認訂單

(6)pay by bank check / by traveler's check/ by T/T / by bank transfer
　　用銀行本票／旅行支票／電匯／銀行匯款付款

(7)send by air parcel post / by courier　請用郵包／快遞寄出

(8)upon receipt of the sample ＝ upon receiving the sample　當一收到樣品

(9)before placing the order ＝ before we place the order　下訂單前

9. 索送樣品或規格確認函常用的主題句／說明句／結論句

主題句(Main Idea)

(1)We confirm the acceptance of your offer dated Jan. 20.

(2)Thanks for your e-mail of Jan. 20 confirming the acceptance of our price and requesting for our sample of ED-12.

說明句(Explanation)

(1)Before placing the order, please send us 3 free samples for approval.

(2)Please kindly send us one sample of model 01 by express for evaluation.

(3)As requested, we will send you the samples by air parcel post tomorrow.

(4)We are sorry for that we have no sample available now and would suggest sending you the similar sample for quality reference first.

(5)Model ED-12 has been out of production for a year. We would suggest that you take model ED-10 as the substitute.

(6)Please pay the sample charge in advance as it is a high-value product.

結論句(Conclusion)

(1)We hope to receive your sample soon.

(2)We will confirm our order upon sample approval.

(3)Please confirm your approval and wait for your order soon.

範例 1　客人接受髮夾(hair clip)報價，要求 3 個免費樣品供確認

From: davidhook@usatrade.com

To: julywang@fancyornament.com

Subject: Hair Clips

Dear July,

We confirm the acceptance of your prices quoted in your quotation of Jan. 20. However, before placing the order, please send us 3 free samples for your model no. 102 , 103 and 105 for the hair clips.

We will place you our formal order upon sample approval.

We hope to receive the above samples within one week, please speed them up.

Best regards

David

文章結構

主題：我們確認接受你們 1/20 報價單上的價錢，但是下訂單以前，請寄三個免費的髮夾樣品給我們，型號 102、103 及 105。

說明：樣品一確認，我們馬上下正式訂單給你們。

結論：我們希望於 1 週內收到，請加速送來。

生字／片語

1. confirm　確認；acceptance (N)；accept (V)　接受
2. quoted in your quotation of Jan. 20 (Adj. Ph.)　你們 1/20 的報價單上所報的
3. free samples ＝ samples free of charge　免費樣品
4. hair clips　髮夾
5. formal order　正式訂單；upon sample approval　當樣品一確認
6. speed up　加速

範例 2 賣方回覆範例 1

From: julywang@fancyornament.com

To: davidhook@usatrade.com

Subject: Hair Clips model 102/103/104

Dear David,

Thanks for your e-mail of Jan. 22 confirming your acceptance of our prices and requesting our above samples.

We are glad to inform you that the samples will be sent tomorrow by express air parcel post for your quality approval.

We are awaiting your order soon.

Best regards

July Wang

文章結構

主題：謝謝您 1/22 來信確認接受我方報價及索樣。

說明：很高興告知樣品可於明日以郵政快遞包裹寄去供確認。

結論：等候貴公司訂單。

生字／片語

1. by express air parcel post　用郵政快遞包裹
2. for your quality approval　供貴公司品質確認

範例 3　客人接受報價，但開模前，要求手製樣品供確認

From: claudewitch@floppyco.com

To: jackli@keyswitch.com

Subject: Your Quotation SQ-52B-01 Membrance Switch

We accept both your unit price and tooling cost shown on your above quotation. However, before starting the tooling, we would like to know if you can make a handmade sample with a membrane switch at the right position for our test first. The appearance is not important.

If the feeling is acceptable, we will decide our tooling order right away.

Please reply the above soon.

文章結構

主題：我們接受你們上述報價單上的單價及模具費，然而開模前，我們希望知道貴
　　　公司是否可手製一個樣品，只要有薄膜開關在正確位置供確認，外觀不重要。

說明：如果觸感可接受，我們將馬上確認模具訂單。

結論：請儘快回覆。

生字 / 片語

1. membrane switch　薄膜開關
2. unit price　單價；tooling cost　模具費
3. before starting the tooling　開模前
4. make a handmade sample　手製一個樣品
5. at the right position　在正確位置
6. appearance　外觀；tooling order　模具訂單
7. If the feeling is acceptable　如果觸感可接受

範例 4　賣方回覆範例 3，要求樣品費

From: jackli@keyswitch.com

To: claudewitch@floppyco.com

Subject: Handmade Sample for Membrance Switch

Thanks for your acceptance of our quotation SQ-52B-01, and we confirm it is no problem to make a handmade sample with a membrane switch at the right position for your test.

However, since the sample charge for the handmade sample is very high, we have to ask you to pay for the sample charge US$50/each. The lead time for sampling will be one week.

Please confirm the above for proceeding.

文章結構

主題：謝謝您接受我們報價單 SQ-52B-01 的報價，並確認可以手製有薄膜開關在正確
　　　位置的樣品供測試。

説明：但是手製樣品的成本很高，我們必須要求每個樣品 50 元美金，做樣品的時間
　　　要一個星期。

結論：請確認以上，以便進行。

生字 / 片語

　　1.　sample charge　樣品費

　　　　handmade (Adj.)　手製的

　　3.　lead time for sampling　做樣品時間

　　4.　for proceeding　以便進行

範例 5　請客人先付樣品費，正式下單時再退還此費

Re: Business Card Holder

Dear Ruth,

Thanks for your quick reply by e-mail of Jan. 25 to our offer sent yesterday. As requested, we will send 50 samples of Business Card Holders via UPS directly to your head office in Canada. However, since the cost is high, we ask you to T/T us in advance the sample charge, which will be deducted from your order or be refunded after you confirm the formal order.
Please confirm the above.

Best regards
Mary Huang

文章結構

主題：感謝 1/25 來信迅速回覆我方昨日報價。

說明：如所要求，我們會將 50 個名片夾的樣品以快遞方式直接寄至貴公司在加拿大
　　　的總部。然而因為成本很高，我們要求貴公司先將樣品費用電匯給我們，此
　　　費用將會從貴公司訂單中扣除，或於貴公司確認正式訂單後退還。

結論：請確認以上。

生字 / 片語

1. As requested ＝ As you requested ＝ As per your request　如所要求
2. Business Card Holders　名片夾
3. via UPS ＝ by express　以快捷方式
　　（UPS、FedExpress、DHL 皆為快遞公司名稱）
4. head office　總部／總公司；sample charge　樣品費
5. T/T (Telegraphic Transfer)　電匯；in advance　事先
6. deducted　扣除；refunded　退還
7. confirm　確認；formal order　正式訂單

範例 6　賣方告知客人正在開模具，開模完成即可送樣

From: timtan@msa.hinet.net

To: estella@hotmail.com

Subject: Samples will be sent after the mold is completed

Dear Estella,

Thank you very much for your interest in our latest PDA Model SP-01, and for your sample order for 2 sets of this model. Since our mold is still under modification and this latest hi-tech product will be launched in the early of next month, we can only send you the relevant samples after the mold is completed. Please kindly wait patiently.

Best regards

Tim Tan

文章結構

主題：感謝您對本公司最新個人數位助理型號 SP-01 的興趣及兩個樣品的訂單。

說明：因為我們的模具仍在修改中，且此最新高科技產品將於下月初才能上市，我們只能於模具完成後再將相關樣品寄給您。

結論：請耐心等候。

生字 / 片語

1. latest ＝ newest ＝ current　最新的

2. PDA (Personal Digital Assistant)　個人數位助理

3. sample order　樣品訂單；hi-tech product　高科技產品

4. mold　模具；under modification　修改中

5. launched ＝ released ＝ entered the market　上市

6. relevant ＝ related ＝ relative　相關的

7. completed　完成；patiently　耐心地

MODERN
BUSINESS
ENGLISH

Part 4 （五、索送樣品或
規格確認函）
商用英文信函寫法

249

範例 7　客戶發 e-mail 要求電子字典(Electronic Dictionary)樣品

Sender: Jkuteo@alltel.net

Receiver: Jessica.ku@telecheck.com

Subject: Sample for Electronic Dictionary

Dear Jessica,

The quotation and catalog of your electronic dictionary interest us very much especially the model no. ED-12A.

Please kindly send us one sample of the above model for evaluation, we will send you the sample charge by bank check upon receipt.

Thanks for your quick reply in advance.

文章結構

主題：貴公司電子字典的報價及目錄令我們極感興趣，特別是型號 ED-12A。

說明：請儘快寄一個上述型號的樣品供評估，收到樣品後，將馬上用銀行本票寄樣品費過去。

結論：先感謝您儘快回覆。

生字 / 片語

1. Electronic Dictionary　電子字典
2. especially　特別地
3. by bank check　用銀行本票
4. upon receipt ＝ as soon as we receive it　當一收到

MODERN BUSINESS ENGLISH

範例 8　賣方回覆範例 7，告知沒有現樣，可否先寄類似型號的樣品

Sender : Jessica.Ku@telecheck.com

Receiver: Jkuteo@alltel.net

Subject: Sample for Electronic Dictionary

Thank you for your e-mail sent on Jan. 17 requesting the sample of our Electronic Dictionary Model No. ED-12A.

We are sorry to inform you that we have no sample available for the above model at this moment, which will be available only after we receive the new orders from other customers. However, we would suggest you taking the similar model no. ED-12B as substitute for quality reference as both quality and functions are the same except there is a little difference in appearance.

Please confirm the above for sending the alternative sample.

文章結構

主題：謝謝您 1/17 的電子郵件，要求型號 ED-12A 電子字典的樣品。

說明：很抱歉目前沒有這個型號的樣品，此樣品要等到其他客戶下新訂單時，才會生產出來；然而，我們建議貴公司先拿類似的替代品型號 ED-12B 參考品質，由於兩種型號的品質及功能都相同，只是外觀略有不同。

結論：請確認以上，以便送樣品。

生字 / 片語

1. We have no sample available　我們沒有樣品可提供
2. at this moment ＝ now　目前；the alternative sample　替代的樣品
3. as substitute　當作替代品；for quality reference　供品質參考
4. both quality and functions are the same　兩者的品質及功能都相同
5. except there is a little difference in appearance　除了外觀上有一點不同

MODERN
BUSINESS
ENGLISH

Part 4 (五、索送樣品或
規格確認函)
商用英文信函寫法

251

練習 16 索送樣品或規格確認函練習

1. 根據下列大意寫一封完整的英文索樣函
   （注意：劃底線部分的形容詞及副詞位置要作調整）

主題：我們　接受　你們手提包(handbags)的　報價，但是下訂單<u>以前</u>，我們希望
　　　收到　<u>型號 HB-20 及 HB-25</u>　的　樣品。

說明：請　<u>儘快</u>　<u>用郵包</u>　寄　樣品　給我們，樣品一確認，我們將　<u>馬上</u>　確認
　　　我們的訂單。

結論：等候貴公司回音。

---

2. 請根據下列大意寫一封完整的英文送樣函

主題：謝謝貴公司來函要求　<u>我們手提包</u>　的　樣品。

說明：我們很高興告知貴公司我們已於　<u>今日</u>　寄出　<u>型號 HB-20 的</u>　樣品，但是
　　　<u>型號 HB-25</u>　的　樣品，要　<u>下週</u>　才能寄去，因為　我們　<u>現在</u>沒有樣品。

結論：請　<u>儘快</u>　告知意見並　確認訂單。

---

3. 自由發揮寫一封索取手錶(Watch)樣品的索樣函

4. 依照中文大意填入正確英文單字／片語：

   (1)我們接受你們的價錢。請寄給我們一個免費的樣品供確認。

   　We _____ your _____ . Please send us one _____ sample for _____.

   (2)收到樣品費後，我們會馬上寄出樣品。

   　We will send you the sample upon receiving the _____ _____,

   (3)樣品一確認，我們將馬上確認我們的訂單。

   　Upon sample approval, we will _____ our order.

六、追蹤／催促函(Follow Up Letters)

　　在各種信函發出後,如果在預期時間內沒有收到對方回音,都可寫追蹤／催促函,例如推銷函、詢問函、報價函、索樣送樣函送出後,催訂單／信用狀、催貨或是追蹤很久沒聯絡的客戶,甚至於展覽後或客戶來訪後等。

　　除了推銷函的追蹤及對很久沒聯絡客戶的追蹤,需要一些技巧禮貌外,其他一般信函催促的寫法都很簡單直接,甚至有時與一些常聯絡的老客戶之間的追蹤催促,連信函也不寫,直接在要催促的舊函上蓋上或寫上 "Follow Up" 字樣,再發送一遍,以省時間。

　　因此追蹤催促函的寫法分下列幾種:

1. 長期合作的熟客戶:直接在要催促的舊函上,蓋上或寫上 "Follow Up" 字樣,再發一遍。或寫上 2nd re-send / 3rd re-send 亦可。

2. 一般信函的追蹤／催促(參範例 1/2/7)

主題:有關某事,自從本公司於××月××日去函之後,至今尚未收到貴公司回音。 說明:因為此事緊急／或我們正等候貴方回覆,才能進行下一個步驟,因此,請速告知是否有收到上述信函,或何時可回覆。 結論:如未收到,請告知,以便重送。已收到,請儘快答覆。謝謝!

3. 對很久沒聯絡的老客戶,為了挽回生意要稍具技巧禮貌(參範例 3)

主題:參照我方記錄,很抱歉有一段時間未與貴公司聯絡,希望貴公司一切順利如意。 說明:從過去聯絡的資料中,我們知道貴公司一向對××產品有興趣,我們很高興在此附上最新的目錄及特惠價,以供參考。 結論:請撥空參看以上資料,並歡迎隨時賜教,我們將很榮幸再為您服務。

4. 展覽後的追蹤函寫法（參範例 4）

主題：首先，感謝貴方能於××時間參觀我們在×××地點的展覽。

說明：(1)按照貴方在會場的要求，我們在此寄上所索取的資料或樣品。

　　　(2)看過我方目錄資料後不知貴方是否還需要任何進一步資料，或對我方產品是否有任何寶貴意見。

結論：期盼能有機會為貴公司服務，或能儘快與貴公司建立商業關係。

5. 客人來訪，回去後的追蹤函，寫法如下：（參範例 8）

主題：感謝您來拜訪本公司，相信您現已平安到家。

說明：有關您拜訪時，所談到的×××某事（例如決定正式訂單／開信用狀／規格研商等），請儘快告知結果或意見。

結論：期盼您儘快的回覆，謝謝！

6. 推銷函寄出後的追蹤函較難寫，因為沒與客戶作過生意，也不清楚客戶目前狀況，如其已有固定良好的配合廠商，要爭取生意就要更多的技巧或有利的誘因，寫法如下：（參範例 5/9）

主題：本公司於×××時間，將本公司最新產品型錄寄去，不知貴公司是否已收到，或已撥空看過。

說明：(1)可再自我介紹，或強調產品特色，或告知市場目前暢銷或正大流行中。

　　　(2)告知目前正有特惠價中，請其把握時機，因為不久即將漲價。

　　　(3)或再補充寄些相關產品的新資料過去，附帶再作追蹤。

結論：表示自己極有誠意要與對方建立商業關係，或極樂意為其服務，因此期盼其回音或任何意見。

7. 追蹤／催促函常用單字／片語

(1)follow up　追蹤；re-send　重送

(2)We have not received your reply.　我們尚未收到貴公司的回音

　　＝ We have received no any reply from you.

(3)Thanks for your assistance and hospitality.　謝謝貴方的協助及款待

(4)urgent　緊急的；patent　專利；file record　檔案紀錄

(5)past communication　過去的聯絡；correspondence　書信往來

(6)Goods are ready for shipment. (＝ ready for dispatch)　貨已準備好可裝運

8. 追蹤／催促函常用的主題句／說明句／結論句

主題句(Main Idea)

(1)With reference to our file record, we have not heard from you for a certain time. (＝ it has been a long time we have not contacted you)

(2)We have not received your reply to our last e-mail sent on Jan. 10.

(3)Thank you for your visit at our booth and showing your interest in our products at the Hanover Fair held last month.

(4)The goods are ready for shipment, but we haven't received your L/C yet.

說明句(Explanation)

(1)From our past communication, we know your interest in our subject products and are glad to enclose our latest catalog for your reference.

(2)We hope you have well received our last e-mail and await your reply soon.

(3)As requested, we are glad to send you our catalog as attached for reference.

(4)Please kindly advise if you have opened the L/C and send us the details.

結論句(Conclusion)

(1)Please kindly reply the above as soon as possible.

(2)We believe that to cooperate with us will bring big profits for you.

(3)We expect to develop long term business relationship with you soon.

(4)Your soonest reply to the above will be highly appreciated.

(5)We will release the shipment upon receiving your payment.

範例 1　進口商催促工廠緊急裝運 20 台儀器的訂單

From: Mary

To: pierre.yves.thiry@logiplus.com; Bernard Boisdequin

Subject: Our Order for 20 sets of flowplus-4th sent-URGENT

Dear Pierre and Bernard,

We haven't received your reply to our last e-mail sent to you on Jan. 3 till now. Referring to our first communication about this order three months ago, you told us the goods will be ready in 4-6 weeks. How about the production situation now? Please kindly advise us when you can have 20 sets of Flowplus ready to ship? Please send us your Pro-forma Invoice(P/I) right away so that we can arrange the T/T payment to you.

We need to know your situation urgently or at least a word from you. Your soonest reply will be highly appreciated.

Best regards

Mary Huang

文章結構

主題：我們至今尚未收到你們回覆我們 1/3 的 e-mail。

說明：參照我們 3 個月前第一次聯絡有關此訂單的往來信函，你們告知貨物可於 4～6 週內準備好。現在的生產情形為何？

結論：請告知何時可裝出 20 組 Flowplus 儀器？並請儘快將預約發票傳來以便於我方安排電匯付款給你們。我們急需知道貴方狀況或至少隻言片語。感謝您儘快回覆。

生字 / 片語

1. communication　聯絡；production situation　生產情形
2. Pro-forma Invoice(P/I)　預約發票；T/T(Telegraphic Transfer)　電匯
3. right away ＝ at once ＝ immediately ＝ as soon as possible　馬上／儘快
4. urgent (Adj.), urgently(Adv.)　緊急的／緊急地
5. at least a word from you　至少隻言片語

範例 2 客人催促「心形物體」(heart module)專利的回覆

BRIGHT GIFT IMPORT CO.

To: Mr. Herb Wu/Unit Gift Mfg.

Re: Heart Module

Dear Herb,

We have received no other communications from you concerning the patent for the heart module.

A sufficient amount of time has elapsed and we would appreciate some type of response.

We are anxious to conclude this matter since our Sales Manager John Bill is planning a business trip to Spain to one of the markets next month.

Best regards

Sun William

文章結構

主題：有關心形物體專利問題，我們沒有收到你們其他聯絡信函。

說明：已有好一段時間過去了，我們將會感謝您任何的回音。

結論：我們急著把這件事做個結論，因為我們業務經理 John Bill 正計劃於下個月到我們國外市場之一的西班牙去出公差。

生字／片語

1. communication　聯絡；anxious　著急

2. patent　專利；heart module　心形物體

3. A sufficient amount of time ＝ a lot of time ＝ much time　很多時間

4. has elapsed ＝ has passed　已經過去

5. some type of response ＝ any reply　任何回覆

6. business trip　商務旅行＝出公差

7. to conclude this matter　把這件事做個結論

範例 3　對很久沒聯絡的老客戶，重新再追蹤

From: carolyen@globintl.com

To: alexhulio@ukimporter.com

Subject: New Product-Glove

With reference to our file records, it has been a long time we have not contacted you. We sincerely wish you to have a prosperous business.

From our past communication, we know of your interest in our subject products and are glad to enclose herewith our new catalog for reference. Please take a look into it and feel free to contact us if any item interests you. We will accordingly offer you the best quotation and the relevant sample right away.

Thank you for your kind attention and look forward to renewing and continuing the good business relationship with you very soon.

Best regards

Carol Yen

文章結構

主題：參照我方檔案記錄，我們已有一段時間未曾與您聯絡了，我們誠心希望貴公司生意興隆。

說明：從過去的聯絡中，我們知道貴公司對上述主題產品有興趣，在此附上新目錄以供參考，請撥空看看，如有任何項目令您感興趣，請隨時告知，我們將馬上提供最好的報價及相關樣品。

結論：謝謝您的注意並期盼很快與貴公司重新繼續良好的生意關係。

生字／片語

1. Glove　手套；with reference to ＝ referring　參照
2. file record　檔案記錄；prosperous business　生意興隆
3. past communication　過去的聯絡；take a look　看看；feel free　隨時
4. relevant sample　相關的樣品；to renewing and continuing　重新繼續
5. good business relationship　良好的生意關係

範例 4　展覽後，追蹤來參觀過的客戶

From: edmondchen@starcoolingfan.com

To: racepress@annindusty.com

Subject: Thanks for your visit

Dear Race,

First of all, we would like to thank you for your visit our booth and showing interest in our products at the Cebit Fair held in Hanover last month.

As inquired, we are glad to send you our catalog of cooling fan for your evaluation. Please specify the items interest you, or send us your detailed specification for making exact quotation.

We look forward to starting business relationship with you soon.

Best regards

Edmond Chen

文章結構

主題：首先，感謝您參觀上個月在德國漢諾威舉辦的 Cebit 展覽我方攤位，並表示對我們產品的興趣。

說明：如所洽詢，我們很高興寄上冷卻風扇的目錄供評估，請指出有興趣的項目或寄來貴公司詳細的規格，以便報正確的價錢。

結論：期盼儘快與貴公司開始生意關係。

生字／片語

1. booth = stand　攤位；cooling fan　冷卻風扇
2. showing interest in = expressing interest in = with interest in　對……感到興趣
3. Cebit Fair held in Hannover　在漢諾威舉行的 Cebit 展覽
4. As inquired = As requested　如所洽詢／要求
5. specify = indicate = point out　指出／指定；detailed(Adj.)　詳細的

範例 5　對新開發客戶的追蹤

From: robertyang@hugesundry.com

To: kimdrill@bluebirds.com

Subject: Sundry Products

Dear Mr. Drill,

On Feb. 1, we have e-mailed you our latest catalog and the best quotation for our Sundry Products. We are wondering if you have well received them since we have received no any response from you till now.

From our catalog and the quotation attached, you will find that we offer many innovation products with the best quality at the lowest prices for the respective customer like you.

Please don't hesitate to let us know your interest and comments.

Best regards

Robert Yang

文章結構

主題：在 2/1 我們寄給貴公司我們雜貨產品最新的目錄及最好的報價，由於我們至今未收到貴方回音，不知貴方是否完好收到以上。

說明：從我們所附的目錄及報價單，貴公司可發現，我們以最低價提供許多高品質的新產品，以便給像貴公司這般令人尊敬的公司。

結論：請不要遲疑讓我們知道您是否感到興趣及意見。

生字／片語

1. Sundry Products　雜貨產品
2. innovation products　創新的產品
3. with the best quality ＝ of the best quality　具有最好的品質
4. at the lowest prices ＝ at the best prices　用最低（好）價錢
5. the respective customer　令人尊敬的客戶

範例 6 賣方催促買方付款

Subject: Follow up for the payment

Dear Michael,

Please kindly be informed that the goods of your order number A-110 are ready for shipment, but to our regret, till now we haven't received your payment yet.

Please kindly advise us if you have remitted us the payment by bank transfer as agreed on your order. If we couldn't receive your payment before the end of this week, we wouldn't be responsible for the delay in shipment. So, please confirm by return when we can expect your payment.

Best regards
Leo Brown

文章結構

主旨：請被告知訂單號碼 A-110 的貨已完成可裝運，但很遺憾我們至今尚未收到貴方付款。

說明：請告知貴公司是否已如訂單上所同意經由銀行匯款給我方。如果我們無法於這週前收到匯款，我們將不負責延遲出貨的問題。

結論：因此，請盡快確認何時我方可收到貴方的付款。

生字 / 片語

1. Goods are ready for shipment. = ready for dispatch 貨已完成可裝運
2. to our regret 很遺憾
3. haven't received your payment yet 尚未收到貴方付款
4. remit 匯出；bank transfer 銀行匯款
5. as agreed 如所同意
6. be responsible for = take the responsibility for 負責
7. delay in shipment 慢出貨
8. by return = soon 盡快

範例 7 報價給客戶後的追蹤

Dear Mr. Aloi,

Since your last meeting with our director Mr. Chung in our office last December in Taipei, we haven't heard from you for a certain time. We hope everything is fine to you and your company.

We appreciate your interest in our various kinds of valves. From our records, we have offered you the prices for our following valves: Air Release Valve, Hydraulic Balancing Valve, DPV etc. Would you please kindly let us know your comments to our above offers or let us know your target prices for re-consideration. Besides, if you have any new inquiry, please don't hesitate to contact us directly for a trial. We appreciate your kind reply to the above soon.

Best regards

Mary Huang / Export Manager

文章結構

主旨：自從去年 12 月在我們台北公司和我們董事長會談後，我們有一段時間沒聽到您的消息了，希望一切安好。

說明：我們很感謝您對我們各種閥類產品感興趣。從之前紀錄上得知，我們已報下列閥類價錢給您：釋壓閥、平衡閥及壓差閥等。請告知您對我方報價的意見，或讓我們知道貴公司的目標價以便重新考慮。此外，如果您有任何新的詢價，也不妨試著直接和我們聯絡。

結論：感謝您能儘快回覆以上。

生字／片語

1. director　董事長；for a certain time　一段時間
2. appreciate your interest　感謝您的興趣
3. Air Release Valve　釋壓閥；Hydraulic Balancing Valve　平衡閥；DPV 壓差閥
4. comments　意見；target price　目標價；re-consideration　重新考慮
5. hesitate　遲疑；for a trail　試看看

範例 8　客人拜訪回去後，追蹤面談時所討論之事

From: bobfritz@rragencies.com
To: ritadin@shunchi.com
Subject: Thanks for your hospitality

Dear Rita,

Firstly I would like to thank you for your kind assistance and hospitality, afforded during our visit. It was great to finally meet you.

Re our meetings, can you please forward your offer for the items discussed during our meetings, including:
(1)the details or breakdown cost about the tooling charge.
(2)the freight cost for safety lamps.

Thank you once again for your assistance, and I look forward to developing a long term business relationship with your company.

Sincerely Yours,
Bob Fritz

文章結構

主題：首先感謝你們在我們拜訪期間的協助及款待，終能見到你們真好。
說明：有關我們的會談，請提供會談時所討論的下列事項的報價：(1)模具費的細目；
　　　(2)安全燈的運費。
結論：再次謝謝您的協助並盼望與貴公司建立長期生意關係。

生字／片語

1. assistance ＝ help　協助；hospitality　款待；afford　給與
2. forward ＝ send ＝ provide　提供
3. breakdown cost　成本細目；tooling cost　模具費；freight cost　運費
4. safety lamps　安全燈
5. long term business relationship　長期生意關係

MODERN BUSINESS ENGLISH

Part 4

（六、追蹤／
催促函）
商用英文信函寫法

263

範例 9　推銷函之後的追蹤

POWER BABYTOY MFG

Exporter & Manufacturer

To: Mr. Walter Ising /LME Co.

Dear Mr. Ising,

<u>Re: New Toy Model KB-01</u>

In keeping the policy of offering our new developed products to the respective customer like you continually, we believe that you should have received our new leaflet sent on Jan. 3.

For your information, we are opening the new mold for the new model KB-01 as per the enclosed photo, which will be finished by the end of this month. Accordingly, we will furnish the relative sample for your evaluation in case you are interested in it.

Persisting the principles of superiority in quality, innovation in products and integrity in business, we have won a very good reputation in the world market.

We believe to cooperate with us will bring considerable profits for you. Please let us know of your interest and comment soon.

Very truly yours,
Power Babytoy Mfg.

Andy Wang
Sales Manager

文章結構

主題：因為是新客戶，用較客套方式作追蹤──持續提供新開發產品給像貴公司這樣的好客戶，我們相信您應已收到我們 1/3 寄去的新單頁目錄。

說明：附帶提供對方一個新資訊，並再強調自己特色

　　(1)提供您一個資訊，我們正在開新型 KB-01 的模具，參如附照片。
　　　此型式將於這個月底完成，如果貴方有興趣，我們將會寄相關樣品給你們評估。

　　(2)秉持優良品質、創新產品及誠信生意原則，我們在世界市場早已贏得很好的名聲。

結論：相信您若與我們合作定會為貴公司帶來可觀的利潤，請儘快讓我們知道您對我們產品是否感到興趣並請提供意見。

生字／片語

1. Toy　玩具；policy　政策
2. respective customer　令人尊敬的客人
3. continually　繼續不斷地
4. new developed products　新開發的產品
5. sent on Jan. 3 (Adj. ph.)＝ which was sent on Jan. 3 (Adj. Cl.)
　形容詞片語修飾名詞 leaflet（單頁目錄）
6. for your information　提供您一個資訊；new mold　新模具
7. as per the enclosed photo　參照如附照片
8. finish　完成；accordingly　根據以上／於是
9. furnish ＝ send ＝ forward ＝ provide　提送
10. relative sample ＝ relevant sample ＝ related sample　相關樣品
11. persisting the principles of　堅持…原則
12. superiority in quality ＝ excellent quality　優良的品質
13. innovation in products　產品的創新
14. integrity in business　誠信的生意
15. good reputation　好的名聲
16. to cooperate with us　與我們合作（名詞片語＝ S）

練習 17　追蹤／催促函練習

1. 根據下列中文大意寫一封完整的追蹤／催促函
（注意：劃底線部分的 Adj. Ph. 及 Adv. Ph. 要調整位置）

主題：本公司　於 9 月 20 日　寄去　我們最新　流行鞋子　的　目錄，並想知道　是否　貴方已經收到它。

說明：因為　這些　都是　今年　最流行的樣式，非常受　消費者　所歡迎，如果　貴公司　有興趣　任何款式，請　儘快　決定　訂單　或　來函索取樣品，我們（是）很樂意　隨時　為您服務。

結論：我們希望得知您的興趣及意見。

生字／片語

(1) 我們最新 流行鞋子 的 目錄→我們的最新的目錄 of fashion shoes

(2) fashion styles　流行的樣式

(3) welcome　歡迎（受歡迎＝被動）

(4) 非常受消費者的歡迎 →非常受歡迎 by 消費者(consumers)

(5) write us for the samples　來函索取樣品

(6) at any time　隨時（放句尾）

(7) to serve you 或 to be of service to you　為您服務

2. Write a letter to follow up the customer who visited at your booth after coming back from Munich Show to see if any possibility to do business with them. (The product can be decided by your own)

3. Write a letter to follow up the buyer to confirm the sample approval after sending samples for approval for a certain time.

七、訂購及接單函(Ordering and Confirmation Letters)

買賣雙方在價錢談妥,樣品／規格也確認後,大都就會直接決定訂單。買方下訂單或賣方確認接單的信函非常簡單,也很簡短,但是要注意下列幾個重點:

1. 訂單上所有的條件與談妥的報價單上的所有條件,要完全一致。

    注意:Quotation(報價單),Purchase Order (P/O)(訂單),Proforma Invoice (P/I)(預付發票),Sales Contract(銷售合約書)或 Sales Confirmation (S/C)(銷售確認書),甚至 Commercial Invoice(出貨發票),內容條件都完全一樣,只是標題不同而已。

2. 訂單屬於合約的一種,如要有法律效用,除了用電報或傳真發送外,最好要求買方將正本,一式兩份,簽好字,用掛號郵寄過來;賣方收到正本訂單後,簽回一份,自己留一份(重點是,訂單一定要正本且具有雙方的正式簽名,才有法律效用,日後如有糾紛發生時,可作為依據)。

    買方發出的英文訂單名稱,常用的有下列幾種:

    Purchase Order / Order Sheet / Indent / Purchase Contract,

    Purchase Agreement / Purchase Confirmation / Purchase note

3. 如果買方沒有寄來正本訂單,或者買方需申請輸入許可證時,賣方要打預付發票(P/I),或銷售確認書(S/C),一樣一式兩份正本,寄給買方,簽回一份,這與正本訂單一樣具有法律效用。

4. 訂單的名稱種類分下列幾種:

    (1)試銷訂單 Trail order ＝樣品訂單 Sample order ＝初步訂單 Initial order ＝小量訂單 Small order

    (2)正式訂單 Formal order ＝大訂單 Large order/big order

    (3)循環訂單 Repeat order

    (4)肯定的訂單 Firm order

    (5)定期採購訂單 Regular order

    (6)最低量訂單 Minimum order

    (7)尚未出貨的訂單 Pending order/Open order

5. 訂單上的數量，按國際公約的規定，可以有＋／－ 10%的誤差，意即賣方可以決定多裝或少裝 10%的數量；因為有時生產時，不良率偏高，無法掌控生產出來的數量會剛好；但也可由買賣雙方自行決定可接受的誤差比率，＋／－ 3%，5%或 15%，只要雙方簽字同意即可。

6. 除了報價單上的條件外，買方也可能在訂單上作下列的特別要求：

(1)指定出貨嘜頭(Shipping Mark)：分主嘜(Main Mark)及側嘜(Side Mark)

(2)指定攬貨公司(Forwarder)或裝運的船公司／航空公司(Shipping Company)

(3)指定檢驗公司(Inspection Company)：要註明檢驗費用是由買方或賣方負擔

(4)要求工廠全檢，以保證品質(100% test/inspection)

(5)要求慢出貨或無法出貨的賠償條款，例如：

Penalty: In case delay shipment happens, the penalty for delay interest will be based on annual rate 15% of the total contract amount, but not over one month, or the order will be cancelled.

（罰款：如果慢出貨，按年利率 15%計算利息，但不得慢超過一個月，否則取消訂單。）

(6)分批裝運(Partial Shipments)：買方如有要求出貨要分批裝運，報價上要註明最多分幾批出貨，因為每批出貨都會有一筆報關手續費及文件製作成本等費用產生。

(7)轉運(Transshipment)：買主指定目的地港口，出口港如無直達的船或飛機可到達，必須到其他港口轉運才能到達的話，最好於報價單上註明。

(8)指定航線／船公司：買方如有指定特殊航線或船公司時，賣方要特別注意遵照。

例如：Shipment is to be made via Panama Cannel and to be effected by American President Line Vessel.

(9)保險(Insurance)：如為 CIF 報價，賣方必須將保險費用計算進去。買方如無特別規定，一般都只投保全險(All Risk)：戰亂地區加保兵險(War Risk)：落後地區則會加保竊盜未能交貨險(TPND ＝ Theft, Pilferage and Non-Delivery)。投保金額按國際標準規定，以 CIF 金額×110%×保險率計算，例如：

① Insurance: Against All Risk for 110% of Invoice Value.

② Insurance: Covering ICC (A) and War Risk for CIF invoice value plus 10%.

7. 訂單雖屬正式合約，但是即使買賣雙方簽字後，仍會有變數，例如：

(1)買方增加訂單數量(add/increase order quantity)（參範例 4）

(2)買方減少訂單數量(reduce order quantity)（參範例 7）

(3)取消訂單(cancel the order)：特別在景氣不好時(Economic Depression)或遇到不可抗力事件(Force Majeure)，例如天災(Acts of Gods)、罷工(Strike)、戰爭(War)或政府法令改變(Governmental Restriction)等。（參範例 7）

如為不可抗力事件造成取消訂單，賣方要自認倒霉；但是，如為景氣關係或其他原因，造成要取消合約，買賣雙方可在訂單上註明罰則：

If the seller or buyer wants to cancel the contract, 5% of total contract value would be charged as penalty to that party.

（如果買方或賣方任何一方想取消合約，要給另一方 5%的解約金）

解約金的百分比，可由買賣雙方自行約定簽認。

8. 與第一次交易的客戶，或有外匯管制的開發中國家的客戶，做生意時，即使訂單已經簽定，也不要下去生產，最好等到信用狀或付款收到，再下去備料生產，因為這些客戶，有可能開不出信用狀或延遲付款時間，那就會造成賣方的風險(risk)或損失(loss)。

按照國際公約規定，信用狀要在簽約後，15 天內開出：

Payment: Buyer shall arrange to establish L/C in favor of Seller with the terms stipulated on the face of the order within 15 days.

9. 訂單(Purchase Order)樣章：參附件一；銷售合約書(Sales Contract)：參附件二；預付發票／預約發票(Proforma Invoice)樣章：參附件三。

10.買方確認訂單(Ordering)的寫法

主題：我們接受貴公司報價單上的價錢及其他條件，並很高興確認我方訂單號碼×
　　　×××，如附。（參範例 1/3/6）

說明：要求賣方要特別注意某些特殊條款要求。（參範例 1/3）

結論：請儘快確認以上，或寄 P/I or S/C 過來，以便安排開信用狀或付款。（參範例
　　　6/8）

11. 賣方回覆確認訂單條款(Confirmation Letter)的寫法

> 主題：先謝謝買方下的訂單，並確認接受訂單上的所有條款。（參範例 2）
>
> 說明：如所要求，附上 P/I 或 S/C 供開信用狀，並請買方將正本簽回一份。（參範例 2）
>
> 結論：當一收到信用狀或付款，馬上安排生產，並會按簽訂的條件執行，表現定當令客人滿意。（參範例 2）

12. 如上所述，訂單簽訂後，買方也有可能要增減數量，例如：

(1)很抱歉景氣不好／計算錯誤，可否減少 1000 個。

We are sorry to ask you to reduce 1000 pcs from O/No.xxx due to economic depression (wrong calculation).

(2)因市場需求增加，我們欲追加 5000 個，請修正訂單數量並確認

Due to high market demand, we would like to increase the order quantity for 5000 pcs (please add 5000 pcs into O/No.xxx).

Please revise the order quantity and confirm by return.

13. 賣方回覆以上，可自行決定是否接受，因為訂單已簽字，要改變需雙方同意才能再做修正，如接受則直接確認；如不接受，則找個理由拒絕。

(1)我們同意接受，請修正訂單數量

We agree to accept your revised quantity, please modify the order quantity.

(2)很抱歉無法接受減少數量，因材料皆已購買／生產已完成

We are sorry we cannot accept your quantity reduction because we have purchased all materials (we have completed the production).

(3)很抱歉無法接受增加的數量，因生產已完成，所增加的數量太少，很難備料或生產

We are sorry we cannot accept your quantity increase because we have completed the production and it is difficult to prepare the material for the increased small quantity.

14.訂單上也可能註明下列其他條款(Other terms and conditions)

(1)The seller hereby agrees to indemnify the purchaser against all costs. expenses, loss or damages incurred or suffered by the purchaser by reason of the delay of shipment or if inferior quality of the goods delivered do not correspond with description set out above.

(2)Force Majeure: Non-delivery of all or any part of the merchandise caused by war, blockage, revolution, insurrection, civil commotions, riots, mobilization, strikes, lockouts, act of God, severe weather, plague or other epidemic, destruction of goods by fire or flood, obstruction of loading by storm or typhoon at the port of delivery, or any other cause beyond the Seller's control before shipment shall operate as a cancellation of the sale to the extent of such non-delivery. However, in case the merchandise has been prepared and ready for shipment before shipment deadline but the shipment could not be effected due to any of the above mentioned causes, the Buyer shall extent the shipping deadline by means of amending relevant L/C or otherwise, upon request of the Seller.

15.訂購及接單函常用單字／片語

(1)Purchase Order (P/O)　訂單

(2)Pro-forma Invoice (P/I)　預約發票

(3)Sales Contract / Sales Confirmation (S/C)　銷售合約書／銷售確認書

(4)shipping mark　出貨嘜頭；forwarder　攬貨公司

(5)accept　接受；confirm the order　確認訂單

(6)Acceptable Quality Level (AQL)　可接受品質標準

(7)formal order　正式訂單；firm order　肯定／確定的訂單

　　trail order　試銷訂單；repeat order　循環訂單

　　regular order　定期採購訂單；minimun order　最低量訂單

　　pending order/open order　尚未裝出（尚未出貨）的訂單

(8)cancel the order　取消訂單

(9)as instructed　如所指示；as stipulated in the order　如訂單上所規定

(10)latest shipping date　最後出貨日

(11)partial shipment　分批裝運；transshipment　轉運

(12)arrange the production　安排生產；ship from stock　從庫存出貨

(13)accept　接受；price　價錢；delivery time　交期

(14)confirm the order　確認訂單；confirm your acceptance　確認接受

(15)as enclosed ＝ as attached　如附；we enclose /we are enclosing our P/I.

(16)for opening L/C ＝ for issuing L/C　以便開信用狀

(17)do 100% test (inspection) before shipment　出貨前做 100%測試

(18)small amount order　小額訂單

(19)sea/ocean shipment　海運出貨；air shipment　空運出貨

(20)Please contact our forwarder.　請和我們的攬貨公司聯絡

　　＝ Please keep in contact with our forwarder

(21)as follows ＝ as below　如下

(22)economic depression ＝ economic recession　經濟不景氣／衰退

(23)have no stock　沒有庫存；urgent (adj)/urgently(adv)　緊急的／地

(24)Please do ship the goods on time.　請務必準時出貨

(25)Please confirm the above.　請確認以上

16.訂購及接單函常用的主題句／說明句／結論句

主題句(Main Idea)

(1)We accept your price and terms and confirm our order no. 001 as attached.

(2)We confirm the acceptance of your order no. 001 and attach our P/I for opening the L/C.

(3)Please confirm if you can supply the following items from stocks.

(4)We will send you our formal order sheet by air mail.

說明句(Explanation)

(1)Please send us your P/I for arranging the payment by T/T.

(2)Before shipment, please do 100% inspection to make sure the quality and do ship the goods on time.

(3)Please confirm your best shipping date.

(4)We have no stock now and the soonest delivery time is in two weeks.

(5)Because this is a small order, please pay by T/T.

(6)As to the sea/air shipment, please contact our forwarder listed below.

結論句(Conclusion)

(1)Please confirm if you accept partial shipments.

(2)Due to economic recession, we hope to cancel this order.

(3)Please confirm the above and send us your P/I soon.

(4)We are enclosing our P/I and awaiting your L/C soon.

(5)We will arrange the production as soon as we receive your L/C.

　　We will arrange the shipment as soon as we receive your T/T payment.

　　（down payment 訂金）。

附件一　訂單樣本(Purchase Order)

MerryBest International Co.

P.O. Box 44-6, Taipei, Taiwan, R.O.C.

〈 PURCHASE ORDER 〉

P/O No: P002-2 Date: Jan. 10, 2012	Shipping Mark:
To: CONTROL SYNERGIE	M.B.
200 Bukit Batok, Singapore	Keelung
Attn: Mr. K. A. Tan/ GM	C/No.1-6
Your Ref: CSBQ-289R4	Made in France

Model No.	Specifications	Qty	U/P	Total Amount
SR6/AR04	Actuator Ex-Proof, 3 Phase Motor 220V, 60Hz	6 sets	FOB French Port FFR 13,470	FFR 80,820

Total Amount: FFR Eighty Thousand Eight Hundred and Twenty only.

Packing: By Standard Export Wooden Case.

Shipment: 12 weeks ARO (After receiving the order)

Payment: by confirmed and irrevocable L/C at sight in favor of:

　　　　ETS L. Bernard S. A., 60 Ave. Du President Wilson, France.

　　　　Bank: Banque Nationale De Paris

Note: Following documents are required with shipment:

　　　(1)Warranty letter to certify the guarantee period: 20 months

　　　(2)Original manufacturer certificate

　　　(3)Material certificate

　　　(4)Inspection and testing report

　　　(5)Operation and maintenance manual (Instruction manual)

　　　(6)Fumigation certificate

The Seller
Control Synergie

The Buyer
Merrybest International Co.

K.A. Tan/General Manager

Daniel Huang/ V.P.

附件二　銷售確認書樣本(Sales Confirmation)

ETS L. BERNARD S.A.

60 AVENUE DU PRESIDENT, WILSON, FRANCE

SALES CONFIRMATION

To: Merry Best International Co.

    P.O. Box 44-6. Taipei, Taiwan

Attn: Mr. Daniel Huang/V.P.

S/C No.1111

Date: Jan. 12, 2012

Your Ref.: P002-2

Model No.	Specifications	Qty	U/P	Total Amount
SR6/AR04	Actuator Ex-Proof, 3 Phase Motor 220V, 60Hz	6 sets	FOB French Port FFR 13,470	FFR 80,820

Total Amount: FFR Eighty Thousand Eight Hundred and Twenty only.

Packing: By Standard Export Wooden Case.

Shipment: 12 weeks ARO (After receiving the order)

Payment : by confirmed and irrevocable L/C at sight in favor of:

       ETS L. Bernard S. A., 60 Ave. Du President Wilson, France.

       Bank: Banque Nationale De Paris

Note: Following documents are required with shipment:

    (1)Warranty letter to certify the guarantee period: 20 months

    (2)Original manufacturer certificate

    (3)Material certificate

    (4)Inspection and testing report

    (5)Operation and maintenance manual (Instruction manual)

    (6)Fumingation certificate

The Seller

ETS L. Bernard S.A.

The Buyer

Merrybest International Co.

_____

K.A. Tan/General Manager

_____

Daniel Huang/ V.P.

附件三　預付發票／預約發票樣本(Proforma Invoice)

ETSL. BERNARD S.A.

60 AVENUE DU PRESIDENT, WILSON, FRANCE

PROFORMA INVOICE

To: Merrybest International Co.　　　　P/I No.1111

　　P.O. Box 44-6. Taipei, Taiwan　　　Date: Jan. 12, 2012

Attn: Mr. Daniel Huang/V. P.　　　　　Your Ref.: P002-2

Model No.	Specifications	Qty	U/P	Total Amount
SR6/AR04	Actuator Ex-Proof, 3 Phase Motor 220V, 60Hz	6 sets	FOB French Port FFR 13,470	FFR 80,820

Total Amount: FFR Eighty Thousand Eight Hundred and Twenty only.

Packing: By Standard Export Wooden Case.

Shipment: 12 weeks ARO (After receiving the order)

Payment: by confirmed and irrevocable L/C at sight in favor of:

　　　　ETS L. Bernard S. A., 60 Ave. Du President Wilson, France.

　　　　Bank: Banque Nationale De Paris

Note: Following documents are required with shipment:

　　　(1)Warranty letter to certify the guarantee period: 20 months

　　　(2)Original manufacturer certificate

　　　(3)Material certificate

　　　(4)Inspection and testing report

　　　(5)Operation and maintenance manual (Instruction manual)

　　　(6)Fumigation certificate

The Seller　　　　　　　　　　　The Buyer

ETS L. Bernard S.A.　　　　　　　Merrybest International Co.

————————————　　　　　————————————

K. A. Tan/General Manager　　　　　Daniel Huang/ V. P.

附件四　一般訂單條款(General Terms and Conditions)

在和大陸買主簽約時，訂單條款背面常會有下列特殊條款要求：

General Terms and Conditions

(1)Unless otherwise specified, this contract is subject to Incoterms 2010.

(2)Fixed Prices: The prices indicated in this contract are fixed, and not subject to any change.

(3)Payment: As per the statement on the front page. All banking charges outside Taiwan shall be for Seller's account. Seller shall not require the letter of Credit be confirmed unless Seller bears the confirmation charges.

(4)Certificate/Inspection: Seller shall send the maker's inspection certificate to Buyer on/before the scheduled shipment date, but Seller shall be still responsible to indemnify Buyer for all and any variances or defects found after delivery.

(5)Guarantee: Seller guarantees that the products furnished shall conform to this order and shall be brand new and of good workmanship and quality, free from any defects.
If any defect, non-conformity or shortage is found by Buyer, replace the rejected products with new ones, or make up the shortage in accordance with the terms and conditions of this order. Seller shall bear the cost, insurance and freight for the replacement.

(6)Warranty: Seller warrants to Buyer that all products are free from defects in material and/or workmanship for a period of 18 months from the date of shipment, or 12 months from the start-up, whichever comes first. If any defect is found within the warranty period hereof, Seller hereby shall agree Buyer to replace or repair in time. If Buyer deems appropriate and shall provide Buyer with all necessary assistance forthwith, and agree to indemnify for loss and damage claimed by Buyer thereafter, and the period of warranty shall be extended to 12 months after the date of replacement or satisfactory completion of the repair work.

(7)Packing: The products covered by this order shall be packed in such a manner as will be adequate for seaborne or airborne export shipment. Such packing must be sufficient to secure safe arrival at destination fully covering such overseas shipping hazards as rough handling and possible collision. For any loss or damage in transit attributable to improper packing, Seller shall take the responsibility for the compensation or replacement of the loss and damage.

(8)Forwarding: If shipment is arranged by Seller, Seller shall be responsible for arrangement of vessel. In arranging the vessel, Seller shall refrain from using the overage vessel or the vessel less than 1,000GRT(gross register tonnage); otherwise, any additional insurance premium thus incurred shall be for Seller's account.

(9)Insurance: If the insurance is effected by Seller, unless otherwise specified, All risks, war, SRCC including unloading port to Buyer's warehouse, shall be covered by Seller. The insured amount shall be invoice value plus 10%.

(10)Shipment Advice: Seller shall advise Buyer of the details of shipment by fax within 48 hours after shipment. Such fax shall contain Buyer's order number, name of vessel, description of goods, ETD, ETA, number of packages, gross weight, invoice value, agent of shipping company in Taiwan, etc. Seller shall airmail directly to Buyer on the same day a full set of non-negotiable shipping documents.

(11)Force Majeure: Neither Buyer nor Seller is responsible for delay or non-performance of this contract if it is caused by war, blockade, revolution, insurrection, civil commotion, riot, mobilization, strikes, lockouts, acts of God , plague or other epidemic, fire, flood, obstruction, government's policy change, etc.

(12)Termination: Provided Seller fails to fulfill any of his obligation under this contract within specified date, Buyer shall have the right to terminate this contract by written notice, and Seller shall be responsible for all the Buyer's losses caused by the termination of the contract.

(13)Arbitration: Any claim or dispute arising from this contract shall be settled amicably. If the parties hereto fail to come to the settlement, the parties hereby agree to be finally settled through arbitration by the Commercial Arbitration Association in Taiwan.

(14) According to Mainland China regulation, effective from 1st Jan. 2006, all goods export to China with wooden materials (including load-bearing, packing, bedding, support and fastening of goods, such as nailed wooden box, wooden crate, wooden pallet, wooden frame, wooden bucket, wooden cask, wooden bearing, chock, sleeper, crosser, etc.) must be heat treated or bromide fumigated. Government Authority's approved IPPC special tag shall stick on an eye-catching spot on the opposite two sides of wooden packing (avoid red or orange color tag) is required. Besides, particularly using as the fixes of the lumber and the pad material, is the key point of the inspection, so must handle according to the wooden rule provision. Any non-compliance of the new regulation on wood packaging

material will be destroyed or returned together with the cargo to the consignments under supervision of China Inspection and Quarantine, the supplier shall bear all the relevant expenses, and if repacking happens on the way to our factory and causing the equipment damage and various losses, the supplier should undertake all the expenses happen.

(15) The supplier shall prepare a statement for non-wooden packing goods, in the event that no wooden material is used. (The definition of non-wooden materials is the deeply – processed wooden packing materials through artificial synthesis or heating and compression, such as peeled core, saw dust, wood and wood shavings.) If any wooden materials being inspected, all the relevant expenses for the treatment of the wood packing materials and the cargo shall be born by the supplier.

(16) The net weight and gross weight for each item on the P/L must be indicated clearly in KG unit (LB is not allowed). The quantity and unit shown on the P/L must be the same as those on the B/L and invoice.

(17) According to the latest announcement No.70 in 2007 of the General Administration of Quality Supervision Inspection and Quarantine, the Ministry of Commerce, the General Administration of Customs of the People's Republic of China, on the Enforcement of Entry and Exit Inspection and Quarantine of Human Food-stuff and Animal Feeding Additive and Raw Materials, the suppliers are requested to print its purpose on the outer package (sticker is not acceptable), and send the pictures with shipping documents to our Custom Affairs Department.

(18) R. O. C. is not allowed to be shown on all shipping documents. If inspected by China customs, the goods will be confiscated and destroyed.

GENERAL TERMS AND CONDITIONS（一般條款中文大意）

(1)除非另有規定，此合約出貨須按照國貿條規 2010 年版(Incoterms 2010)。

(2)固定價錢：此合約所列價錢為固定，不得調漲。

(3)付款：參照首頁上所敘述或在此所附的補充條件，所有離開台灣的銀行的費用由賣方負擔，賣方不得要求保兌信用狀除非保兌費用由賣方負擔。

(4)Certificate/Inspection 檢驗證明：賣方須於出貨前或出貨時提供工廠的檢驗證明給買方，但於出貨後發現有任何差異或不良，賣方仍須負責賠償。

(5)Guarantee（保證）：賣方在此保證所提供的設備／材料符合此訂單規格，並為全新產品及具良好的手工及品質，不能有任何不良。 如有任何不良，不符合規格或短裝被買方發現，賣方於收到買方通知同時，要馬上於買方規定時間內用新品更換不良品，或按訂單條款補足短少部份。賣方要負擔所更換新品的成本、保險及海空運運費，以及不良品退回賣方的所有費用。

(6)Warranty（保固）：除非另有規定，賣方向買方保證所有設備／材料的原料和手工的品質完好並保固從出貨日算起 18 個月，或從啟用算起 12 個月，看哪個優先。保固期間如有任何不良，賣方同意更換新品或即時修復。

如果買方認為適當，賣方須提供買方所有必須的協助，並同意賠償買方索賠的損失及損壞。保固時間要從賠貨後或令人滿意的修復後延長 12 個月。

(7)Packing（包裝）：此訂單包括的設備／材料的包裝，須為適當海空運外銷出貨的包裝，要夠安全且完好抵達目的地，並要能承受所有海外裝運危險、粗糙處理及碰撞。如因不適當的包裝造成任何損失或損壞，買方可自行決定要求賣方按照第五點所提規定賠貨。

(8)Forwarding（運送）：如果賣方安排出貨，賣方應負責安排船，但不得使用船齡過老的船或低於 1000GRT（噸位）的船，否則因此產生的額外保險費用要由賣方負擔。

(9)Insurance（保險）：如由賣方投保，除非另有規定，賣方要保全險、戰爭險、罷工暴動內亂險(SRCC)包括從卸貨港到買方倉庫。保額為出貨發票金額加 10%。

(10)Shipping Advice（出貨通知）：出貨同時，賣方要馬上用傳真(fax)通知買方出貨明細，包括買方合約或訂單號碼、船名、裝運貨的品名、裝運港名及時間、預計離港日及到港日、包裝箱數、毛重、發票金額、台灣的運輸公司代理名稱。賣方於裝運當天直接郵寄一套出貨文件給買方，此發出的傳真(fax)或郵寄(airmail)文件要

註明收件人 attention of PURCHASING DIVISION（採購部）。

賣方如未在規定時間內提供以上，所有產生的額外費用由賣方負擔。

(11)Force Majeure（不可抗力事件）：如因官方確認的戰爭、封鎖、革命、暴動、內亂、總動員、罷工、停工、天災、瘟疫、流行病、火災、洪水、障礙、政府法令變更等狀況，造成延遲出貨或無法履行合約，買賣雙方都不須負責。

(12)Termination（終止）：如果賣方無法在規定日內履行合約上的責任，買方有權以書面通知終止此合約，賣方須負責因終止此合約造成買方所產生的所有損失。

(13)Arbitration（仲裁）：從此合約所產生的任何索賠或爭執，雙方須盡快和平解決，如雙方無法解決，則經由在台灣的商業仲裁協會來仲裁。

(14)按照中國大陸規定，從 2006 年 1 月生效，所有含木頭材料外銷到大陸貨物（包括承載、包裝、鋪蓋、支稱、及貨物的固定物，例如打釘的木箱、木板條箱、木頭墊板、木框、木桶、木軸、止滑塊、枕木、橫木等）都須被熱處理或溴化煙燻消毒，政府授權確認的 IPPC 貼條，被要求要貼在木頭包裝外箱兩側看得見的地方（避免紅色或橘色貼條）。此外，特別用來當墊料的固定物是檢驗的重點，因此必需按木頭法令規定處理，如未按木頭包裝材料的新規定，在中國檢驗隔離規定下，將被銷毀或和貨物一起退回。供應商需負擔所有相關費用。如重新包裝運至我方工廠和造成設備損壞及各種損失，皆由供應商負責所有發生的費用。

(15)如沒有使用木頭包裝的話，供應商須準備一份非木頭包裝貨物的聲明（非木頭材料的定義是，經由人工合成或熱壓的加工處理的木頭包裝材料，例如層壓膠合板、薄板、纖維板等；以及木頭材料同等品或厚度低於 6mm：例如去皮殼心、鋸灰及木削）。如有任何木製材料被驗出，所有相關木頭包裝材料和貨物的處理費用須由供應商負擔。

(16)在包裝表上要清楚註明每項產品的淨重和毛重，以公斤(KG)標示（不可以英磅 LB）。包裝表上的數量和單位須和提單和發票上的一致。

(17)按照 2007 年中華人民共和國海關行政商業部門，品質監督檢疫一般行政的最新公告，加強在人類填充食物、動物食物添加物、及原物料的進出口檢疫。供應商被要求在外箱包裝上印出其目的／意圖（不可用貼條），並將其照片和出貨文件一起寄至我方的海關事務部門(Custom Affairs Department)。

(18)所有出貨文件上不可顯示"R.O.C"字樣，如被中國海關查到，貨將被沒收及銷毀。

範例 1 買方下新訂單

From: pennyturner@hushfax.com
To: eunicechang@sbco.com
Subject: New Order No. HK-010

Dear Eunice,

Thank you for your quotation SQ-50. We accept your price and terms and would like to confirm our firm order no. HK-010 as per the copy attached. The formal order sheet will be followed by post.

Please do 100% test before shipment to make sure the quality and ship the goods on time as instructed.

Please confirm your acceptance by return and send us your P/I for opening L/C.

Best regards
Penny Turner

文章結構

主題：謝謝你們的報價單SQ-50，我們接受貴方價錢及條件，並確認我們的正式訂單 HK-010，影本如附，正本將郵寄過去。

說明：出貨前，請做 100%測試以確保品質，並按指示準時出貨。

結論：請速確認以上，並傳來貴公司的預付發票，以便開狀。

生字 / 片語

1. firm order 肯定的訂單；formal order sheet 正式訂單

2. will be followed by post 隨後以郵件寄去

3. Please do 100% test before shipment 出貨前，請做 100%測試

4. as instructed 如所指示

5. for opening L/C 以便開信用狀

範例 2　賣方回覆範例 1 確認接受新訂單

From: eunicechang@sbco.com

To: pennyturner@hushfax.com

Subject: Your Order No. HK-010

Dear Penny,

Many thanks for your e-mail of 15-01 and confirm the acceptance of your order no. HK-010. We also confirm that we will do 100% test before shipment and will proceed as stipulated in the order.

The duly signed P.O. is appended herewith. As requested, we also attach our Proforma Invoice for opening L/C. We will arrange the production upon receiving your L/C.

We believe our performance will certainly meet your full satisfaction and look forward to fostering and strengthening our future business cooperation.

Best regards

Eunice Chang

文章結構

主題：謝謝您 1/15 來函，並確認接受貴公司訂單 HK-010，同時也確認於出貨前將做 100% 測試，並將遵照訂單上所規定的執行。

說明：在此附上已簽名的訂單。如所要求，我們也附上 P/I 供開 L/C 用。當一收到信用狀，我們將馬上安排生產。

結論：我們相信我方表現定當令貴公司滿意，並期盼助成加強雙方未來的生意合作。

生字 / 片語

1. proceed　進行；as stipulated　如所規定；duly　適當的
2. appended ＝ enclosed/attached　附上；herewith ＝ herein　在此
3. As requested ＝ As you requested　如所要求
4. arrange the production　安排生產；performance　表現
5. meet your full satisfaction　令貴公司滿意
6. fostering　助成；strengthening　加強
7. look forward to　期盼（注意：to ＝介系詞＋ N 或 V-ing）

範例 3　買方下訂單並指定空運公司及要求隨貨發票做成本一半之金額

DDC COMPUTER CO.

Importer & Distributer

From : Andy Muller

To: Mr. Rain Woo/Taiwan Computer Mfg.

Re: New order no. NB-020

Thanks for your today's e-mail and the fine prices you made us for the computer systems. We confirm the order as enclosed and we will have the goods delivered to Associated Cargo Express Ltd. latest Feb. 1st. Our freight prices are as follows:

1-45kg	US$6.00/kg
46-99kg	US$5.50/kg
100kg up	US$5.00/kg

We will pay the freight to you so you can put it on the invoices. Please send us a Proforma Invoice as soon as possible and advise which account we should remit the money to.

As usual, we hope you will make us a half price invoice (50% under value) which has to be sent with the goods, and send the invoice with correct amount to our Accounting Dept. Thanks.

Best regards

Andy Muller

the Buyer

文章結構

> 主題：謝謝您今日的電子郵件及你們給我們電腦系統之最好價錢。我們確認訂單如
> 　　　附，我們要求貴方將貨於最慢 2/1 前，交給 Associate 空運貨運公司，我們的
> 　　　運費報價如下：
>
> 　　　　1 − 45 公斤　　　　每公斤 US$6.00
>
> 　　　　46 − 99 公斤　　　　每公斤 US$5.50
>
> 　　　　100 公斤以上　　　　每公斤 US$5.00
>
> 說明：我們將付運費給你們，因此你們可加到發票上，請儘快寄 P/I 給我們並告知我
> 　　　們應該把錢匯到那個帳戶。
>
> 結論：如以前一樣，我們希望你們以成本一半的金額來製作隨貨發票，正確金額那
> 　　　份發票，直接寄到我們會計部門，謝謝。

生字／片語

1. the fine price　好價錢

2. computer systems　電腦系統

3. delivered ＝ despatched ＝ shipped　被裝至

4. Cargo　空運貨物

5. freight prices　運費報價

6. which account we should remit the money to　我們應把錢匯至那個帳戶

7. as usual　如以前一樣

8. a half price invoice ＝ an Invoice with 50% under value　低於成本一半的發票

9. Invoice with correct amount　正確金額的發票

10. Accounting Dept.　會計部門

範例 4　客人下訂單後，又追加項目

From: andrewsmile@tuttline.com

To: lindalin@swanexport.com

Subject: Order No. 87111

Thank you very much for your e-mail of Feb. 27 and the relevant Proforma Invoice for our order # 871111.

Since our customer just increased following four items in this order, please re-send us your Proforma Invoice with best prices and have all goods ready within two weeks. I will pay you the money upon my visit on March 16 or by T/T — will let you know the decision later on.

Please confirm the above soon.

文章結構

主題：謝謝您 2/27 的電子郵件及給我們訂單#871111 的相關預付發票。

説明：由於我們的客戶剛增加下列四個項目於此訂單，請重傳最好價錢之 P/I，並於兩週內出貨，我將於 3/16 拜訪時付款給你們或用電匯付款——隨後會讓你們知道決定。

結論：請儘快確認以上。

生字 / 片語

1. relevant　相關的
2. increased　增加；items　項目
3. re-fax　重新傳真
4. have all goods ready within two weeks　在兩週內出貨
5. upon my visit ＝ when I visit　當我拜訪時

範例 5　客人追加訂單並要求提早出貨

From: claude@laubtech.com

To: samyen@snowball.com

Subject: PO17003

(1)Currently, quantity on the order is 2,000 pcs, our customer needs to increase to 4,000 pcs.

(2)Your shipping date of 2/26 is not good enough, the customer needs all 4,000 pcs much sooner. What can you do to improve the shipping date? We need all 4,000 pcs via air.

Please review and advise your very best shipping date by return. Thanks.

文章結構

主題：有關訂單 17003

說明：(1)目前訂單上的數量是 2,000 個，我們的客戶要增加到 4,000 個。

　　　(2)你們的出貨日 2/26 不夠好，客人更早需要此 4,000 個，你們將如何改進交期？我們要全部 4,000 個用空運。

結論：請重查看並立即告知你們最好的出貨日，謝謝！

生字／片語

1. currently ＝ presently　目前地

2. increase　增加

3. shipping date ＝ shipment time　出貨日

4. much sooner　更早

5. improve the shipping date　改進出貨日

6. review　重新查看

MODERN
BUSINESS
ENGLISH

Part 4 (七、訂購及
接單函)
商用英文信函寫法

287

範例 6　客人下新訂單並指定海運船務公司

From: alanwood@cusby.com

To: candywang@epiccomputers.com

Subject: Order No.7738

Dear Candy,

Herewith we are pleased to order from you:

25,000 sets Keytops SW-302, 12 Keys, at US$0.21/set FOB Taiwan

25,000pcs Rubber Pad KB-501 　　　at US$0.18/pc

Delivery time: before end of March

Shipment by sea. For shipping operation , please contact:

HBA Transport (Taiwan) Ltd.

Phone: 02-22798888 fax: 02-22799999

Please send us your P/I and inform us about your best delivery time.

Best regards

文章結構

主題：很高興在此下給你們下列訂單：

說明：25,000 組字鍵 SW-302，12 鍵，每組 US$0.21 的 FOB 台灣價

　　　25,000 片橡膠片 KB-501，每片 US$0.18

　　　交期：3 月底以前

　　　出貨用海運，有關出貨安排，請洽：

　　　HBA 台灣運輸公司，電話：02-22798888　傳真：02-22799999

結論：請寄 P/I 給我們並告知你們的最快交期。

生字 / 片語

1. Herewith ＝ Herein　在此
2. Keytops　字鍵；Rubber Pad　橡膠片
3. For shipping operation　有關出貨安排
4. Transport　運輸

範例 7　客人下單後，要減少數量及取消部分訂單

From: dickdonus@uselectronics.com

To: dinhu@royalresistor.com

Subject: Order No. 16620 Resistors

We have found that we are overstocked on some items. Please advise if you will accept the following reductions / cancellations:

Zero ohm 1/4W 500 Mpcs-Cancel

120 ohm 1/4W -reduce from 100M to 75M pcs

Please review and advise by return.

If you cannot accept to cancel the quantity listed above, could you accept a partial reduction? Please advise.

文章結構

主題：有關訂單 16620 電阻

　　　我們發現某些項目，庫存過多，請告知是否可接受下列的減量／取消量：

說明：零歐姆 1/4 瓦 500,000 個——取消

　　　120 歐姆 1/4 瓦——從 100,000 個減至 75,000 個

結論：請重新查看並即刻告知

　　　如果無法接受上面所列數量的取消，能否接受部分減少？請告知。

生字／片語

1. resistors　電阻
2. overstocked　庫存過多；on some items　在某些項目上
3. reductions　減少之量；cancellations　取消之量
4. 100M ＝ 100,000 pcs（M ＝ 1,000 或 K ＝ 1,000）
5. partial reduction　部分減少

MODERN
BUSINESS
ENGLISH
Part 4 （七、訂購及接單函）
商用英文信函寫法
289

範例 8　客人下訂單前，又殺價

From: shwu@camelstarmedical.com

To: chilo@eelcuk.com

Subject: Order for PFC2

Dear Mr. Chilo,

Thanks for your e-mail of March 9. We expect to place you an order for 1 set of PFC 2. However, to be competitive, we need the price at FOB London Airport Euros 5,200.- which is 15% less than your offer.

Please confirm your acceptance by return and send us relevant Proforma Invoice immediately.

Best regards

S. H. Wu

文章結構

主題：謝謝你們 3/9 的 e-mail，我們希望下一組 PFC2 的訂單給你們。

說明：然而，為了要有競爭性，我們需要 FOB 倫敦機場價歐元 5,200，比你們的報價低 15%。

結論：請即確認接受並馬上寄來相關之預付發票。

生字／片語

1. to be competitive　為了要有競爭性

2. Euros　歐元

3. which is 15% less than your offer　比你們報價低 15%

範例 9　賣方確認訂單並詢問是否要分批裝運

From: gracechen@poolingtrade.com

To: smithkio@horsetube.com

Subject: Your order no.9801

Re: Your Order #9801

Thanks for your above order and please be advised that all items are in stock except one item 35LD, which will be ready early April.

Please confirm if you want us to ship the stock first or ship all items together until early April.

Please let us know your decision as soon as possibe.

Best regards

文章結構

主題：謝謝你們上述的訂單，請被告知所有項目都有庫存，除了 35LD 這一項，要到
　　　4 月初才有貨。

說明：請確認是否您要我們先出庫存貨，或者全部一起等到 4 月初再一起出貨。

結論：請儘快告知決定。

生字／片語

1. all items are in stock　所有項目都有庫存

2. to ship the stock first　先裝庫存的貨

3. ship all items together　所有項目一起出貨

範例 10 賣方回覆不接受訂單上的價錢，因爲數量太少

From: Peter Lee

To: bradlois@obatrading.com

Subject: Your P.O. #B1852

We received your above order today, in which you confirmed us the quantity of two containers only.

As offered in our last correspondence, we would give you 1.5% discount only based on the order for total 3 containers. So, please re-confirm if you want to increase the quantity to 3 containers to get the 1.5% discount, or only order 2 containers without any discount.

We are waiting for your reply soon.

文章結構

主題：我們今日收到貴公司上述的訂單，在訂單上您只確認 2 個貨櫃量。

說明：如我們上封書信中所報價，我們願意提供 1.5%折扣係根據訂單量為 3 個貨櫃。因此，請再確認是否貴公司要增加數量為 3 個貨櫃以獲取 1.5%折扣，或照原價只購買 2 個貨櫃。

結論：我們等候貴方儘快回覆。

生字／片語

1. container　貨櫃；as offered　如所報價

2. in our last correspondence　在我們上封書信中

3. discount　折扣；based on　根據

4. total　全部；re-confirm　重新確認

5. without any discount　沒有任何折扣

練習 18 訂購及接單函練習

1. 將下列句子翻譯成英文

(1)我們接受你們的價錢及交期,並確認我們的訂單 AB-012 如附件

(2)請確認貴方的接受,並寄給我們貴方的預付發票,以便開信用狀

(3)我們將 用郵件 寄 我們的正式訂單給你們(by mail 放句尾)

(4)我們確認 出貨前 將做 100%測試(before shipment 放句尾)

(5)因為這是一張小額訂單,請 用電匯 付款(by T/T 放句尾)

(6)請告知最好的出貨日

(7)有關海運出貨,請聯絡我們的報關行(forwarder) 如下:

(8)由於經濟不景氣(economic depression),我們希望取消此訂單

(9)我們 目前 沒有庫存(stock),最快交期是 2 個星期

(10)請告知是否接受分批出貨(partial shipments)

2. 請依下列大意,寫一封完整的買方下單函

主題:我們接受貴公司 報價單上 的 價錢及交期,並很高興附上 我們的新訂單號碼
　　　AB-012。

說明:請務必要 準時 出貨,因為我們要趕聖誕節銷售。

結論:請確認以上,並傳預付發票給我們,以便開信用狀。

生字／片語

　　1. 務必＝ do,放動詞前,加強語氣

　　2. 趕聖誕節銷售　to catch the Christmas sale

　　3. 以便開信用狀　for opening the L/C

3. Write a letter to confirm the acceptance of the above order no. AB-012 and attach the relevant P/I for opening the L/C.

4. 克漏(填入適當的英文生字／片語)

(1)我們接受貴公司的報價並確認我方訂單號碼 001 如附。

　　We _____ your offer and _____ our order number 001 as _____.

(2)請將尚未出貨部分的訂單於這個月底前全部裝出。

Please ship all _____ orders before the end of this month.

(3)我們急需此貨。請告知可否從庫存出貨？

We need the goods _____. Can you ship them from _____?

(4)附上我方的預約發票以供開信用狀。

We are _____ our P/I for _____ L/C.

(5)當一收到貴公司的信用狀，我們將馬上安排生產。

We will _____ the _____ upon receiving your L/C.

5. 將下列句子翻譯成中文

(1)We are pleased to place you an order for 100,000 sets of MP3 players. Please do ship them on time as we have to catch the sales season here. Any delay in shipment will not be accepted.

(2)Please produce according to the attached drawing and specification and send us some shipping samples for approval before shipment. We will remit money to you by T/T as soon as the samples are approved.

(3)Thanks for your order and attach herewith our P/I for arranging the remittance. We will make the shipment upon the receipt of your T/T payment.

(4)Due to the typhoon, there is no vessel available this week and the soonest available vessel will be in next week. Please kindly accept our delay in shipment and confirm your acceptance by return e-mail.

(5)We just received the additional order 5,000pcs for O/No. 100 from the customer. Please adjust the order quantity and confirm if you can ship all together at the end of June.

八、信用狀及其他付款條件討論函
(L/C & Other Payment Terms Discussion Letters)

1. 當買賣雙方正式簽定訂單或銷售合約書後，按「國際條約」規定：「買方須於簽約後 15 天內開出信用狀，或安排付款。」因此，對於新客戶或第一次交易的客戶就會嚴格要求；但是對老客戶多半會給予通融，只要在出貨前 1～2 週內收到 L/C 即可，但是因為銀行作業慢，或有些國家有外匯管制，也會造成信用狀延誤開出，賣方就要注意去追蹤或催促，以免造成損失。

   注意：所有貿易流程中最大的風險(risk)，即在付款延誤或沒收到付款。

2. 一般的付款條件除了信用狀(L/C)以外，常用的有：
   D/P（付款交單）、D/A（承兌交單）、O/A（月結／記帳）、COD（交貨付現）、CWO（下單付現）、Installment（分期付款）及 On Consignment（寄售）等。(CAD: Cash against Documents = D/P)
   有關付款條件(Payment Terms)及付款方法(Payment Methods)的細節，請參看「三、報價及回覆函」章節，在此就不再重複。

3. 因為信用狀的銀行費用較高，一般多用於訂單金額超過 5,000 美金以上的交易，否則多半採用其他付款方法。

4. 信用狀上的基本內容
   (1)申請人(50: Applicant/Drawee)：買主公司名稱／地址
   (2)受益人(59: Beneficiary/Drawer)：賣方公司名稱／地址
   (3)開狀銀行(Opening Bank)：買方的往來銀行（大都在信頭處）
   (4)通知銀行(57D: Advising Bank)：開狀行的分行或往來銀行
   (5)押匯銀行(41D: Negotiation Bank)：賣方的往來銀行（大都不限押）
   (6)保兌銀行(49: Confirming Bank)：視需要再指定
   (7)補償／清算銀行(Reimbursement Bank)：代付款行（視需要而定）
   (8)付款銀行(42D: Drawee: Issuing Bank)：大都為開狀行
   (9)開狀日期(31C: Issuing Date/Date of Issue)
   (10)信用狀號碼(20: L/C No./Documentary Credit Number)

(11)金額及幣別(32B: Currency Code, Amount)

(12)最後出貨日(44C: Latest Shipping Date/Latest Date of Shipment)

(13)有效期(31D: Expiry Date/Date and Place of Expiry)

(14)提示押匯時間(48: Drafts Presented Time for Negotiation)

(15)啟運港(44A:Loading On Board From…/Port of Loading)

(16)目的地港(44B: Port of Discharge)

(17)出貨品名／數量／單價及貿易條件(45A:Shipment of Goods..CFR..)

(18)分批裝運(43P: Partial Shipment: Allowed or Prohibited)

(19)轉運(43T: Transhipment:Permitted or Forbidden)

(20)信用狀格式及種類：

　　① 40A: Irrevocable 不可撤銷

　　② 42C: at sight 即期 或 at xxx days sight (usance)遠期

　　③ transferable 可轉讓：做轉口或三角貿易時使用

　　④ back to back 遮蓋式轉讓：可改變開狀人名稱／金額及最後出貨日

　　⑤ revolving L/C 循環信用狀：在一定金額及時間內可重複使用

　　⑥ standby L/C 擔保信用狀＝ Performance Bond 保證金（賣方開給買方）

(21) 46A: Documents Required 文件／單據要求：

　　下列三項為必備：

　　①商業發票 Commercial Invoice

　　②包裝表 Packing List

　　③提單 Clean on Board Ocean Bill of Lading（海運）；

　　　Air Way Bill（空運）；Forwarder's Receipt（攬貨人收據）

　　　注意：L/C 上會規定在提單上要註明下列四點：

　　　　　　・運費誰付：FOB/FCA 出貨：Freight Collect（運費待收）

　　　　　　　CFR/CIF/CPT/CIP：Freight Prepaid（運費預付）

　　　　　　・收貨人／貨主(Consignee)：客戶名稱或 Consigned to the order of xxx

　　　　　　　bank/To order of shipper（＝貨主是銀行）

　　　　　　・通知人(Notify)：買主或其進口報關行名

　　　　　　・賣方於提單背面空白處背書：Blank Endorse

　　一般正本提單為三張，L/C 上有時要求三張正本（3/3 of B/L 或 Full set of B/L）

　　都送入銀行，有時要求兩張正本(2/3)送入銀行押匯，一張正本(1/3)直接寄給買主。例如：Full set clean on board ocean bill of lading indicating freight collect, blank endorse, consigned to the order of Banco de Chile. Notify: Buyer.

以下視需要或客戶要求，再提示

④保單　Insurance Policy or Certificate（CIF 出貨才需要）

⑤產地證明　Certificate of Origin──由商會發出

⑥海關發票　Special Custom Invoice (SCI)

⑦其他文件：檢驗報告／品質證明／操作手冊／出貨通知等或其他特別指示，視客戶要求照做即可。

5. 信用狀是按照訂單條款開出，包括以上基本內容，所以收到 L/C 後，要馬上核兌以上各點，看有無錯誤或矛盾之處，如果有錯誤，最好請客戶儘快修改(amend L/C)，否則就有瑕疵，押匯時會被對方拒付(unpaid)；如果有些老客戶為了省修改費(admendment charge)或時間來不及修改，同意接受瑕疵，一定要要求客戶發一封公司正式同意函來，送入銀行，當作押匯文件之一。

注意： 如此雖可免遭馬上拒付的危機，但仍會被銀行扣瑕疵費。現在有些國家例如澳洲法律已規定銀行拒付瑕疵押匯。

6. 信用狀開法有三種

(1)郵寄信用狀(Airmail L/C)＝正本

(2)短電信用狀(Short cable L/C)＝通知＋正本會另外郵寄

(3)全電信用狀(Full cable L/C)＝正本，但要注意 L/C 上要註明：This cable is the operative instruction（可押匯）或 This credit subject to UCP 600（按統一慣例條款）

7. 信用狀附件

附件一　信用狀申請書樣本 (Application for Opening L/C)

附件二　全電信用狀樣本 (Full Cable L/C / SWIFT)

附件三　信用狀上常見的文件要求及額外條件

　　　　(Documents Required & Additional Conditions)

8. 買方告知信用狀已開出寫法（參範例 1）

We are glad to inform you that we have opened L/C No.×××on×××.

Please proceed the order and confirm the receipt.

（我們很高興於××時間已開出 L/C 號碼×××，請進行訂單並於收到時確認）

9. 賣方確認收到信用狀寫法（參範例 2）

We confirm the receipt of your L/C No.×××with thanks.

（我們感謝收到貴公司信用狀號碼×××）

Thank you for your L/C No.×××which was received on××××.

（謝謝貴公司信用狀，已於×××時間收到）

10.賣方催信用狀或催付款（參範例 4）

Re: Your order No.×××, goods are ready for shipment, please confirm

when you opened the L/C. (or if you have remitted us the payment by T/T)

（有關訂單×××，貨已完成，請告知何時開 L/C，或是否已電匯付款）

11. 賣方收到信用狀，發現有錯誤，請買方修改（參範例 5）

主題：我們確認已收到貴方的信用狀，但是發現有下列的錯誤：

(We confirm that we have received your L/C, but we found following errors.)

說明：(1)品名金額不對

(The description of the goods and the total amount are not correct.)

(2)日期太短，來不及裝運

(Time is too short to ship before the latest shipping date.)

(3)價錢條件錯誤，例如 FOB 打成 CIF

(The price terms shoud be FOB not CIF.)

(4)分批轉運限制錯誤，應改為允許

(Partial shipment and transhipment should be allowed.)

(5)或以上基本內容任何一項出錯，例如將提單改為攬貨者收據

(Please change AWB to Forwarder's Receipt.)

結論：請儘快修改以上，如為買方過失，買方付修改費，如為賣方的錯，賣方付修改費。

(Please amend the above as soon as possible and pay the amendment charge at your side as the errors were caused by you.)

12.買方回覆已修改 L/C 或將接受瑕疵押匯

> 主題：我們收到貴方×××的 e-mail 要求修改信用狀
>
>   (We received your e-mail dated ××× asking for the L/C amendment.)
>
> 說明：(1)如所要求，已修改。(We have amended it as requested.)
>
>   (2)因時間來不及，我們接受貴方瑕疵押匯。
>
>   (We will accept your discrepant documents as there is no enough time to amend it.)
>
> 結論：(1)因為不是我方過錯，修改費由貴方負擔。
>
>   (Because it is not our fault, the amendment charge is for your account.)
>
>   (2)請於收到信用狀修改書後，儘快裝運。
>
>   (Please arrange the shipment as soon as you receive the L/C amendment.)

13.其他付款方式，討論內容：

  (1)貨款已電匯或已用支票寄出

    (The payment has been made by T/T or by bank check.)

  (2)附上對帳單，請儘快付清

    (The statement is enclosed. Please settle it soon.)

  (3)買主將直接開 L/C 給工廠，請其保留佣金或將差額匯至×××帳戶

    (The buyer will directly issue L/C to the factory. Please reserve our commission or remit the difference to our account.)

14.因信用狀的開狀手續費較高，且文件要求嚴謹，如稍有錯誤，就會遭銀行拒付 (unpaid)，所以除非金額非常巨大，付款條件現在大都會要求客戶於出貨前以電匯方式付款(Payment terms: By T/T before shipment)。但是要特別注意電匯(T/T)是付款方式(Payment method)，不是付款條件(Payment terms)，不要只打 Payment by T/T，這將很容易造成付款糾紛，因為客戶也許會拖延付款時間，甚至於貨物裝出後一段時間都尚未將貨款電匯出來。所以如要求客戶以 T/T 付款，一定要記得註明電匯的時間是出貨前或出貨後多久。例如：

(1)Payment: By T/T before shipment.

(2)Payment: 1/2 by T/T upon order confirmed; 1/2 by T/T before shipment.

(3)Payment: by T/T within 30 days after shipment.

15.信用狀及其他付款條件討論函常用單字／片語

(1)applicant　申請人；beneficiary　受益人

(2)latest date of shipment　最後出貨日；date of expiry　有效期

(3)partial shipment　分批出貨；transhipment　轉運

(4)permitted/allowed　允許

(5)not permitted/not allowed/prohibitted/forbidden　不允許

(6)irrvocable　不可撤銷；transferable　可轉讓；at sight　即期；confirmed　保兌；
in our/your favor　受益人為我方／貴方

(7)standby L/C　擔保信用狀；revolving L/C　循環信用狀

(8)freight collect　運費待收／運費到付；freight prepaid　運費預付

(9)Shipping Documents　出貨文件：Commercial Invoice　商業發票；Packing List　包
裝表；Weight List　重量表；Insurance Policy or Certificate　保單；Certificate of
Origin　產地證明；Ocean Bill of Lading　海運提單；Air Way Bill　空運提單；
Warranty Letter　保固書；Certificate of Original Manufacturer　原廠證明書；
Inspection Report　檢驗報告；Instruction Manual　操作手冊；Certificate of Quality
and Quantity　品質數量證明書；Fumigation Certificate　煙燻證明書；Shipping
Advice　出貨通知

(10)unit price　單價；total amount　總價

(11)net weight　淨重；gross weight　毛重；freight charge　運費

(12)Payment terms　付款條件：D/P　付款交單；D/A　承兌交單；O/A　記帳／月
結；CWO　下單付現；COD　交貨付現；Installment　分期付款；On
consignment　寄售

(13)L/C amendment/amend L/C　信用狀修改；amendment charge　修改費

(14)AQL (Acceptable Quality Level)　可接受品質標準

(15)inspect at random　抽驗；major defect　主缺點；minor defect　次缺點

(16)receipt　收據；wire ＝ remit ＝ T/T　匯款／電匯

(17)non-sufficient funds　存款不足；overdrawn　透支

(18)money order　郵政匯票；outstanding payement　欠款

(19)Forfaiting　遠期 L/C 賣斷；Factoring　應收帳款管理；OBU　境外金融

16.信用狀及其他付款條件討論函常用的主題句／説明句／結論句

主題句(Main Idea)

(1)We are glad to inform you that we have issued an irrevocable and transfererable L/C at sight in your favor on 5/20.

(2)Regarding O/No. 001, we have opened you the full cable L/C last week. Please refer the copy attached and arrange the shipment as soon as possible.

(3)We confirm the receipt of your L/C Number xxx and will ship the goods as scheduled.

(4)We received your L/C but found following errors: Partial shipment and Transshipment should be allowed.

説明句(Explanation)

(1)Please arrange the production as soon as you receive the L/C and do 100% inspection before shipment.

(2)We will settle the payment by Bank Draft for P/O 001.

(3)We will arrange the shipment upon receiving your T/T payment.

(4)Please remit our commission to our bank account upon receipt of the payment from the customer.

(5)Debit Note 010 is enclosed. Please settle it before 6/30 as agreed.

(6)The amendment charge should be at your accont as the error was caused by you.

結論句(Conclusion)

(1)Please confirm the receipt of our remittance and send us shipping details soon.

(2)The shipment will be made before the latest shipping date as stipulated in your L/C.

(3)We will send you the relevant shipping advice upon shipment.

(4)Please feel free to contact us if you still have any question.

(5)We are waiting for your shipping information soon.

範例 1　買方通知信用狀已開

From: richchild@earthlink.net

To: mikelord@ms67.hinet.net

Subject: L/C for P.O. 179

Dear Mike,

Please be informed that we have opened L/C No. 202775 on Jan. 15 through "Amsterdam Bank, Taipei" to your bank by cable.

Please check and send us shipping details ASAP. Thanks.

Best regards

Rich Child

文章結構

主題：有關訂單 179 的信用狀

說明：請被告知本公司已於 1/15 經由「台北阿姆斯特丹銀行」開出 cable 信用狀號碼 202775 到貴公司的銀行。

結論：請查看並儘快發出貨明細給我們，謝謝！

生字／片語

1. have opened L/C ＝ have issued L/C　已開出信用狀

2. Amsterdam Bank, Taipei　台北阿姆斯特丹銀行

3. Shipping details ＝ shipping advice ＝ shipping information　出貨明細／出貨通知

4. ASAP ＝ as soon as possible ＝ soon ＝ immediately　儘快

範例 2 回覆範例 1，賣方確認收到信用狀

From: mikelord@ms67.hinet.net

To: richchild@earthlink.net

Subject: Confirm receiving the L/C for P.O.179

Dear Rich,

We confirm with thanks the receipt of your cable L/C No.202775 for your order no.179 today, and have started production already. The shipment will be made before the latest shipping date as stipulated in your L/C, and the relevant shipping advice will be followed upon shipment. Will keep you informed.

Best regards

Mike Lord

文章結構

主題：我們帶著感謝確認於今日收到貴公司訂單 0179 的 cable 信用狀，同時也已開始生產。

說明：將於信用狀規定的最後出貨日前出貨，相關的出貨通知亦會於出貨同時發出。

結論：將會與您保持聯絡。

生字／片語

1. confirm　確認；cable L/C　電報方式開出的信用狀

2. production(N)　生產；produce (V)

3. shipment will be made ＝ goods will be shipped　出貨

4. latest shipping date　最後出貨日

5. as stipulated　如所規定

6. relevant　相關的；shipping advice　出貨通知

7. upon shipment　出貨同時

8. (We) will keep you informed ＝ will keep you posted　將會與您保持聯絡

範例 3　買方要求賣方重傳更新的 P/I 以便開 L/C 及告知付款

Dear Win,

Re: P/O 543 P/I 886-2

With regard to the additional order, are you going to send us a revised P/I, or are you going to re-bill on a new P/I for applying import licence and opening the L/C?

Please let us know so we know how to proceed the L/C.

Re: Payment for O/No 525

Please be informed that our bank has remitted US$12,000 to your bank on Jan. 9. Reference no. is 202350.

Please check and confirm upon receipt.

Best regards, Ciao

文章結構

主題：有關訂單 543 P/I 886-2 增訂部分。

說明：你們要寄更新的 P/I 給我們或另開一張新的 P/I 以便申請輸入許可證及開 L/C？

結論：請讓我們知道以便進行開信用狀。

主題：有關訂單 525 的付款。

說明：請被告知我們銀行已於 1/9 匯款美金 12,000 元到貴方銀行，參考號碼是 202350。

結論：請查看並於收到後確認。

生字 / 片語

1. additional order　增加的訂單
2. revised　修正的；re-bill　重做；proceed　進行
3. have remitted　已經匯款；remitance　匯款(N)
4. reference no.　參考號碼；upon receipt　當一收到

範例 4　回覆範例 3，賣方催信用狀及催付款

Dear Ciao,

Re: P/O 543

We have sent you our revised P/I as requested last Monday, but we have not received your L/C till now. Since the goods are ready to ship, please kindly advise when you opened the L/C and the L/C details.

Re: Payment for O/No.525

Sorry to inform you that we have not received your T/T payment US$12,000 till now. Please check with your bank to see if they did remit the money on 1/9 as you mentioned in your last mail. Please advise by return.

Best regards

Win

文章結構

主題：有關訂單 543。

說明：上週一已按要求將修改的 P/I 寄去，但至今尚未收到貴公司的信用狀。

結論：因貨已完成準備裝運，請告知何時開出 L/C 及其明細。

主題：有關訂單 525 的付款。

說明：抱歉通知您我們至今尚未收到貴公司 US$12,000 的電匯付款。請與銀行查看
　　　是否真的已於您上封信中所提的 1/9 匯出。

結論：請速告知。

生字 / 片語

1. goods are ready to ship (for shipment)　貨已完成準備裝運
2. L/C details　信用狀明細
3. T/T payment　電匯付款
4. remit the money　匯錢
5. as you mentioned　如您所提及
6. by return ＝ right away ＝ soon ＝ immediately　馬上／即刻

MODERN
BUSINESS
ENGLISH
Part 4 （八、信用狀及其他
付款條件討論函）
商用英文信函寫法 305

範例 5　賣方收到信用狀，發現有錯誤，請買方修改

From: stonelin@ms33.hinet.net

To: petrovic@earthlink.net

Subject: Please ament L/C ASL-0120

Dear Petrovic,

We confirm receiving your L/C No. ASL-0120, but we are sorry to inform you that we found following errors:

1. The trade terms should be CFR New York by sea freight, not CIP New York by air fright.

2. The partial shipment and transhipment should be allowed.

Please kindly amend the above as soon as possible. The amendment charge is to be at your account as the errors were caused by you.

Best regards

Stone Lin

文章結構

主題：我們確認收到貴公司信用狀 ASL-0120，但抱歉告知發現下列錯誤。

說明：1. 價錢條件應該是 CFR 海運到紐約，而非 CIP 空運到紐約。

　　　2. 分批出貨及轉運應該要允許。

結論：請盡快修改以上。因此過錯為貴方所造成，修改費用由貴公司負擔。

生字／片語

1. errors　錯誤；trade terms　貿易條件
2. sea freight　海運；air fright　空運
3. partial shipment　分批出貨；transhipment　轉運
4. amend　修改；amendment charge　修改費用
5. at your account　由貴公司負擔
6. caused by you　為貴方所造成

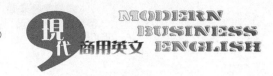

範例 6　賣方告知出貨價錢中已含配額費用(Quota Charge)

From: allenpan@textiledyeing.com

To: carolbrain@usamerchandise.com

Subject: Shipment of L/C No. AA12345

Please note we have included the quota charge into the unit price of the merchandise shipped under above L/C no.

The detail is as follows: US$10.50 per piece and total amount is US$25,000.

Thank you for your attention to the above. Please feel free to contact us if you still have any question.

Best regards

Allen Pan

文章結構

主題：有關信用狀 AA12345 的出貨。

　　　請注意我們已將配額費用包括在上述信用狀的出貨單價中。

説明：細節如下：每個美金 10.50，總金額美金 25,000。

結論：謝謝您對以上的注意，如有疑問，請隨時和我們聯絡。

生字／片語

1. quota charge　配額費用；quota　配額
2. unit price　單價；total amount　總金額
3. merchandise　商品；per piece　每一個
4. Please feel free to contact us　請隨時和我們聯絡

範例 7　買方告知將用電匯(T/T)付款

To: Mr. Muller /ALMAC

From: Alex Wang / Yi-Ming Trade Co.

Sorry for having not replied you earlier because I was terribly busy these few days for the Exhibition.

Regarding O/No. 8701, we will settle the payment by T/T in next two days. However, please also include two pieces of 34HC into this order and have them ready and shipped upon receipt of our payment.

Thanks for your cooperation.

文章結構

主題：很抱歉未能儘早回覆，因為近日為了展覽特別忙。

説明：有關訂單 8701，我們過兩天會用電匯付清貨款，然而，請加入兩個 34HC 在此訂單，並準備好於收到貨款時出貨。

結論：謝謝您的合作。

生字 / 片語

1. terribly busy　極忙碌；these few days　這幾天
2. for the Exhibition　為了展覽；by T/T　用電匯
3. settle the payment　付清貨款；next two days　過兩天
4. upon receipt of our payment　當一收到我方付款
   ＝ upon receiving our payment

範例 8　進口代理告知其買主要直接付款給工廠並要求賣方保留其佣金

From: johnsmith@tradeagency.com.

To: phli@shoesmfg.com

Subject: Direct Payment by our Customers

Dear Mr. Li,

As expected, the business between us is getting better and better. However, we are facing the problem that most of our customers would like to pay direct to your company.

Therefore, we must find a best way to settle our commission: either to reserve at your account or to remit to our account for which we are going to open an account at the same bank as yours during this visit.

Please advise us your best suggestion and comments.

Best regards

John Smith/President

文章結構

主題：如所預期我們之間的生意是愈來愈好。然而我們正面臨一個問題，那就是我們大部分的客人都希望直接付款給貴公司。

說明：因此，我們必須找出一個最好的方法來支付我方佣金：是保留在貴公司的帳上或匯入我方帳戶，有關此點，我準備在這次的拜訪期間，在貴公司同一銀行內開一個戶頭。

結論：請告知貴公司的最佳意見。

生字／片語

1. direct payment　直接付款；as expected　如所預期
2. is getting better and better　愈來愈好
3. facing　面臨；settle our commission　來支付我方佣金
4. reserve at your account　保留在貴公司的帳上
5. remit to our account　匯入我方帳戶；open an account　開一個戶頭
6. during this visit　在此次拜訪期間

範例 9　買方欲修改信用狀，問出貨狀況

From: Edmund Din

To: mikewolf@rpmpump.com

Subject: L/C No. 7AMAH1

Please confirm by return the followings:

(1)When will the shipment be ready?

(2)What is the latest shipping date you would like us to amend on L/C?

(3)Please advise the total gross weight for two sets of pumps. We need this information to calculate the air freight cost, which will be deducted from the L/C amount.

We need to get the above information before tomorrow.

文章結構

主題：請即告知下列事項：

說明：(1)這批貨何時可裝出？

　　　(2)你們希望修改 L/C 上的最後出貨日至何時？

　　　(3)請告知這兩組幫浦的總毛重，我們需此資料計算空運費，以便從信用狀的金額中扣除。

結論：我們需要於明日前得知上述資料。

生字 / 片語

1. latest shipping date　最後出貨日；L/C amount　信用狀金額

2. amend (V), amendment (N)　修改；air freight cost　空運費

3. total gross weight (GW)　總毛重；net weight (NW)　淨重

4. two sets of pumps　兩組幫浦；deduct　扣除

範例 10　買方告知以電匯支付貨款並要求收據

From: jefflin@evatechnology.com

To: martin@sctco.com

Subject: T/T payment US$4,987

We wired the sum of US$4,987 to your account at Bank of America this afternoon as the payment for our P/O #3026 & P/O # 3027.　We believe that it will reach you in a couple of days.　Please send us a receipt when you get the money.

文章結構

主題：我們今日下午電匯美金 4,987 元的金額至貴公司美國銀行的帳戶，當作訂單
　　　3026 及 3027 的付款。

說明：我們相信此匯款一兩天內會入貴帳戶內。

結論：請於收到錢時寄收據給我們。

生字／片語

1.　sum = amount　金額

2.　wired = remit by T/T　電匯

3.　as the payment　當作付款

4.　in a couple of days　一兩天內

5.　receipt　收據

MODERN
BUSINESS
ENGLISH

Part 4 (八、信用狀及其他
付款條件討論函)
商用英文信函寫法

311

範例 11 回覆範例 10，賣方確認收到電匯，發出收據

SCT COMPANY

P.O. BOX 2216, CA 9104, USA

E-mail address: sct@earthlink.com.

To: EVA Technology Co.

Attn: Mr. Jeff Lin   Feb. 2, 2012

RECEIPT

Dear Mr. Lin,

We confirm with thanks the receipt of your T/T payment total amount US$4,987 covering following orders:

P/O # 3026	US$3,039.00
P/O # 3027	US$1,948.00
Total Amount Paid	US$4,987.00

Thanks & best regards
SCT Company

文章結構

主題：收據

說明：我們確認收到貴方電匯總金額美金 4,987 元支付訂單 3026 貨款 US$3,039 及訂
單 3027 貨款 US$1,948。

結論：謝謝。

範例 12　賣方要求付款改為可轉讓信用狀

Subject: Changing the Payment Terms

Dear Anton,

Please kindly be informed that we will ask for the payment by irrevocable and transferable L/C from now on, or if it is possible, we would prefer you pay us by T/T before shipment.

Due to the world market depression, it is more and more difficult to run the business, and many manufacturers of textile here in Taiwan either shut down or moved to Mainland China, Thailand or other Asian countries. To survive and to be more competitive in the market, we could not but change our business style by ordering the material from our associated companies in China and making the shipment from Hong Kong instead of from Taiwan. That is why we need the payment by a transferable L/C or T/T in advance.

Anyway, we will discuss more detailed cooperation with you next month when we attend the Fair in Hannover.

Thanks for your attention to the above and look forward to seeing you soon.

Best regards

Shock Lin

文章結構

主題：請被告知從現在開始，我們將要求付款更改為不可撤銷可轉讓的信用狀，或者，如果可能的話，我們更喜歡你們在出貨前電匯給我們。

說明：由於世界市場不景氣，愈來愈難做生意，而許多台灣的紡織工廠不是關閉就是搬遷至中國大陸、泰國或其他亞洲國家。為了生存及能在市場上競爭，我們不得不改變我方的生意型態，從大陸的關係企業購料，再從香港出貨，取代以往的在台生產，那也是為何我們需要信用狀要能轉讓或付款能事先電匯之原因。

結論：(1)無論如何，我們在下個月參加漢諾威展覽時，會和貴公司再討論細節。
　　　(2)感謝您對以上之注意並盼望很快和您見面。

生字／片語

1. by irrevocable and transferable L/C　用不可撤銷可轉讓信用狀

2. from now on　從現在開始；associated companies　關係企業

3. before shipment　出貨前；textile　紡織

4. Due to the world market depression　由於世界市場不景氣；economic depression　經濟不景氣

5. to run the business　經營生意

6. shut down　關閉；move to　搬遷至

7. Mainland China　中國大陸；Thailand　泰國；Asian Countries　亞洲國家

8. To survive　為了生存；business style　生意型態

9. attend the Fair in Hannover　參加漢諾威（德國）的展覽

範例 13　賣方要求修改信用狀並討論驗貨標準

From：Howard Chen <howardchen@sohotech.com>

To：Adrian Boosh

Subject：L/C 900718 for Towball Shipment

Dear Adrian,

We regret to inform you that we can only ship above goods at the end of Feb. if we can receive your following amendment by cable before this week:

(1)Please change the trade terms to FOB Shanghai, China instead of FOB Hong Kong and shipment from China to Australia or New Zealand via Hong Kong.

(2)Transshipment is to be allowed.

(3)Extend the Latest Shipping Date to March 5 and the Validity (Expiry Date) to March 15.

We will proceed the packing and inspection only upon receipt of your above amendment. To show our sincerity, we will pay for the amendment charge and will return you the freight cost from China to Hong Kong after shipment is made.

The goods will be inspected according to the International Inspection Standard as per AQL Level II: Major 0.65, Minor 1.5 - please refer the attached Data.

Please move faster on this amendment to avoid the delay in shipment. Please understand that our bank can only make the transferable L/C to our factory upon receipt of your correction.

We are really sorry for the trouble caused and appreciate your great help on this matter.

Best regards

Howard Chen

文章結構

主題：有關滑鼠拖球的信用狀 900718，很抱歉告知我們只能在二月底出貨，但要於本週收到貴方下列電報修改。

說明：(1)請將貿易條件由 FOB 香港改為 FOB 大陸上海，而出貨將由大陸經香港到澳洲或紐西蘭。

(2)轉運要改為允許。

(3)延長最後出貨日至 3/5，有效期至 3/15。

當一收到上述修改，我們馬上進行包裝與檢驗，為了表示誠意，我們將負擔修改費，並於出貨後，退還大陸到香港的運費。出貨將按國際檢驗標準 AQL Level II，主缺點 0.65，次缺點 1.5 來檢驗——請參照附表。

結論：請快速修改，以免延遲出貨，請了解我們銀行只能於收到貴方的修改書後，才能開出轉讓 L/C 給工廠。我們為了所造成的麻煩致歉，並感謝貴公司在這件事上的大力幫忙。

生字／片語

1. towball　電腦滑鼠中的拖球；freight cost　運費
2. trade terms　貿易條件；amendment charge　修改費
3. Australia　澳洲；New Zealand　紐西蘭；via　經由
4. transshipment　轉運；Latest Shipping Date　最後出貨日
5. extend　延長；Validity (Expiry Date)　有效期
6. to show our sincerity ＝ to show our faith　為了表示誠意
7. International Inspection Standard　國際檢驗標準
8. AQL ＝ Acceptable Quality Level　可接受品質標準
9. Major (defect)　主缺點；Minor (defect)　次缺點
10. to avoid the delay in shipment　為了避免延遲出貨
11. move faster　快速行動
12. make the transferable L/C　開出轉讓信用狀

範例 14　賣方告知買方銀行存款不足，被退票

From：Tony Lau <tonylau@pacificnational.com>

To：WH Chang<whchang@hitech.com>

Subject：Your Check #838-097

Dear Mr. Chang,

We deposited your above check for US$537.70 into the bank, but it was returned by your bank "Chase Manhattan Bank" for the reason of "Non-Sufficient Funds" Your account is overdrawn US$313.08.

Please deposit the above amount to your bank to clear the overdraft, or advise how you will settle the above payment.

Best regards

Tony Lau

文章結構

主題：我們將貴公司上述支票 US$537.70 存入我方銀行，但卻遭到貴方銀行 "Chase Manhattan" 銀行退票。

說明：理由是「存款不足」，貴方帳戶透支 US$313.08。

結論：請存入銀行上述不足的透支，或告知將如何支付上述欠款。

生字／片語

1. deposit　存入

2. Non-Sufficient Funds　存款不足

3. overdrawn(V)　透支；to clear the overdraft　結清透支

範例 15　買方回覆範例 14，將另寄郵政匯票付款

From：WH Chang<whchang@hitech.com>

To：Tony Lau <tonylau@pacificnational.com>

Subject：Our Check #838-097

Dear Tony,

Thanks for your e-mail of 1/31 and sorry for the trouble we caused to you.

We are sending you today the money order US$537.70 to clear the outstanding payment.

Please confirm upon receipt.

Best regards

W.H. Chang

文章結構

主題：感謝貴公司 1/31 來信，並對我方造成的麻煩道歉。

說明：我們今日寄上郵政匯票 US$537.70 來結清欠款。

結論：請於收到後確認。

生字／片語

1. money order　郵政匯票

2. outstanding payment　未付款／欠款

MODERN BUSINESS ENGLISH
現代商用英文

附件一　信用狀申請書樣本(Application for Opening L/C)

開　發　信　用　狀　申　請　書
APPLICATION FOR IRREVOCABLE LETTER OF CREDIT

致：彰化銀行　　　　　　　　　　　　　　　申請日期 Date:＿＿＿＿＿＿
TO : CHANG HWA COMMERCIAL BANK, LTD.
茲請　貴行按下列條件以全文電報/簡略電報/航郵開發不可撤銷信用狀：　　L/C No. ＿＿＿＿＿＿＿＿
THE UNDERSIGNED HEREBY REQUESTS YOU TO OPEN BY □ DETAIL
CABLE □ BRIEF CABLE □ AIRMAIL YOUR IRREVOCABLE LETTER OF　□三角貿易□當日開出□可以次日開出
CREDIT ON THE FOLLOWING TERMS AND CONDITIONS：
本信用狀適用之信用狀統一慣例，以開狀當時國際商會公佈之最新版本為準
The credit opened under this application is subject to the prevailing Uniform Customs and Practice for Documentary Credits
published by the International Chamber of Commerce at time of L/C issuance.

(57D) Advising Bank 通知銀行（如需指定通知銀行時請填列）：

	SWIFT Code:
	Reimbursing Bank（由本行填寫）＿＿＿＿

(31D) 本信用狀單據提示期限(有效期限)為　　年　月　日於＿＿＿＿＿＿＿＿，
This Credit is valid until ＿＿＿＿＿＿ at ＿＿＿

(50) Applicant申請人（Name, Address）：ID營利事業統一編號＿＿＿＿＿；□申請人地址對外不加註"R.O.C."

(59) Beneficiary（Name, Address）受益人

(32B) Amount Not Exceeding幣別□USD□JPY□EUR□其他(請註明)＿＿＿；金額（小寫）
金額（大寫）

以下中英文併列處，擇一填寫即可

(42C) 匯票期限：請洽兌受益人所簽發以　貴行或　貴行代理行為付款人之匯票 available by negotiation of draft drawn on you
or your correspondent

□ 即期：□即期無融資 / □對外開發即期信用狀，惟對內向　貴行融資＿＿＿＿天。
□ Sight：□ sight □without / □with financing from you for ＿＿＿＿ days.
□ 遠期：□對外開發受益人負擔□見票日起/ □裝運日起/ □＿＿＿＿＿＿＿ 起＿＿＿ 天利息之遠
　　　　 期信用狀，另自匯票到期日起，向　貴行融資 ＿＿＿ 天。
　　　　 承兌費用由 □申請人(買方) □受益人(賣方)負擔。
　　　　 貼現息費用由 □申請人(買方) □受益人(賣方)負擔。
□ Usance at ＿＿＿＿＿ days after □sight □shipment date □＿＿＿＿＿＿＿with financing from you for
　　 days after maturity of drafts.
　　 Acceptance commissions are for □ Applicant's □ Beneficiary's account.
　　 Discount charges are for □ Applicant's □ Beneficiary's account
□ Deferred Payment ＿＿＿＿days

(43P) Partial Shipments 分批裝運□Permitted 准許□Prohibited不准許（如未註明，概以准許分批裝運開發）
(43T) Transhipment 貨物轉運□Permitted 准許□Prohibited 不准許（如未註明，概以准許轉運開發）
(44C) Latest shipment date最後裝船日＿＿＿＿年(YYYY)＿＿＿月(MM)＿＿＿日(DD)

申請人願遵守本申請書（共三頁）背面所列以及/或有關開發信用狀條的所訂各條款 The applicant duly abides by the terms and conditions on the reverse hereof and , if any, those of the relative contract for letter of credit.	申請人簽章For and on behalf of the applicant （請蓋原留印鑑）Authorized Signature TEL:＿＿＿＿＿ 分機：＿＿＿ FAX:＿＿＿＿＿	營業單位 驗印及初審記章： 負責人： 經辦： 分機：	外匯指定單位 負責人： 複　核： 謄　打： 審　核： 收　件：

（申請人加蓋騎縫章）　　　　　　第1頁，共3頁　　　　（國連 68 新）96.6.1000 本 210×297 箱 8K60P(盛)

附件二　全電信用狀樣本(Full Cable L/C/SWIFT)

CABLE L/C（SWIFT）編號及內容（中英文）：

第 1 行　開狀銀行(ISSUING BANK/SENDER)

第 2 行　通知銀行(RECEIVER)

27　　頁數(SEQUENCE OF TOTAL)

40A　信用狀格式(FORM OF DOCUMENTARY CREDIT)：不可撤銷(IRREVOCALBE)

20/21 信用狀號碼(DOCUMENTARY CREDIT NUMBER)

31C　開狀日(DATE OF ISSUE)

31D　有效期(DATE AND PLACE OF EXPIRY)

40E　應用條款(APPLICABLE RULES) UCPURR LATEST VERSION

50/51 申請人(APPLICANT)

59/60 受益人(BENEFICIARY)

32B　金額及幣別(CURRENCY CODE, AMOUNT)

39A　金額誤差百分比(PERCENTAGE CREDIT AMOUNT TOLERANCE)

41D　押匯銀行(AVAILABLE WITH..ANY BANK BY NEGOTIATION)：大都不限押

42C/42A　即期或遠期(DRAFT AT SIGHT../DRAFT AT 90 DAYS SIGHT..)

42D　付款行(DRAWEE: ISSUING BANK)：大都為開狀銀行

43P　分批裝運(PARTIAL SHIPMENT)：ALLOWED 允許／NOT ALLOWED 不允許

43T　轉運(TRANSHIPMENT)：ALLOWED 允許／NOT ALLOWED 不允許

44A　裝運港(LOADING ON BOARD FROM..)：海運出口港

44B　目的地港(FOR TRANSPORTATION TO..)：海運目的地港口

44E　裝運港(PORT OF LOADING/AIRPORT OF..)：空運出口機場

44F　目的地港(PORT OF DISCHARGE/AIRPORT OF DESTINATION)：空運機場

44C　最後出貨日(LATEST DATE OF SHIPMENT)

45A　出貨品名／數量／單價及貿易條件(SHIPMENT OF GOODS.. CFR..)

46A　文件要求(DOCUMENTS REQUIRED)

47A　額外條款(ADDITIONAL CONDITIONS)

　　　（瑕疵扣款費用／可接受數量誤差／第三者文件可否接受..等）

48/50 提示押匯時間：（國際規定最慢出貨後 21 天內）＝ 31D 日期－ 44C 日期

49/51 保兌指示(CONFIRMAITON INSTRUCTIONS)：WITHOUT 不須

57D　通知銀行(ADVISE THROUGH BANK..)

71B　費用(CHARGES)：所有開狀行國外所有銀行費用都由受益人負擔

72/73　銀行對銀行指示(BANK TO BANK INFORMATION)：例如：文件用快遞寄出

71　　銀行對補償銀行指示(BK TO BK INFORMATION-REIMBURSEMENT SUBJECT TO ICC URR 526)

78＊　銀行對付款／承兌或讓購銀行之指示(INSTRUCT TO PAY/ACCEPT/NEGOT BANK)

BANK OF TAIWAN

Head office International Operations Department

Beneficiary	Date: June 09, 2012
King-Tech Industrial Inc. 1F, No.20,	Advising No. 5200112020
Shingsheng N. Rd.. Taipei, Taiwan, R.O.C.	

Dear Sirs,

We wish to inform you that we have received a SWIFT message from BANK DE CHILE, which reads as follows: MT 700 ISSUE OF A DOCUMENTARY CREDIT

SEQUENCE OF TOTAL	27:	1/1
FORM OF DOC. CREDIT	40A:	IRREVOCABLE
DOC. CREDIT NUMBER	20:	B6E2-210568900
DATE OF ISSUE	31C:	120610
APPLICABLE RULES	40E:	UCP LATEST VERSION
EXPIRY	31D:	DATE 120730 PLACE IN THE BENEFICIARY'S COUNTRY
APPLICANT	50:	YXL WIRELESS S.A. MACHENNA 4812 EDIFICIO 512, CHILE
BENEFICIARY	59 :	KING-TECH INDUSTRIAL INC. 1F, No.20, SHINGSHENG N. RD., TAIPEI, TAIWAN, R.O.C.
AMOUNT	32B:	CURRENCY USD AMOUNT 50,000.00
AVAILABLE WITH/BY	41D:	ANY BANK IN THE BENEFICIARY'S COUNTRY BY NEGOTIATION.
DRAFTS AT⋯	42C:	SIGHT FOR 100PCT OF INVOICE VALUE
DRAWEE	42D:	ISSURING BANK
PARTIAL SHIPMENTS	43P:	PROHIBITED
TRANSSHIPMENT	43T:	PERMITTED
PORT OF LOADING	44E:	ANY TAIWAN PORT
PORT OF DISCHARGE	44F:	SANTIAGO
LATEST DATE OF SHIPMENT	44C:	120715
SHIPMENT OF GOODS	45A:	VALVES AS PER P/I NO. KT-011 TRADE TERMS:CIF SANTIAGO

（續上頁）

DOCUMENTS REQUIRED

46A: 1. SIGNED COMMERCIAL INVOICE IN QUADRUPLICATE

2. PACKING LIST IN TRIPLICATE

3. FULL SET CLEAN ON BOARD BILL OF LADING ISSUED TO ORDER OF BANK DE CHILE MARKED FREIGHT PREPAID AND THIS CREDIT NUMBER NOTIFYING APPLICANT.

4. INSURANCE POLICIES OR CERTIFICATES IN DUPLICATE ISSUED OR ENDORSED TO OUR ORDER COVERING ALL MARINE INSTITUTE (CARGO)CLAUSES (A) AND INSTITUTE THEFT PILFRAGE AND NONDELIVERY CLAUSES (CARGO) ALL IRRESPECTIVE OF PERCENTAGE FROM WAREHOUSE TO BUYER'S WAREHOUSE.

5. BENEFICIARY'S CERTIFICATE IN DUPLICATE STATING THAT ONE SET OF NON-NEGOTIABLE DOCUMENTS HAVE BEEN SENT DIRECTLY TO THE APPLICANT BY REGISTERED AIRMAIL WITHIN THREE DAYS AFTER SHIPMENT.

ADDITIONAL COND.

47A: 1. INVOICES EXCEEDING THIS CREDIT AMOUNT ARE NOT ACCEPTABLE

2. ALL DOCUMENTS SHOULD BE DATED ON OR AFTER THE DATE OF ISSUE OF THIS DOCUMENTARY CREDIT.

3. L/C NUMBER SHOULD BE MENTIONED IN ALL DOCUMENTS.

4. A DISCREPANCY DOCUMENT FEE OF USD120.00 OR EQUIVALENT IS TO BE DEDUCTED FROM BENEFICIARY'S ACCOUNT FOR EACH PRESENTATION OF DISCREPANT DOCUMENTS UNDER THIS L/C.

5. THIRD PARTY DOCUMENTS ARE NOT ACCEPTABLE

DETAILS OF CHARGES　71B: ALL BANKING CHARGE AND STAMP DUTY OUTSIDE CHILE ARE FOR BENEFICIARY'S ACCOUNT.

PRESENTATION PERIOD　48: DOCUMENTS MUST BE PRESENTED FOR NEGOTIATION WITHIN 15 DAYS AFTER THE DATE OF SHIPMENT BUT NOT LATER THAN THE VALIDITY OF THIS CREDIT.

CONFIRMATION INSTRUCTION　49: WITHOUT

INSTRUCTIONS　78: ALL REQUIRED DOCUMENTS MUST BE SENT DIRECTLY TO US IN ONE LOT BY REGISTERED AIRMAIL.

TRAILER:　　　　NAA:96854AB

附件三　信用狀上常見的文件要求及額外條件
(Documents Required & Additional Conditions)

46A : DOCUMENTS REQUIRED

1. SIGNED COMMERCIAL INVOICE IN TRIPLICATE SHOWING COUNTRY OF ORIGIN IS CHINA.

2. PACKING LIST IN QUADRUPLICATE SHOWING N/W, G/W OF EACH CARTON AND TOTAL CBM.

3. FULL SET (3/3 SET) OF ORIGNAL CLEAN ON BOARD OCEAN BILL OF LADING CONSIGNED TO APPLICANT (MADE OUT TO ORDER), MARKED "FREIGHT PREPAID", BLANK ENDORSE (ENDORSED IN BLANK), NOTIFY APPLICANT AND "TELEX RELEASE"

4. MARINE INSURANCE POLICY OR CERTIFICATE IN DUPLICATE FOR FULL CIF INVOICE VALUE PLUS 10 PERCENT COVERING INSTITUTE CARGO CLAUSES (A), INSTITUTE WAR CLAUSES(CARGO)AND INSTITUTE STRIKES CLAUSES (CARGO)EVIDENCING CLAIMS PAYABLE IN HONG KONG IN THE CURRENCY OF THE DRAFT.

5. BENEFICIARY'S CERTIFICATE CERTIFYING THAT:

   (A) ONE SET OF NON-NEGOTIABLE DOCUMENTS (AND 1/3 SET OF CLEAN B/L WITH BLANK ENDORSED) HAVE BEEN SENT TO APPLICANT BY COURIER WITHIN THREE DAYS AFTER SHIPMENT.

   (B)BENEFICIARY'S CERTIFICATE CERTIFYING THAT WOOD PACKING MATERIALS HAVE BEEN TREADED AND MARKED WITH "IPPC" MARK IF GOODS PACKED IN WOOD MATERIALS.

   (C)BENEFICIARY'S DECLARATION DECLARES THAT NO-WOOD PACKING MATERIAL IS USED.

   (D)BENEFICIARY'S CERTIFICATE CERTIFYING THAT SHIPPING ADVICE HAS BEEN SENT TO THE APPLICANT BY FAX WITHIN 48 HOURS AFTER SHIPMENT.

6. BENEFICIARY'S CERTIFICATE CERTIFYING THAT QUALITY AND QUANTITY CONFORM TO THE ORDER.

7. CERTIFICATE OF ORIGINAL MANUFACTURER

47A : ADDITIONAL CONDITIONS

1. INVOICE EXCEEDING THIS CREDIT AMOUNT ARE NOT ACCEPTABLE.

2. 3 PERCENT MORE OR LESS IN QUANTITY AND AMOUNT ARE ACCEPTABLE

3. ALL BANKING CHARGES OUTSIDE ISSURING BANK ARE FOR THE ACCOUNT OF BENEFICIARY.

4. A DISCREPANCY DOCUMENT FEE OF USD120 OR EQUIVALENT PLUS ALL RELATIVE CABLE CHARGES WILL BE DEDUCTED FROM THE REIMBURSEMENT CLAIM(BENEFICIARY'S ACCOUNT) FOR EACH PRESENTATION OF DISCREPANT DOCUMENTS UNDER THIS DOCUMENTARY CREDIT.

5. THIRD PARTY DOCUMENTS ARE NOT ACCEPTABLE. (THIRD PARTY DOCUMENTS ARE ACCEPTABLE.)

6. FORWARDER'S CARGO RECEIPT IS ALSO ACCEPTABLE.

46A：文件要求

1. 簽名的商業發票三份，註明產地為中國。

2. 包裝表四份，註明每箱的淨重、毛重、及總材積。

3. 整套（三張正本）清潔海運提單，註明收貨人為申請人（為 TO ORDER ＝銀行），運費已付，背面空白處背書，通知人為申請人及電放。

4. 海運保單兩份，保額為出貨發票金額加 10%，投保全險、戰爭險、及罷工險，註明索賠地在香港，按照信用狀上的幣別。

5. 受益人證明書證明

   (A)一套副本出貨文件（及一張正本清潔海運提單背面空白處背書）已於出貨後三天內用快遞寄給申請人。

   (B) 出貨如有木頭材料的包裝，受益人須出具證明書證明木頭包裝材料已被處理過並註明 "IPPC" 標誌。

   (C)如出貨無木頭包裝，受益人須出具聲明書聲明沒有使用木頭包裝材料。

(D) 受益人證明書證明出貨通知已於出貨後 48 小時內用傳真傳給申請人。

6. 受益人出具品質數量證明書證明數量品質符合訂單規定。

7. 原廠證明書。

47A：額外條件

1. 發票金額超過此信用狀金額將不被接受。

2. 數量及金額可接受+/- 3%。

3. 所有離開開狀行以外的銀行費用皆由受益人負擔。

4. 在此信用狀下每筆押匯的瑕疵文件，須由受益人帳上扣除瑕疵費用USD120 或相近費用加上所有相關電報費。

5. 第三者押匯文件不接受（第三者押匯文件可接受）。

6. 攬貨公司的收據可接受（出貨如經由FORWARDER時L/C上須要求註明此條件）。

練習 19　信用狀及其他付款條件討論函練習

1-1.依據前面附件二信用狀(L/C NO. B6E2-210568900)回答下列問題：

通知銀行？（英文作答）	
信用狀號碼？（英文作答）	
受益人？（英文作答）	
貿易條件？（英文作答）	
保險種類？（中文作答）	
裝運港？（英文作答）	
商業發票份數？（中文作答）	
提單上的收貨人？（英文作答）	
目的地港？（英文作答）	
瑕疵費用扣款多少？（英文作答）	
可否超押？	☐可　　　☐否
可否接受第三者文件？	☐可　　　☐否
可否分批出貨？	☐可　　　☐否

1-2. 將下列翻成中文：

(1) 2/3 set of clean on board marine bills of lading made out to order blank endorsed marked freight collect indicating this credit number 009-80 and notify applicant.

(2) Beneficiary certificate certifying that 1 set of non-negotiable documents and 1/3 set of clean B/L with blank endorsed have been sent to applicant by courier within three days after shipment.

(3) One copy of declaration of no-wood packing material issued by beneficiary. Beneficiary certificate certifying that wood packing materials have been treated and marked with 'ippc' mark if goods packed in wood material.

2. 根據下列中文大意寫一封完整的賣方催 L/C 函

> 主題：有關貴公司的訂單號碼 9901
>
> 說明：我們很高興告知貨已準備好，但是我們尚未收到貴公司的信用狀。
>
> 結論：請告知你們是否已經開出信用狀，如是，請告知信用狀明細(L/C details)或傳送影本給我，謝謝！

3. 根據下列中文大意寫一封完整的買方通知已開 L/C

> 主題：有關訂單 9901，我們很高興告知我們已於昨日開出 L/C No. 880128。
>
> 說明：請儘快進行此訂單，並務必準時出貨。
>
> 結論：請於收到時確認（upon receipt 放句尾）。

4. 根據下列中文大意寫一封賣方確認收到 L/C，並要求修改函

> 主題：我們感謝收到 L/C No. 880128，但是我們發現下列錯誤：(following mistakes)
>
> 說明：(1) 價錢條件應該是 FOB Hong Kong。
>
> (2) 轉運應該要允許。
>
> (3) 最後出貨日應該是 10 月 30 日。
>
> 結論：請儘快修改以上，並速確認。

5. Write a reply letter to the above letter concerning L/C No.880128 telling the seller that the buyer will accept the discrepant documents instead of amending L/C due to no enough time for amendment, and also push for the shipping information.

九、包裝/生產/出貨/驗貨討論函
(Packing/Production/Shipment/Inspection Letters)

當訂單簽定或客人開出信用狀/安排付款後,仍會有許多後續問題會產生或需要討論處理,例如:

1. 包裝(packing)及出貨嘜頭(shipping mark)

如果訂單上沒有註明詳細包裝或出貨嘜頭時,按照「國際公約」規定:「買方需於出貨前一段合理時間內,告知賣方其指定的包裝及出貨嘜頭,如未能及時告知,賣方可按其工廠標準包裝及嘜頭來作業,買方必須接受。」

包裝常用材料有:塑膠袋(plastic bag)、紙盒(box)、紙箱(carton)、木箱(wooden case)、板條箱(crate)、墊板(pallet)、普麗龍(foamed polystyrene)、氣泡布(air bubbles plastic sheet)及貨櫃(container)等(等詳細可參附表十三)

注意:因為怕有細菌蟲卵寄生問題,現在大部分國家都禁止進口貨品使用木箱包裝;但如考慮成本一定要用木箱包裝,都會要求所使用的木箱要事先做熱處理(ippc)或煙燻處理(fumigation treatment),並出具相關證明文件。

包裝方式有:散裝(bulk packing)、小包裝(small packing)、個別包裝(individual packing)

出貨嘜頭分主嘜(main mark)與側嘜(side mark):
主嘜為出口海關檢驗部分,至少要寫出:(1)目的地港(destination);(2)箱數(carton number);(3)產地(country of origin),有時第一行會加打客人公司名稱,以方便其提貨時辨認,例如:Main Mark

側嘜為客人指定,客人如無特別規定,至少要寫:
(1)數量(Quantity);(2)淨重(Net Weight);(3)毛重(Gross Weight)。此外,亦可加註品名規格,例如:Side Mark

```
586 Notebook Computer
QTY：1 set
N. W.：10 kgs
G.W.：12 kgs
```

2. 出貨方式／裝運指示(Shipping Method)

FOB 出貨時，如果訂單上沒有註明出貨方式(by air or by sea freight)，出貨前，賣方一定要問清楚買方是否有指定運輸公司(Shipping Company)及攬貨公司／報關行(forwarder)，因為有些客人為了節省運費及報關手續費，和某些運輸公司或攬貨公司簽約，請其併貨出口，可享有較好折扣或優待。至於 CFR 或 CIF 出貨，因運費為賣方計算支付，所以由賣方自行決定運輸公司及報關行。

3. 品質管制(Quality Control)及驗貨問題(Inspection)

一般出貨檢驗有四個狀況：

(1)工廠自行檢驗擔保品質

(Quality Certificate issued by Manufacturer)

(2)由政府的商品檢驗局來驗(Quality Certificate issued by Bureau of Commodity Inspection & Quarantine)

(3)由私人檢驗公司／公證行來驗

（Quality Certificate issued by an independent public survey 或客人指定的檢驗公司名稱）

(4)由代理商來驗(Quality Certificate issued by our Agent)

以上以工廠自行擔保品質最好，因為工廠本身可做 100%測試，或工廠本身如有獲國際標準組織 ISO 品質認證通過，或產品有通過歐美安全規定 UL, CE, FCC 等實驗室測試通過，產品品質的穩定性就較有保障。

如果是商檢局／公證行或代理，都只能按照國際檢驗標準(International Inspection Standard)的品質可接受標準(AQL = Acceptable Quality Level)抽驗，AQL 是按照抽樣比率採樣驗貨，大約看一下品質、內容及手工，並無法保證 100%沒問題，真有問題時仍需由工廠負責。

AQL 抽驗方式如下：

例如出貨數量：10,001～35,000 個

抽驗標準：AQL Level II，則任意採樣 315 個

主缺點 Major Defect（多為功能不良）：

　　如果按 Major：0.65, 315 個當中，只能允許壞 5 個；如果壞 6 個，則判退貨（即 5 收 6 退）。

次缺點 Minor Defect（多為外觀不良，像刮傷／破損／顏色等問題）：

　　如按 Minor：1.5, 315 個當中，總共只能壞 10 個；如壞 11 個，則判退貨（即 10 收 11 退）。

抽樣檢驗的不良率如在允收標準內，則判接受(Accept)，如超出，則判退貨(Reject)，但也有時因超出的範圍很小，客人又急著要貨，也會加驗後，判特例接受(waive)。檢驗報告(Inspection report)及 AQL 樣章，請參照附件一、二。

驗貨的步驟如下：

(1)準備好客人的訂單／信用狀及確認的樣品或規格圖面

(2)請工廠按訂單規定，做出正確的發票／包裝表

(3)驗貨的順序：

　　①先核對文件：核對發票／包裝表和訂單規定是否一致

　　②再核對嘜頭：出貨文件／外箱和訂單規定是否一致

　　③清點箱數

　　④按 AQL 標準，隨意開箱抽點數量

　　⑤按 AQL 標準，抽驗出貨的品質（比照確認樣品或確認的圖面規格來驗），客人如無指定驗貨標準，一般產品都按 MIL-105D, AQL Level II, Major defect: 0.65, Minor defect: 1.5 來抽驗。

4. 生產(Production)及出貨(Shipment)問題

　　正式生產時，也會遇到許多問題，例如：

　　(1)生產不順，不良率太高(defective rate too high)

　　(2)模具有問題(tooling problem)

　　(3)原料短缺(material shortage)

(4)人工短缺(labor shortage)

(5)機器故障(machine broken)

(6)樣品無法通過確認(sample can't be approved)

(7)賣方報價計算錯誤(wrong offer)

(8)客人信用狀慢開(L/C delay)

(9)買方訂錯貨(wrong ordering)

(10)買方多訂(overbooking)

(11)市場不景氣(market depression / economic depression)

(12)政府法令變更(government rule changed)

(13)天災／罷工／戰爭／意外等的不可抗力事件

    (force majeure- typhoon/strike/ war/accident)

(14)沒有船位(no shipping space)

(15)船位難訂(It is difficult to book the shipping space)

(16)假日太多(too many holidays)

遇到以上任何狀況，都有可能造成兩個結果：

(1)慢出貨(delay the shipment)

(2)取消訂單(cancel the order)

慢出貨的責任歸屬有三：

(1)買方過失（例如上述問題(8)）——賣方可不負責，買方要：

    ①負責修改信用狀，並接受延遲出貨

      (amend the L/C and accept the late shipment)

    ②買方自行負擔空運費(air freight)

    ③買方付加班費(overtime charge)給賣方，請其趕工

(2)賣方過失（例如上述問題(1)～(7)）——賣方要負責，而買方可能會：

    ①取消訂單(cancel the order)

    ②將海運(by sea)改為空運(by air)，並要求賣方負擔運費差額(to pay the difference between air freight and sea freight)

    ③買方要求賣方賠償因慢出貨所造成的損失，例如負擔其銀行利息或生產損失，視訂單上簽約時所訂定。

(3)雙方都無責任（例如上述問題(12)～(14)）——此時，解決方法：

　①取消訂單(cancel the order)

　②將海運改空運，雙方各付一半運費差額(to share half of the difference of freight cost)

如遇到上述問題(9)～(11)，買方可能要求減少訂單量或取消訂單，此時，賣方可有下列四個選擇：

(1)不接受(not accept)：如果已生產完成或將完成時，就不要接受

(2)無條件接受(accept)：可告知客人是特別幫其忙

(3)接受，但請客人付取消費用(pay cancelling charge)

(4)不接受取消，但同意延緩出貨(accept to delay the shipment for few months)

MODERN
BUSINESS
ENGLISH

Part 4 (九、包裝／生產／
出貨／驗貨討論函)
商用英文信函寫法

331

5. 包裝／生產／出貨驗貨討論函常用單字／片語

(1)plastic bag 塑膠袋；inner box 內盒；carton 紙箱；wooden case 木箱；crate 板條箱；
　　pallet 墊板；foamed polystyrene 普利龍；air bubbles plastivc seet 氣泡布；container
　　貨櫃

(2)net weight (NW)淨重；gross weight(GW)毛重；measurement 材積

(3)metric ton (M/T)公噸；cubit meter (CBM)立方公尺

　　(1 M/T = 2204.6 LBS; 1CBM = 35.315 cubit feet)

(4)broker 報關行；forwarder 攬貨公司

(5)inspect at random 抽驗

(6)inspection standard 檢驗標準

(7)AQL (Acceptable Quality Level)可接受品質標準

(8)QC(quality control)品管；QA (quality assurance)品保

(9)IQC (incoming quality control)進料品管

(10)IPQC (in process quality control)線上品管

(11)major defect 主缺點；minor defect 次缺點

(12)accept 接受；reject 退貨；waive 特例接受

(13)main mark 主嘜；side mark 側嘜

(14)delay the shipment (= delay in shipment)慢出貨

(15)cancel the order 取消訂單

(16)daily capacity 日產能；monthly capacity(= monthly output)月產量

(17)shipping details 出貨明細；shipping information 出貨資訊

(18)shipping advice 出貨通知

(19)labor shortage 人工短缺

(20)defective rate 不良率

(21)make the clearance from the customs 清關 (= clean the goods from the customs)

(22)in house = in the factory 在廠內

6. 包裝 / 生產 / 出貨驗貨討論函常用的主題句 / 說明句 / 結論句

主題句(Main Idea)

(1)Please note that the packing we request is 10pcs in an inner box and 4 boxes in an export carton.

(2)Please do 100% inspection before shipment.

(3)Please ask QA department to inspect the shipment at random according to AQL level II, major defect 0.40 and minor defect 1.0.

(4)We are sorry for the delay in shipment as there is no vessel available this week.

(5)We urgently need 10,000pcs of P/N 100 in house by this Friday, please advise us you very best shipping date by return.

說明句(Explanation)

(1)As we are in urgent situation, please ship this order by direct air.

(2)Please e-mail us flight details right away so that we can trace the goods and make the clearnance form the customs sooner.

(3)You are requested to make a single shipment for this order and pay the difference for the delay shipment.

(4)We will release the shipment as soon as we receive your T/T payment.

(5)When the goods are ready, please phone our forwarder to pick them up.

(6)Due to material shortage, we are sorry for that we can't ship before the latest date of shipment shown on L/C.

結論句(Conclusion)

(1)Please follow above packing instruction.

(2)Please confirm the above soon.

(3)Please send us the shipping advice prior to the shipment.

(4)Since all goods are ready for shipment, please confirm if you have the specified forwarder or shipping company.

(6)We agree to extend the latest shipping date to the end of next month, but no further delay will be accepted.

範例 1　買方告知出貨包裝及出貨方式

From: brainrata@antgsnc.com

To: judyyin@hpc.com

Subject: P/O 89801

Dear Judy,

Please note that the packing we request is 6pcs in an inner box and 6 boxes in an export carton.

Also, please ship this order as same as previous shipments by direct air to us, and use air authorized #91019.

Please follow above packing instruction and advise shipping details soon.

Best regards

Brain Rata

文章結構

主題：請注意我們要求的包裝是 6 個裝一個內盒，6 盒裝一個外箱。

說明：請像以前出貨一樣用直飛方式空運給我們，並使用空運授權編號 91019。

結論：請按照上述包裝指示並儘快告知出貨明細。

生字／片語

1. inner box　內盒；export carton　外銷紙箱
2. as same as previous shipment　像以前出貨一樣
3. by direct air　直飛；packing instruction　包裝指示
4. air authorized #　空運授權編號

範例 2 買方要求全檢並按國際檢驗標準抽驗

From: dickdon@dncimport.com
To: jamesho@swellcomponents.com
Subject: L/C 072-2

Dear Mr. Ho,

Our customer's concern is that these products will be big runners for them and they want to make sure everything is perfect before they get into production.

So, please do 100% inspection before packing and ask the QA department to inspect the shipment at random according to AQL level II major 0.40 and minor 1.0.

Please confirm the above by return.

Best regards
Dick Don

文章結構

主題：我們的客戶主要關心的是此產品為他們的大生意，他們必須確定生產前一切完好。

說明：因此請於包裝前做 100%檢驗，出貨前請品保部門按 AQL Level II 主缺點 0.40、次缺點 1.0 抽驗。

結論：請儘快確認以上。

生字／片語

1. Customer's concern 客戶的關心；big runners 大生意
2. before they get into the production 在他們進入生產前
3. do 100% inspection 做 100%檢驗
4. before packing 包裝前；QA ＝ quality assurance 品保
5. at random 抽驗；major 主缺點；minor 次缺點
6. AQL ＝ acceptable quality level 可接受品質標準

範例 3　買方催貨

From: luannyun@freedomtech.com
To: shipping@marchmore.com
Subject: Order No.4017-Top Urgent

Dear March,

Would you please confirm if you have shipped the goods of the above order to our Cargo forwarder as instructed in our L/C. If so, please kindly advise us of the flight details including Flight No., AWB No., & estimated time of arrival at Taoyuan airport, so that we can trace the goods and make the clearance from the customs sooner.

We are in urgent situation because we have to deliver the goods to our customer before the deadline shown on the L/C.

Please send us the shipping advice prior to the shipment.

Thanks & best regards
Luann Yun

文章結構

主題：請告知是否已按 L/C 指示將貨裝給我們指定的空運攬貨公司，如是，請告知班機明細包括班機號碼、提單號碼及預計到桃園機場時間，以便我們追蹤並提早清關。

說明：我方狀況緊急因為我們必須在信用狀上規定之最後期限前交貨給我們的客戶。

結論：請於出貨前傳送出貨通知給我們。

生字／片語

1. Cargo forwarder　空運攬貨公司；as instructed　如所指示
2. flight details　班機明細；Flight No.　班機號碼
3. AWB = airway bill　空運提單；Taoyuan airport　桃園機場
4. estimated time of arrival = ETA　預計到達日；deadline　最後期限
5. trace　追蹤；make the clearance from the customs　清關

MODERN BUSINESS ENGLISH
現代商用英文

範例 4 因賣方慢出貨造成無法併貨，買方要求賣方負擔差額

From: brigittahobor@rochdiago.com

To: tomshu@modenintl.com

Subject: Our Order 294B

Dear Mr. Shu,

With reference to your e-mail of 17.01, the consolidated shipment would arrive at Munich too late.So, you are requested to make a single shipment for this order and pay the difference:

  Air freight for Group shipment EUR 18.05

  Air freight for Single shipment EUR213.30

    Difference EUR195.25

We will send a Debit Note to you for the above amount.

Kind regards

Brigitta Hobor

文章結構

主題：參照貴公司 1 月 17 日來函，併貨到慕尼黑將太慢。

說明：因此，貴方被要求單獨裝此訂單並支付差額：

  併貨空運費   EUR 18.05

  單獨出貨空運費 EUR213.30

    差額 EUR195.25

結論：我們將寄一張上述金額的應收帳單給貴公司。

生字／片語

1. Consolidated shipment 併貨；Munich 慕尼黑
2. single shipment 單獨出貨；difference 差額
3. air freight 空運費；EUR 歐元
4. Debit Note 應收帳單；Credit Note 應付帳單

範例 5　賣方告知無法提早於 30 天出貨

Angelantoni Scientifica

Attn: Mr. Chang/ CAM Tech.

Dear Sirs,

With reference to your e-mail of today, we must regretfully inform you that we cannot deliver the equipment in 30 days as stated in your order. The soonest delivery time is 40 working days as stated in our proforma invoice. Please confirm.

Yours Sincerely
Angelantoni Scientifica

文章結構

主題：參照貴方今日 e-mail，我們必須遺憾地告知貴公司我們無法於貴公司訂單所述的 30 天內出貨。

說明：如我們在 P/I 上所述，最快交期要 40 個工作天。

結論：請確認。

生字／片語

1. regretfully　遺憾地
2. equipment　設備
3. as stated　如所敘述
4. working days　工作天

範例 6　買方回覆範例 5 將延長 L/C，但要求賣方儘快出貨

CAM TECHNOLOGY CO.

To: Angelantoni Scientifica

Dear Mr. Angelantoni,

We are sorry to hear from your e-mail of 1/18 that you cannot ship the equipment in 30 days. Since we have phoned you and told you that we need this equipment for the Exhibition here by Feb. 18, we thought you should have started the production at that time.　Anyway, we will extend the L/C to end of Feb., but still please try your best to effect the shipment ASAP.

Best regards

Paul Chang

文章結構

主題：很遺憾由貴方 1/18 的 e-mail 聽到貴公司無法於 30 天內裝出此設備。

說明：因為我們已經打過電話並已告知我方需要於 2/18 前收到此設備以便參展，我們以為貴方在那時應已開始生產了。

結論：無論如何，我們將延長 L/C 到二月底，但是仍請盡力儘快出貨。

生字／片語

1. equipment　設備；at that time　在那時

2. for the Exhibition　為了參展

3. have started the production　已經開始生產

4. try your best　盡力；ASAP ＝ as soon as possible　儘快

5. effect the shipment ＝ make the shipment　出貨

範例 7　買方要求出清剩餘訂單

From: ericbracel@wwimport.com

To: carrychen@santo.com

Subject: Model 001 Recorders

Dear Carry,

As we do not need any more this model for a few months, we do not need to place a further order at this moment.

As a result, could you please send the 128 sets ready ones and then consider the order closed.

Thanks and await your shipping information.

Best regards

Eric Bracel

文章結構

主題：有關錄音機型號 001，由於我們在幾個月內將不再需要此型號，我們目前將不
　　　需進一步訂購。

說明：因此，請將剩餘的 128 台完成品裝出，此訂單就此結束。

結論：謝謝並等候貴公司的出貨資料。

生字／片語

1. Recorders　錄音機
2. further order　進一步的訂單；at this moment　目前
3. ready ones　完成品
4. consider the order closed　此訂單視為結束
5. shipping information ＝ shipping advice　出貨資料

範例 8　由快遞公司直接發給買方的出貨通知

Dear Sirs,

TNT is pleased to advise you that Holoms ab has arranged for a shipment to be collected from them on 23-June, and delivered to King-Tech Inc.

The shipment has a TNT CONSIGNMENT NOTE NUMBER: 632542912

To be able to check the status of the shipment, simply click here >> Track my Shipment and more detailed information on the consignment will be available.

If the sender of your shipment has provided a reference, this reference can be used to track the consignment.

Best regards

文章結構

主題：很高興通知 Holoms ab 公司已於 6/23 安排一批出貨，運費到收，收貨人為 King-Tech Inc.。

說明：此批出貨的 TNT 編號為 632542912。請按此處可追蹤出貨狀況>>追蹤我方出貨並可得知此出貨更多詳細資訊。

結論：如此貨的寄件者已提供參考資料，此參考可用來追蹤此託運之貨。

生字 / 片語

1. TNT 快遞公司
2. has arranged for a shipment 已安排一批出貨
3. to be collected 運費到收（運費待收）
4. check the status of the shipment 查看出貨狀況
5. click here 按此處
6. detailed information 詳細的資訊；track the consignment 追蹤託運之貨

範例 9　買方要求賣方負擔漏裝部分的出貨費用

Subject: P/O T98009

Dear Tim,

Thank you for the shipping information. However, regarding the balance of 19 pcs, since you did not ship all together, you must be responsible for the additional shipping and handling cost for the balance of 19 pcs.

We have advised you many many times not to split a shipment unless you have been instructed by us. It's your fault that you mixed up the material. Please confirm you will pay the above.

Best regards
Victory

文章結構

主題：謝謝你們的出貨通知，然而有關剩餘 19 個，因為你們沒有一起裝出，你們必須負責這 19 個額外的出貨手續費用。

說明：我們已經告訴過你們很多很多次，沒有我方指示，不得分批出貨，此次為貴方自己混錯料所造成之錯誤。

結論：請確認貴公司將負擔以上費用。

生字／片語

1. the balance of 19 pcs　剩餘的 19 個
2. you must be responsible for　貴方需負責
3. additional shipping and handling cost　額外之出貨手續費
4. split a shipment ＝ make partial shipments　分批裝貨
5. fault　過失；mixed up the material　混料

範例 10　客人催模具及交貨

Subject: P/O SW0701

Dear Mr. Din,

Please do what you can to improve tooling time in order to support our manufacturing schedule. We will need minimum of 5,000 pcs shipped via air by this Friday.

Can you complete the tool by 1/18 and produce first article ASAP? What is daily capacity of tool based on an 8 hours shift? Please update us tomorrow.

Best regards
Fred

文章結構

主題：為了支持我們生產交期，請盡全力改進開模時間，我們需要於本週五前，至少空運 5,000 個。

說明：你們可否於 1/18 前完成模具並儘快做第一次生產的貨？根據 8 小時一班的生產，此模具日產量是多少？

結論：請於明日給我們最新狀況。

生字／片語

1. tooling time　開模時間；support　支持；complete　完成
2. manufacturing schedule　生產交期；minimun　最少
3. first article　第一次生產的貨；based on　根據
4. daily capacity　日產能／日產量；8 hours shift　8 小時一班
5. update　最新狀況

範例 11 買方願意付加班費，要求賣方趕貨

From: vancebrown@longwellimport.com

To: slimwei@yokeelectronic.com

Subject: P/O 15744

Dear Slim,

We urgently need 10,000pcs of P/N 7580 in house by this Friday or we will lose our contract. We will pay whatever overtime expediting etc. in order to get these parts immediately. We realize that you have many holidays coming up, but we really need your help.

Please review very carefully and advise your very very very best shipping date tomorrow without fail!

Thanks & best regards
Vance Brown

文章結構

主題：我們緊急需要 10,000 個 P/N 7580 於本週五前到達本廠內，否則我們將會丟掉此筆生意。

說明：我們願意付任何加班趕工等費用，只要能馬上拿到這些零件。我們了解你們即將有很多假日，但是我們真的需要你們的幫忙。

結論：請仔細重新查看此事並於明日告知你們最快最快的出貨日，不得延誤！

生字／片語

1. urgently (Adv.)，urgent (Adj.) 緊急地；lose our contract 丟掉此筆生意
2. in house = in our factory 到達本廠內
3. whatever overtime expediting etc. 任何加班趕工等
4. many holidays coming up 許多即將來臨的假日
5. review 重新查看；without fail 不得延誤

範例 12　因賣方調價，買方欲取消訂單

From: ToddWilse@hhdistribution.com

To: williamyeh@keyswitch.com

Subject: P/O 2055 and 2056

Dear Mr. Yeh,

Re: P.O. 2055

We found you increased the price to US$3.75/M, which is higher than what we can get in local. So, please put the order on hold until next Monday because we may want to cancel this order.

Re：P.O. 2056

Please explain why you can not supply this switch assembly now.

Perhaps you can propose an alternative, please advise.

Best regards

Todd Wilse

文章結構

有關訂單 2055

我們發現你們把價錢漲到每 1,000 個 US$3.75，那比我們國內可買到的價錢高，因此請保留此訂單到下週一，因為我們可能要取消此訂單。

有關訂單 2056

請解釋為何現在無法供應這種開關的裝配？或許你們可以建議一個替代品，請告知。

生字／片語

1. increased the price　漲價；in local　在國內；M = 1,000
2. put the order on hold　保留此訂單
3. cancel the order　取消訂單；switch assembly　開關裝配
4. perhaps　也許；propose　建議；alternative　替代品

範例 13　賣方催買方修改信用狀及其他資料

From: BillyWang@maxcomputer.com

To: LynnLorsch@bbccomputer.com

Subject: Shipment of L/C 1178M21

Dear Lynn,

We are sorry for the delay in the shipment under your above L/C, as there is no vessel available this week. The soonest vessel will be early next week. So, please kindly amend/extend the latest shipping date to 2/15 and the expiry date to 2/25.

Also, please e-mail us your inspection standard and confirm if every piece packed in a blister is OK.

Since all goods are ready to ship, please rush to reply the above.

Best regards

Billy Wang

文章結構

主題：很抱歉上述信用狀的出貨要慢裝，因為本週沒有船，最快的船要下週初。

說明：因此請修改／延長最後出貨日到 2/15，有效期到 2/25，同時，請 e-mail 你們的檢驗標準給我們，並確認是否可接受每個各別包裝於透明罩內。

結論：由於所有貨已完成可出貨，請速回覆以上。

生字／片語

1. delay in the shipment　延遲出貨

2. there is no vessel available this week　本週沒船

3. amend　修改；extend　延長；rush to reply　速回覆

4. inspection standard　檢驗標準

5. every piece packed in a blister　每個各別包裝於透明罩內

附件一　檢驗報告

INSPECTION REPORT									
P/O NO: 51K-1878-01		TTL ORD Q'TY:				QC NO:			
P/N: XZAM-21R-B60		RECEIVE Q'TY: 3,000PCS				DATE REC'D:			
DWG NO: 40083412708		BALANCE Q'TY:				DATE REQ'D:			
USED NO:		DESCRIPTION & SPEC.				VENDOR:			
DEPT:		檢驗項目及標準參如附 Rubber 21K				TEL:			
CUST CODE:						CVTS OF TOOL:			
QC SPECS: □MIL-STD-105D NORMAL INSPECTION-AQL II　　□LTPD									
SAMPLE SIZE: 125	CRITICAL		MAJOR	0.65	MINOR	1.5	MECHANICAL & ELECTRICAL TEST		
	A/R	0	A/R	2/3	A/R	5/6			
SAMPLE DEF: 1	CRITICAL		MAJOR	0	MANOR	1	DIMENSION		
ITEM	DEF/SAMP	DEFECT DESCRIPTION		CRIT.	MAJ.	MIN.	n	c	d
1	1/125	印刷不良 PRINTING DEFECT				✓	24	0	24
							FIT OR ASS'Y		
							n	c	d
							A/F(GRAM)		
							MAX.	NOM.	MIN.
							196	150±30gs	112
							RESISTANCE(Ω)		
							MAX	NOM.	MIN.

REMARKS:
　　摩擦試驗 10PCS 數據如下：52, 46, 48, 51, 47, 55, 46, 52, 47, 55
　　VIBRATION TEST FOR 10PCS WHICH READINGS ARE: 52, 46, 48, 51, 47, 55, 46, 52, 47, 55

PREPARED BY:		CHECKED BY:		☑ACCEPT　　□REJECT			
APPROVED BY:		DISPOSITION				COMMENTS	
	WAIVE	COND. ACPT.	REJECT	SORTING	SCRAP		
ENG:							
MFG:							
PUR:							
QC/QA.							
Q'TY ACCEPTED:		Q'TY REJECTED:		Q'TY DESTROYED:		INVOICE NR:	

附件二　AQL 抽樣表樣本（MIL-STD-105D 計數抽樣表）

樣本代碼

批量 N （每批出貨數量）	特殊檢驗水準 （抽樣數量代碼）				普通檢驗水準 （抽樣數量代碼）		
	S-1	S-2	S-3	S-4	I	II	III
2-8	A	A	A	A	A	A	B
9-15	A	A	A	A	A	B	C
16-25	A	A	B	B	B	C	D
26-50	A	B	B	C	C	D	E
51-90	B	B	C	C	C	E	F
91-150	B	B	C	D	D	F	G
151-280	B	C	D	E	E	G	H
281-500	B	C	D	E	F	H	J
501-1200	C	C	E	F	G	J	K
1201-3200	C	D	E	G	H	K	L
3201-10000	C	D	F	G	J	L	M
10001-35000	C	D	F	H	K	M	N
35001-150000	D	E	G	J	L	N	P
150001-500000	D	E	G	J	M	P	Q
500001 以上	D	E	H	K	N	Q	R

正常檢驗單〉次抽樣（主抽樣表）

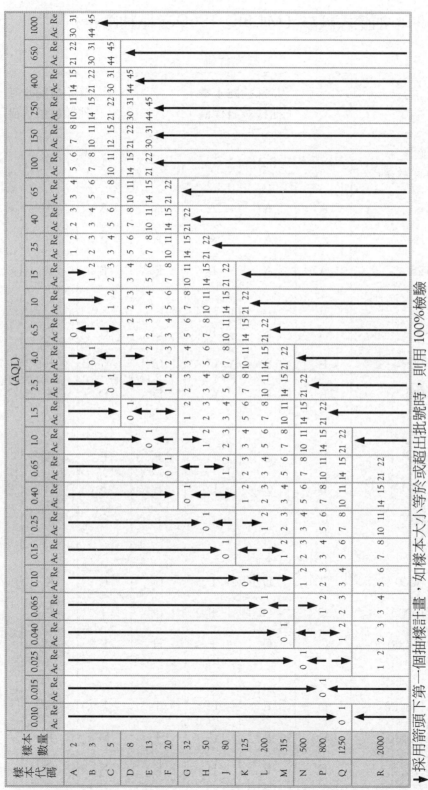

(AQL)

樣本代碼	樣本數量	0.010 Ac Re	0.015 Ac Re	0.025 Ac Re	0.040 Ac Re	0.065 Ac Re	0.10 Ac Re	0.15 Ac Re	0.25 Ac Re	0.40 Ac Re	0.65 Ac Re	1.0 Ac Re	1.5 Ac Re	2.5 Ac Re	4.0 Ac Re	6.5 Ac Re	10 Ac Re	15 Ac Re	25 Ac Re	40 Ac Re	65 Ac Re	100 Ac Re	150 Ac Re	250 Ac Re	400 Ac Re	650 Ac Re	1000 Ac Re

↗ 採用箭頭下第一個抽樣計畫，如樣本大小等於或超出批號時，則用 100%檢驗
↖ 採用箭頭上第一個抽樣計畫

允許數 Ac ＝ Accept
拒收數 Re ＝ Reject

練習 20　包裝／生產／出貨／驗貨討論函練習

1. 將下列翻譯成英文

(1)外銷紙箱

(2)內盒

(3)貨櫃

(4)散裝

(5)淨重

(6)毛重

(7)運輸公司

(8)攬貨公司／報關行

(9)品質管制

(10)檢驗報告

(11)人工短缺

(12)罷工

(13)沒有船位

(14)抽驗

(15)清關

(16)日產能

(17)併貨

(18)取消訂單

(19)請延長最後出貨日到 3/31。

　　Please extend the ＿＿＿＿＿＿＿ date of shipment to 3/31.

(20)訂單 001 的貨已完成可出貨。

　　The goods of O/N 001 are ready for ＿＿＿＿＿＿＿.

(21)信用狀上週已開出，請盡快發出貨通知給我們。

　　The L/C has been ＿＿＿＿＿＿＿ last week.

　　Please send us shipping ＿＿＿＿＿＿＿ soon.

(22)請告知是否有指定的攬貨公司。

　　Please advise if you have a specified ＿＿＿＿＿＿＿.

2. 根據下列中文大意寫一封出貨包裝指示函

> 主題：有關 L/C NO.AB02113，請注意包裝必須 5 個裝入一個內盒，10 盒裝一個外銷
> 紙箱，6 箱打一個墊板。
>
> 說明：當貨準備好時，請打電話給我們的報關行，他們會去你們的工廠提貨。
>
> 結論：請準時出貨不得延誤。

生字／片語

(1) 6 箱打一個墊板　6 cartons in a pallet

(2) 當貨準備好時　when the goods are ready

(3) 提貨　pick up the goods

(4) 不得延誤　without delay

3. Write a letter to tell the buyer that the shipment of L/C No. 7728 can't be shipped before 10/30 due to material shortage, and ask the buyer to extend the latest date of shipment to 11/20.

4. 根據下列大意寫一封買方對上封信的回覆函

> 主題：有關 L/C7728，很遺憾聽到貴公司要延遲出貨。
>
> 說明：我們同意延長最後出貨日到 11/20，但是請付修改費，因為這是貴方的過錯。
>
> 結論：請確認並不得再延後。

生字／片語

(1) 遺憾　regret

(2) 延遲出貨　delay the shipment

(3) 延長　extend

(4) 修改費　amendment charge

(5) 不得再延後　won't delay again

十、出貨通知及押匯文件(Shipping Advice & Shipping Documents)

　　當生產完成，信用狀或付款都沒問題時，就要安排出貨，出貨的工作很單純但很瑣碎，步驟如下：

1. 先跟工廠要來正確的實際出貨包裝明細表。

2. 按信用狀上規定的條款，做好出貨的發票(Invoice)與包裝表(Packing List)。

3. 找適當的船或班機：可查船期版廣告，打電話問船公司或委託出口報關行代查。

4. 將發票、包裝表及信用狀影本交給出口報關行，到船公司或空運公司訂機位或船位（Shipping Order = S/O 裝貨單，出口通關報關用）並做出口報單。

5. 聯絡卡車將貨送至機場貨運站或指定的貨櫃場（不同的船公司，有不同的貨櫃場），請報關行的人在現場等貨，進行通關／報關的手續。

6. 注意一定要在訂單或信用狀上規定的最後出貨日之前，把貨裝出去。按國際條約規定：「以提單上的日期視為實際出貨的日期」(Date of B/L shall be accepted as a conclusive date of shipment.)。
   另外要注意，提單上所蓋的實際出貨日期＝空運的放行日（不是進倉日）＝海運的開航日（不是結關日）。

7. 台灣現在進出口金額超過 US$5,000 以上，已不須簽輸出入許可證，改以進出口報單代之。

8. 貨裝出後，如為 CFR 或 CIF 出貨，要拿 S/O 去船公司繳清運費才可換領到正本提單(B/L)（一式三份）。如為 FOB/FCA 出貨則請報關行拿 S/O 直接去船公司換提單。

9. 一般報關行也可代為做出所有押匯文件，或至少代為申請做出：提單　(B/L)、海關發票(SCI)、產地證明(Certificate of Origin)及出口報單，其他文件由出口商自行做出。

10. 出貨當天或最慢隔天（參照 L/C 上規定），以電報發出出貨通知(Shipping Advice)給客戶。（參範例 2/3）Shipping Advice 內容至少包括下列幾項：

   (1)Order No.　訂單號碼

   (2)L/C No.　信用狀號碼

   (3)Description of goods　品名規格

   (4)Total Quantity　總數量

   (5)Total Amount　總金額

   (6)Total Gross Weight　總毛重

   (7)Name of Vessel　船名，或 Flight No.　班機號碼

   (8)ETD (Estimated time of departure)　預計離港日

   (9)ETA (Estimated time of arrival)　預計到港日

   (10)B/L No.或 AWB No.　提單號碼

11. 按照信用狀上文件的要求，逐項做出所有的出貨文件。

   一般最簡單的L/C，至少會要求三樣文件：發票(Invoice)、包裝表(Packing List)及提單（B/L 或 Air Way Bill）。

   如為 CIF 出貨，要向保險公司投保，做出保單(Insurance Policy or Certificate)，其他文件視 L/C 要求，再照做即可。

12. 文件整理好，填好押匯銀行的出口押匯申請書，附上 L/C 上要求的整套正副本押匯文件及正本信用狀，送入銀行押匯(Negotiation with the Bank)。

13. 另外將整套押匯文件，用掛號或快捷方式（視 L/C 要求）寄一套副本押匯文件給客人。

14. 銀行檢視過押匯文件，如果沒有問題，則馬上付款。

   除了押匯手續費、郵電費外，一般銀行會扣 10～14 天的預付利息費用。（從送文件押匯到銀行付款，最快 1 天，最慢 7 天，視公司與押匯銀行的關係，以及文件有無錯誤而定）

15.出貨通知及押匯文件樣本參下列附件：

附件一　出貨通知(Shipping Advice)

附件二　水單(Bill of Exchange)

附件三　出口押匯申請書(Application for Negotiation of Drafts Under L/C)

附件四　出口報單（出口商使用）

附件五　進口報單（進口商使用）

附件六　出貨發票(Commercial Invoice)

附件七　包裝明細表(Packing List)

附件八　產地證明書(Certificate of Origin)

附件九　海運提單(B/L)

附件十　保險單(Insurance Policy)

附件十一　原產地證明書(Certificate of Origin for ECFA)

附件十二　煙燻證明書(Fumigation Certificate)

附件十三　原廠證明書(Certificate of Original Manufacturer)

附件十四　校正證明書(Certificate of Calibration)

附件十五　稽核報告(Audit Report)

16.進口商／買方收到賣方的出貨文件後，開始辦理進口事宜：

(1)先去銀行贖單或蓋章做擔保提貨。

(2)再請進口報關行拿提單(B/L)去船公司換小提單(Delivery Order ＝ D/O)，如為 FOB 出貨，進口商要先繳清運費，才可換領小提單（如果銀行正本出貨文件傳送太慢，出口商郵寄的副本也尚未收到，但貨已到達，則請出口商將出貨文件中的提單／發票／包裝表傳真或 e-mail 過來，可用副本進行上面兩項作業）。

(3)進口商將出貨文件中的發票／包裝表／小提單及進口的產品型錄交給進口報關行，進行報關手續。

(4)海關先將進口貨物以電腦分類，分為 C1（免審免驗）、C2（審核文件）及 C3（驗貨）三種方式，然後進行查驗。

(5)繳清進口關稅及相關海關費用。

(6)提貨，可自行叫車或請報關行安排叫卡車，將貨送至指定地點。

範例 1　買方通知賣方已匯出貨款，並催出貨通知

RE: Please advise when you shipped 20 sets of Flowplus

Dear Pierre,

We have remittted you the money by T/T for 20 sets of Flowplus Test Instruments on January 5, and e-mailed the relevant bank receipt of the remittance to you at the same time. Please confirm if you have got the money and if you have shipped the above goods to us.
If so, please kindly advise us the shipping details.
Thanks for your assistance and we are waiting for you shipping advice.

Best regards
Mary Huang/King-Tech Valve

文章結構

主題：我們已經於 1/5 電匯 20 台 Flowplus 測試儀器的貨款給貴公司，並同時 e-mail
　　　相關銀行匯款收據給你們。

説明：請確認貴公司是否已收到錢並已出貨給我們。如是，請告知出貨明細。

結論：感謝協助並等候貴公司的出貨通知。

生字 / 片語

1. remit (V)　匯款；remittance (N)
2. relevant bank receipt of the remittance　相關銀行匯款收據
3. at the same time　同時；test instruments　測試儀器
4. shipping details　出貨明細；shipping advice　出貨通知
5. assistance　協助

範例 2　賣方通知買方貨已裝出

From: pierre.thiry@logiplus.com

To: sb.wow@msa.hinet.net

Subject: Shipping Advice

Dear Mary,

We are glad to inform you that the shipment of 20 sets Flowplus Test Instruments has been made on Jan. 14 via CX-01 from Brussels. ETA Taipei will be on about Jan. 16.

Attached please find the relevant shipping documents including Invoice, Packing List and Air Way Bill. The original shipping documents have been sent to you via registered mail today. We believe that the goods will arrive at your side in good condition, and thanks for your patronage.

Best regards

Pierre Thiry

文章結構

主題：我們很高興通知貴公司 20 台 Flowplus 測試儀器已於 1/14 經由 CX-01 班機由
　　　布魯塞爾裝出，預計到達台北時間為 1/16。

說明：附上相關出貨文件包括發票、包裝表及空運提單。正本出貨文件於今日以掛
　　　號寄去貴公司。

結論：我們相信貨將狀況良好抵達並謝謝惠顧。

生字 / 片語

1. shipping advice　出貨通知；test instruments　測試儀器
2. original shipping documents　正本出貨文件
3. Invoice　發票；Packing List　包裝表；Air Way Bill (AWB)　空運提單
4. registered mail　掛號郵件；patronage　惠顧
5. ETA (Estimated time of arrival)　預計到港日
6. in good condition　狀況良好；Brussels　布魯塞爾（比利時）

範例 3　賣方通知買方貨已海運裝出

From: info@samwellg.com

To: Tony Smith

Subject: Shipping Advice for P/O 8097

Dear Tony,

We have shipped 200M pcs of nuts under your order number 8097 via s.s. "Phillips V-201".
Total amount is US$100,000 and the total gross weight is 500kgs. ETD Keelung is on about
January 31 and ETA London will be on about February 25. B/L No. is KEELND6666. Please
insure at your side.

Attached please find one set of non-negotiable documents for customs clearance use. Thank
you for your order and hope to receive further new orders from you soon.

Best regards

Eva Hu

文章結構

主題：我們已經裝出貴公司訂單 8097 的貨 200,000 個螺帽。船名為 Phillips V-201，
總金額是 10 萬美元，總毛重是 500 公斤，預計離開基隆時間是 1/31，預計到
達倫敦時間為 2/25，提單號碼是 KEELND6666。

說明：請自行投保。如附是一套副本出貨文件供清關用。

結論：謝謝惠顧並希望很快會收到貴公司進一步的新訂單。

生字／片語

1. 200Mpcs of nuts (M = 1,000)= 200,000　個螺帽

2. s.s.= steam ship　汽船："Phillips V-201"　船名及航次

3. total amount　總金額；total gross weight　總毛重

4. ETD (Etimated Time of Departure) Keelung　預計離開基隆時間

5. ETA (Etimated Time of Arrival) London　預計到達倫敦時間

6. B/L = Bill of Lading　（海運）提單

7. Please insure at your side　請自行投保

8. non-negotiable documents = copy shipping documents　副本出貨文件

9. for customs clearance use　供清關用；further　進一步的

MODERN
BUSINESS
ENGLISH

Part 4 （十、出貨通知及
押匯文件）
商用英文信函寫法   357

附件一　出貨通知樣本(Shipping Advice)

Dear Sirs,

<u>RE: SHIPPING ADVICE</u>

We are glad to inform you that we have dispatched the following goods to you:

L/C No.:8856412

P/O: No. AB010

Description of Goods: Screws

Total Quantity: 200Mpcs

Total Amount: US$100,000

Total Gross Weight: 5 M/T.

Shipped via: (1) s.s. "Evergreen V-100" , B/L No. EVE8888

(2) flight no. _____ ; AWB No. _____

(3) courier ____ (ex. UPS/DHL/TNT); Receipt # _____

(4) Air parcel post

ETD Keelung is on about July 30, 2012

ETA L.A. will be on about August 30, 2012.

Attached please find one set of non-negotiable documents including Commercial Invoice, Packing List, Bill of Lading, Insurance Policy and Certificate of Origin for your customs clearance use.

We hope the above goods will arrive at your side in good condition and thank you for your patronage and hope to receive your further new order soon.

Best regards

附件二　水單樣本(Bill of Exchange)

Pay to the order of

For **TAIPEIBANK**

....
Authorized signature

Taipei, Taiwan ....

Draft No.

Exchange for

At .... sight of this SECOND of Exchange (First unpaid) Pay to the order of

The sum of

TAIPEIBANK

Drawn under Letter of Credit No. .... dated ....

Issued by

To

Value received

....
Authorized signature

TAIPEIBANK
ANY BANK, BANKER OR TRUST CO.

Pay to the order of

11-104 (89. 9. 500本)(光)

附件三　出口押匯申請書樣本

(Application for Negotiation of Drafts under L/C)

台北銀行 公鑒　　　　出口押匯申請書　　　　客戶代號：

日期：＿＿＿＿＿＿＿＿＿＿

輸出許可證號碼：＿＿＿＿＿＿＿＿＿＿　收件號碼

茲檢送本公司根據＿＿＿＿＿＿＿＿＿＿＿＿＿＿銀行第＿＿＿＿＿＿號信狀
所簽發之匯票及(或)下列各項單據金額＿＿＿＿＿＿＿＿＿＿請准予辦理押匯，並按照外匯管理之有關規定結付。
本公司負責保證　貴行於押匯後十二天內或承兌到期日收妥貨款，絕不使　貴行因押匯而遭受任何損害。上項匯票及(或)下列各
項單據如發生退票、拒付等情事，不論為押匯金額全數或一部，本公司於接獲　貴行通知後，願立即如數以原幣加息償還及負擔
一切因此而支出之費用，並願依照本公司另立之「押匯總質權書」所列條款履行責任。本公司同意因單據上之欠缺、瑕疵或因單
據正由　貴行審核中，致不能及時辦理押匯而使本公司因匯率之變動而蒙受損失時，由本公司負一切責任，概與　貴行無關。
本筆押匯款項，請依下列方式處理：
　　□一、請以新台幣墊付
　　　　□1.請扣還對　貴行＿＿＿＿＿＿部／分行之貸款。
　　　　□2.押匯款請存入　貴行＿＿＿＿＿部／分行＿＿＿＿存＿＿＿＿＿號 戶名：＿＿＿＿＿＿＿＿
　　□二、請存入外匯存款戶　帳號
　　□三、本筆請以預售遠期外匯契約書號碼＿＿＿＿＿匯率＿＿＿＿＿折換新台幣。

Draft	Invoice	B/L	Conc./ Cust. Invoice	Cert. of Origin	Insurance Cert./ Policy	Packing/ Weight/ List	Insp./ Survey Cert	Benefi/ Shipper Cert.	Assign. Letter	Statement	Carrier Cert	Cable	Postal Receipt	Insurance Co. Cert

電話　　　　　　　　　申請人

（以下各欄請勿填寫）　　　　　　　　　　　　　　　　公司及負責人簽章

CHECK MEMO

Latest Dates: Shipment ＿＿＿＿＿＿＿＿ Expiration ＿＿＿＿＿＿＿ Credit Balance ＿＿＿＿＿＿

□Sight Draft
□Usance Draft＿＿＿Days＿＿Sight/B/L date

DISCREPANCIES AND ACTION TAKEN

Buyer's Account/Shipper's Account

Amount of Draft	Foreign Currency
Master L/C	
Local L/C	
Agent Commission	
Insurance Premium	
Other(s)	
Postage/ Cable Charge	
Loan & Interest	
Discount Charges	

Covering Shipment of

From ＿＿＿＿＿ To ＿＿＿＿＿
Transhipped at ＿＿＿ Via ＿＿＿
Per S.S. ＿＿＿＿ B/L Date ＿＿＿
Reimbursement Method Remarks (Form A/Form B)

□ 轉押匯 ＿＿＿＿＿ Bank.
　Charges: Post. NT$＿＿＿ Tlx. NT$＿＿＿

□ 洽放款 ＿＿＿＿＿ 部／分行

□ Discrepancies Show On The Cover:

本筆出口押匯除上述所列
押匯單據皆合乎L／C要求可否付款
呈請核示
截至／該戶額度美金＿＿＿千元。
自／止未銷帳餘額美金＿＿＿千元。
至／已押美金＿＿＿千元。

Approved By:　　　　Checked By:　　　　First Check

附件四　出口報單樣本（出口商使用）

關 01002

出口報單

類別代號及名稱 G5 國貨出口		聯別	共 1 頁 收單 第 1 頁
報單(收單關別 出口關別 民國年度 給或關代號 發貨單或收序號)號碼 (7)	AW / 101 / 4766 / 8197	收單編號或登錄單號碼 (13) NIL	

報關人名稱、簽章 華茂報關 有限公司	專責人員姓名、簽章 江丕泉 70AA469	統一編號 (8) 12974246	運關監管編號 (9)	託 1 運 (10)	埠單編號	
		貨物出口人(中、英文)名稱、地址 寶閥精密工業股份有限公司 KING-TECH VALVE PRECISION INDUSTRY INC. 台北市中山區新生北路2段127巷47號1F (11)			報關日期(民國) (14) 101 年 1月 02日	出口口岸 (15) TWKEL KEELUNG
(:)	(2)				離岸價格 (16)	金額
469	00469				TWD	108,666
檢附文件字號 (3)		買方統一編號(及海關監管編號) (12) 名稱、地址 BELLS MARKETING SDN BHD.			幣別 FOB Value USD	3,550.00
					運費 (17) USD	100.00
貨物存放處所 (4) 001A1031 中國貨櫃	運輸方式(5) 2				保險費 (18) USD	20.00
					應 加 (19) 減 費用 (20)	

申請審驗方式 (21) N	買方國家及代碼 (22) MY MALAYSIA	目的地國家及代碼 (23) MYPKG PORT KELANG	出口船(機)名及呼號(班次) (24) S6AV5 WAN HAI 315 S052	外幣匯率 30.61	

項次 (27)	貨物名稱、品質、規格 製造商等 (28)	商標	輸出許可證號碼一項次 (29) 商品標準分類號列 (30) 稅則號別 統計號別 (主管機關指定代號)	淨重(公斤) (31) 數量(單位)(32) (統計用) (33)	簽審機關 專用欄	離岸價格 (34) FOB Value (新台幣)	統計方式 (35)
	BRONZE HYDRAULIC BALANCING VALVE, MODEL:41-05 SCREWED END BSPT, WORKING PRESSURE :PN25						
1.3/4"		"NO BRAND"	NIL 8481.10.00.00-0	105 150 PCE 150 PCE		108,666	02

總件數 (25) 7 CTN	單位	總毛重(公斤) (26) 116	海關簽註事項			商港建設費	
BMSB PORT KLANG O/NO.BM-S/2758/10/10 C/NO.1-7 MADE IN TAIWAN						推廣貿易服務費	
			查檔	補檔			
						台　計	
			分估計費	放行		後	
						補 紀	
						錄	
其他申報事項 委任書:46900283 至1011231日止			核發准單	電腦審核			
						證明文件核發	聯別 份數 核發紀錄
			通關方式	(申請)查驗方式			

C-8197 C

出口

附件五　進口報單樣本（進口商使用）

關 01001

進口報單

類別代號及名稱(7) 五堵關　二課一股
G1 外貨進口

聯別

共 3 頁　第 1 頁

收單

報單號碼(8)：收單關別 AW / 轉自關別 / 民國年度 01 / 船或關代號 0553 / 裝單或收序號 0196

理單編號

報關人名稱、簽章	華茂報關有限公司
(1)	469
專責人員姓名、簽章	江丕泉
(2)	00469
提單號數(3)	SNLSHTL0290207C1
貨物存放處所(4)	001A1070 環球貨櫃
運輸方式(5)	2
起運口岸及代碼(6)	SHANGHAI CNSHA
統一編號(9)	12974246
海關監管編號(10)	
繳(11)	3
納稅義務人（中、英文）名稱、地址	寶閥精密工業股份有限公司 KING-TECH VALVE PRECISION INDUSTRY INC. 台北市中山區新生北路2段127巷47號1F
案號(12)	AAB100000042
特(13)	N
賣方國家代碼、統一編號、海關監管編號	CN　KNKH
名稱、地址(14)	KUNSHAN KING-TECH VALVESIFANG RD,CHENGBEI KUNSHAN CITY JIANGSU PROVINCE CHINA
進口船（機）名及呼號（班次）(15)	DONG FANG FU 1208S　BVKS
進口日期（民國）(16)	101年 02月 28日
報關日期（民國）(17)	101年 02月 29日
離岸價格(18) FOB Value	幣別 USD　金額 38,155.76
運費(19)	USD 120.00
保險費(20)	USD 172.24
應加減費用(21)(22)	
起岸價格(23)	USD 38,448.00
CIF Value	TWD 1,136,523
國外出口日期（民國）(24)	101 年02月 27日
外幣匯率	29.56

項次(27)	貨物名稱、牌名、規格等(28)	生產國別(29)	輸入許可證號碼－項次(30) / 商品標準分類號列(31) / 稅則號別 / 統計號別 / (主管機關指定代號)	檢查號碼	單價(32) 條件、幣別 / 金額	淨重（公斤）(33) / 數量（單位）(34) / （統計用）(35)	完稅價格 數量(36)	進口稅率(37) 從價 / 從量	納稅辦法(38) / 貨物稅率(39)
1.	AIR RELEASE VALVE 1" FC-SB	CHINA CN	NIL 8481.20.00.00-8 ( )		CIF USD 19	283.8 60 SET (60 PCE	33,699	3%	31
2.	AIR RELEASE VALVE 1/2" SUS304 -SB	"	NIL " ( )		CIF USD 200	1.4 1 SET (1 PCE	5,913	3%	31
3.	AIR RELEASE VALVE 2"*150LB SUS304-SB	"	NIL " ( )		CIF USD 160	14.8 2 SET (2 PCE	9,459	3%	31
4.	AIR RELEASE VALVE 3/4" FC-SB	"	NIL " ( )		CIF USD 18	283.8 60 SET (60 PCE	31,925	3%	31

總件數(25) 15　單位 WDC　總毛重（公斤）(26) 8,392

海關簽註事項

標記及貨櫃號碼：
KING TECH SB
C\NO.1#-15#
GESU5982902 4500 (2)FCL/LCL

收單建檔補檔　核發稅單

分估計稅銷證　稅款登錄

分估複核　放行

通關方式　（申請）查驗方式

進口稅	44,072
商港建設費	
推廣貿易服務費	454
營業稅	59,029
稅費合計	103,555
營業稅稅基	1,180,595
滯納金（日）	

其他申報事項：委任書:46900283自990101至1011231日止

101IL02004　101/02/29 15:25 101IL02004

進口

MODERN BUSINESS ENGLISH
現代 商用英文

附件六　出貨發票／商業發票樣本 (Commercial Invoice)

KUNSHAN KING-TECH VALVE

SIFANG RD,CHENGBEI KUNSHAN CITY JIANGSU PROVINCE CHINA

COMMERCIAL INVOICE

NO.SB-120226   DATE: FEB. 22, 2012

Invoice of

Shipped in good order and comdit   BY SEA   sailing on or about

from   CHINA   to   KEELUNG, TAIWAN   for account and risk of

Messrs.   KING-TECH VALVE PRECISON INDUSTRY INC

1FL NO.47 LANE 127 SEC 2 SINSHENG N.RD. JHONGSHAN DISTRICT TAIPEI 104 TAIWAN

TEL:886-2-25631098   FAX:886-2-25712152

Marks & No.	Description	Q'TY	Unit Price	Amount
	CIF KEELUNG	set	USD	USD
	AIR RELEASE VALVE 1" FC-SB	60	19.00	1,140.00
	AIR RELEASE VALVE 1/2" SUS304-SB	1	200.00	200.00
	AIR RELEASE VALVE 2"*150LB SUS304-SB	2	160.00	320.00
	AIR RELEASE VALVE 3/4" FC-SB	60	18.00	1,080.00
	AIR RELEASE VALVE 3/4" SUS304-SB	24	39.00	936.00
	DIRECT-ACTIVATED PRESSURE SUSTAINING VALVE 1/2"*150LB RF SUS316-SB	2	86.00	172.00
	HYDRAULIC BALANCING VALVE 4"*10K FCD-SB	7	140.00	980.00
	HYDRAULIC BALANCING VALVE 8"*PN16 FCD-WELL	1	425.00	425.00
	MULTI-FUNCTION VALVE 3"*10K FCD-FLUX	5	126.00	630.00
KING TECH	MULTI-FUNCTION VALVE 4"*10K FCD-FLUX	3	165.00	495.00
SB	SUCTION DIFFUSER VALVE 16"*12" IN:10K OUT:150LB FCD-SB	7	1,200.00	8,400.00
C\NO.1#-15#	SUCTION DIFFUSER VALVE 18"*12" IN:10K OUT:150LB FCD-SB	7	1,300.00	9,100.00
	SUCTION DIFFUSER VALVE 20"*14" IN:10K OUT:150LB FCD-SB	5	2,000.00	10,000.00
	SUCTION DIFFUSER VALVE 6"*6" 10K FCD-SB	10	210.00	2,100.00
	FLOAT VALVE 2"*150LB SUS304-SB	1	270.00	270.00
	PRESSURE SUSTAINING VALVE 4"*10K FCD-SB	2	170.00	340.00
	PRESSURE RELIEF 3"*10K FCD-SB	1	140.00	140.00
	PRESSURE RELIEF 6"*10K FCD-SB	2	350.00	700.00
	NON-SLAM CHECK VALVE 3"*10K FCD-SB	1	140.00	140.00
	NON-SLAM CHECK VALVE 4"*10K FCD-SB	1	180.00	180.00
	NON-SLAM CHECK VALVE 6"*10K FCD-SB	2	350.00	700.00
	TOTAL	204		38,448.00

VVVVVVVV

SAY TOTAL US DOLLARS THIRTY-EIGHT THOUSAND FOUR HUNDRED AND FORTY-EIGHT ONLY

KUNSHAN KING-TECH VALVE INDUSTRY CO., LTD.

Kingsman Jong

附件七 包裝明細表樣本(Packing List)

KUNSHAN KING-TECH VALVE

SIFANG RD,CHENGBEI KUNSHAN CITY JIANGSU PROVINCE CHINA

PACKING LIST

NO.SB-120226   DATE: FEB. 22, 2012

Packing list of

Shipped in good order and comdit   BY SEA   sailing on or about

from   CHINA   to   KEELUNG, TAIWAN   for account and risk of

Messrs.   KING-TECH VALVE PRECISON INDUSTRY INC

         1FL NO.47 LANE 127 SEC 2 SINSHENG N. RD. JHONGSHAN DISTRICT TAIPEI 104 TAIWAN

         TEL:886-2-25631098   FAX:886-2-25712152

Marks & No.	Description	Quantity set	kg/set	NW	GW
NO.1	SUCTION DIFFUSER VALVE 16"*12" IN:10K OUT:150LB FCD-SB	2	253.00	506.00	549.00
NO.2	SUCTION DIFFUSER VALVE 16"*12" IN:10K OUT:150LB FCD-SB	2	253.00	506.00	549.00
NO.3	SUCTION DIFFUSER VALVE 16"*12" IN:10K OUT:150LB FCD-SB	2	253.00	506.00	549.00
NO.4	SUCTION DIFFUSER VALVE 16"*12" IN:10K OUT:150LB FCD-SB	1	253.00	253.00	506.00
	SUCTION DIFFUSER VALVE 18"*12" IN:10K OUT:150LB FCD-SB	1	206.00	206.00	
NO.5	SUCTION DIFFUSER VALVE 18"*12" IN:10K OUT:150LB FCD-SB	2	206.00	412.00	458.00
NO.6	SUCTION DIFFUSER VALVE 18"*12" IN:10K OUT:150LB FCD-SB	2	206.00	412.00	457.00
NO.7	SUCTION DIFFUSER VALVE 18"*12" IN:10K OUT:150LB FCD-SB	2	206.00	412.00	457.00
NO.8	SUCTION DIFFUSER VALVE 20"*14" IN:10K OUT:150LB FCD-SB	1	550.00	550.00	596.00
NO.9	SUCTION DIFFUSER VALVE 20"*14" IN:10K OUT:150LB FCD-SB	1	550.00	550.00	597.00
NO.10	SUCTION DIFFUSER VALVE 20"*14" IN:10K OUT:150LB FCD-SB	1	550.00	550.00	596.00
NO.11	SUCTION DIFFUSER VALVE 20"*14" IN:10K OUT:150LB FCD-SB	1	550.00	550.00	597.00
NO.12	SUCTION DIFFUSER VALVE 20"*14" IN:10K OUT:150LB FCD-SB	1	550.00	550.00	597.00
NO.13	AIR RELEASE VALVE 3/4" FC-SB	60	4.73	283.80	615.10
	AIR RELEASE VALVE 1" FC-SB	60	4.73	283.80	
NO.14	SUCTION DIFFUSER VALVE 6"*6" 10K FCD-SB	8	60.30	482.40	647.90
	HYDRAULIC BALANCING VALVE 4"*10K FCD-SB	4	28.00	112.00	
NO.15	HYDRAULIC BALANCING VALVE 8"*PN16 FCD-WELL	1	90.00	90.00	621.00
	SUCTION DIFFUSER VALVE 6"*6" 10K FCD-SB	2	60.30	120.60	
	HYDRAULIC BALANCING VALVE 4"*10K FCD-SB	3	28.00	84.00	
	AIR RELEASE VALVE 3/4" SUS304-SB	24	1.30	31.20	
	AIR RELEASE VALVE 1/2" SUS304-SB	1	1.40	1.40	
	DIRECT-ACTIVATED PRESSURE SUSTAINING VALVE 1/2"*150LB RF SUS316-SB	2	1.70	3.40	
	MULTI-FUNCTION VALVE 3"*10K FCD-FLUX	5	16.70	83.50	
	MULTI-FUNCTION VALVE 4"*10K FCD-FLUX	3	25.50	76.50	
	AIR RELEASE VALVE 2"*150LB SUS304-SB	2	7.40	14.80	
	FLOAT VALVE 2"*150LB SUS304-SB	1	3.00	3.00	
	PRESSURE SUSTAINING VALVE 4"*10K FCD-SB	2	5.00	10.00	
	PRESSURE RELIEF 3"*10K FCD-SB	1	4.00	4.00	
	PRESSURE RELIEF 6"*10K FCD-SB	2	10.00	20.00	
	NON-SLAM CHECK VALVE 3"*10K FCD-SB	1	5.00	5.00	
	NON-SLAM CHECK VALVE 4"*10K FCD-SB	1	8.00	8.00	
	NON-SLAM CHECK VALVE 6"*10K FCD-SB	2	12.00	24.00	
	TOTAL	204		7704.40	8392.00

SAY TOTAL FIFTEEN (15) WOODEN FRAMES ONLY

KUNSHAN KING-TECH VALVE INDUSTRY CO., LTD.

Kingsman Tong

MODERN BUSINESS ENGLISH
現代商用英文

附件八 產地證明書樣本(Certificate of Origin)

839959017508

臺 灣 省 商 業 會
TAIWAN CHAMBER OF COMMERCE

4F., 158 SUNG CHIANG RD.
TAIPEI 10430,TAIWAN R.O.C.
台北市松江路158號4F
E-mail:tcoc@tcoc.org.tw
TEL:(02)25385455 EXT.106
(02)25817039
FAX:(02)25211980,25679375

產 地 證 明 書
CERTIFICATE OF ORIGIN

日期
Date ___May 19,2012___

1. 出 口 商 名 稱
   Exporter/Manufacturer: ___SAMWELL INTERNATIONAL, INC.___   登 記 號 碼 Registration No.___

   地 址
   Address ___12F-3, NO.5, SEC.2, TUN HWA S. RD., TAIPEI,TAIWAN, R.O.C.___

2. 外國進口商或提貨人姓名
   Importer/Consignee ___ADIMPEX S.R.L.___

   地 址
   Address ___VIA ADRIATICA N.17/C 60022 CASTELFIDARDO (AN) ITALY___

3. 茲證明本證書內所列之產品確係在台灣區生產／加工／製造特給予證明
   This is to certify that the merchandise described is grown/processed/manufactured in Taiwan Origin.

   Loaded on ___AIR FREIGHT___   leaving ___C.K.S. AIR PORT___   on/about___
   (輪船或飛機名稱Name of the carrier)   (臺灣港口名Name of Taiwan port)   (日期Date)

   Destined for ___ITALY___   through___   B/L No.___
   (目的地港口名Name of the destination port)   (港口名Name of port)   持有在臺灣所發提單號碼

包 裝 標 誌 Marks & Numbers	貨 品 名 稱 Description of Goods	數 量 Quantity	備 註 Remarks
SW P/O:300148/04/E P/N:390060032 ITEM:KEYPAD ANCONA C/NO.1 MADE IN TAIWAN R.O.C. INVOICE NO.:SWI-PCL-7731	S9292-ADISRL RUBBER PADS COND. RUBBER KIT P/N 390060032 S9292-ADISRL (275A) SAY TOTAL ONE(1) CARTON ONLY.	100.00PCS ================ 100.00 PCS	May 19,2004

4. Invalid without official seal or with unauthorized alteration. 未加蓋本會印信及校對章者無效。

簽發單位：
Issued by : **TAIWAN CHAMBER OF COMMERCE**   證 號
Certification No.

Authorized Signature

ED 0 4 IA 0 4 4 2 8

HELEN

03-5-27;11:01PM;

附件九 海運提單樣本(B/L)

B/L NO. KUSSEKS112020819

SHAKELC20280

ADP

上海亞東國際貨運有限公司

SHANGHAI ASIAN DEVELOPMENT INTERNATIONAL TRANSPORTATION PU DONG CO., LTD.

上海
SHANGHAI

COMBINED TRANSPORT BILL OF LADING

COPY NON-NEGOTIABLE

托運人 Shipper	
KUNSHAN KING-TECH VALVE INDUSTRY CO.,LTD	
ADD:SIFANG RD CHENGBEI KUNSHAN CITY JIANGSU PROVINCE CHINA	
TEL: 0512-57782089	
FAX: 0512-57779036	

收貨人 Consignee or order	
KING-TECH VALVE PRECISION INDUSTRY INC	
ADD: 1FL,NO.47,LANE 127,SEC.2,SINSHENG N.RD., JHONGSHAN DIST.TAIPEI 104,TAIWAN	
TEL: 886-2-25631098	
FAX: 886-2-25712152	

通知人 Notify address
SAME AS CONSIGNEE

前段運輸 Pre-carriage by	收貨地點 Place of receipt
海運船隻 Ocean vessel DONG FANG FU V.1208S	裝貨港 Port of loading SHANGHAI
卸貨港 Port of discharge KEELUNG, TAIWAN	交貨地點 Place of delivery KEELUNG,TAIWAN

最終目的地 Final destination

正本提單份數 Number of original B/L
ONE

集裝箱號 Container No.	封號及嘜頭 Seal No. Marks and Numbers	箱數或件數 No. of Containers or PKG	包裝種類;貨名 Kind of packages;description of goods	毛重 Gross Weight KGS	尺碼 Measurement CBM
		15 WOODEN CASES		8392	15.79
			OTHER VALVE PARTS OF VALVES		
	KING TECH SB C NO.1#-15#				
		CFS-CFS	FREIGHT PREPAID		

ON BOARD A. D. P.
SHANGHAI(3)

SURRENDER!!

SAY TOTAL FIFTEEN WOODEN CASES ONLY.

總箱數或總件數(大寫)
Total No. of Containers or Packages (in words)

運費和費用 Freight and charges	運費噸 Revenue Tons	費率 Rate	預付 Prepaid	到付 Collect
OCEAN FREIGHT			AS ARRANGED	

C/DOLPHIN LOGISTICS CO., LTD.
7F,NO.267,SEC.3,NANKING, E.RD.,TAIPEI TAIWAN
TEL:00886-02-25459900
FAX:00886-02-25460550
ATTN:LOBO LAI

運費支付地
Freight payable at

INWITNESS where of the number of original Bills of Lading stated above have been signed,one of which being accomplished,the other(s) to be bevoild.

簽單地點和日期
Place and date of issue
SHANGHAI 2012-2-27
SHANGHAI ASIAN DEVELOPMENT INY'L TRANS PUDONG CO.,LTD.
(3)
AS AGENT(S)
AGENT FOR THE CARRIER NAMED ABOVE

TERMS AND CONDITIONS AS PER ORIGINAL,BILL OF LADING

MODERN BUSINESS ENGLISH

現代 商用英文

附件十　保險單樣本(Insurance Policy)

泰安產物保險股份有限公司
TAIAN INSURANCE CO., LTD.
總公司：台北市100館前路59號
Tel:(02)2381-9678
免費申訴電話：0800-012-080
傳真電話：(02)2311-8456、(02)23754002
96.09.14事宜字(96)轉企字第111號函備查
83年台財保第831522372號函核准(公會版)

「本商品經本公司合格簽署人員檢視其內容業已符合保險精算
原則及保險法令，惟為確保權益，基於保險與消費者間平對
原則，消費者仍應評加保險單條款與相關文件，審慎選
擇保險商品。本商品如有虛偽不實或誤導情事，應由本公司及
負責人依法負責」要個人可透過免費申訴電話(0800-012-080)或
本公司網站(http://www.taian.com.tw)、總公司、分公司、及通訊
處查閱及索取電腦資訊公開說明文件

AIR/MARINE CARGO POLICY
NO.　02199062421

KING-TECH VALVE PRECISION INDUSTRY INC.

ASSURED MESSRS.

CLAIM, if any, payable in　USD　Currency
at　PORT KELANG, MALAYSIA

Amount Insured　USD4,037.00
U.S. DOLLARS FOUR THOUSAND THIRTY SEVEN ONLY

in case of loss or damage claimable hereunder,
immediate APPLICATION FOR SURVEY must
be made to our claim agents as specified hereon:

SUBJECT-MATTER INSURED (WARRANTED ALL BRAND-NEW UNLESS OTHERWISE SPECIFIED)

INVOICE NO : SB-101104
150 PCS OF MODEL 41-05 BRONZE HYDRAULIC BALANCING VALVE
PACKED IN 7 CTNS

Links Survey (M) Sdn Bhd AS
AGENCY
No 41C Lorong Cungah
42000 PORT KLANG
SELANGOR DARUL EHSAN
MALAYSIA
TEL : (60)331676441
　　　(60)33167002
FAX : (60)331676442

Per　S.S. 'WAN HAI 315' V-S052

Sailing on or about　NOV. 04, 2010

From　KEELUNG, TAIWAN TO PORT KELANG, MALAYSIA

Conditions- Subject to the following and other concerned Clauses
as per back hereof:

INSTITUTE CARGO CLAUSES (A) 1/1/82
FROM SELLER'S WAREHOUSE TO BUYER'S WAREHOUSE

Marks and Numbers as per Invoice No. Specified above.
Valued at the same as Amount Insured

IMPORTANT
PROCEDURE IN THE EVENT OF LOSS OR DAMAGE FOR
WHICH UNDERWRITERS MAY BE LIABLE LIABILITY
OF CARRIERS, BAILEES OR OTHER THIRD PARTIES
96.09.28參委 (96) 積企字第188號山商定

Signed in　TAIPEI　on NOV. 01, 2010　Issued in　DUPLICATE

It is the duty of the Assured and their Agents, in all cases, to take such measures
as may be reasonable for the purpose of averting or minimising a loss and to ensure
that all rights against Carriers, Bailees or other third parties are properly preserved
and exercised. In particular, the Assured or their Agents are required.
1. To claim immediately on the Carriers, Port Authorities or other Bailees for any
missing packages.
2. In no circumstances, except under written protest, to give clean receipts where
goods are in doubtful condition.
3. When delivery is made by Container, to ensure that the Container and its seals are
examined immediately by their responsible official. If the Container is delivered
damaged or with seals broken or missing or with seals other than as stated in the
shipping documents, to claim the delivery receipt accordingly and retain all
defective or irregular seals for subsequent identification.
4. To apply immediately for survey by Carriers' or other Bailees' Representatives if any loss or damage be apparent and claim on the Carriers or other Bailees
for any actual loss or damage found at such time.
5. To give notice in writing to the Carriers or other Bailees within 3 days of delivery
if the loss or damage was not apparent at the time of taking delivery.
NOTE: The Consignees or their Agents are recommended to make themselves
familiar with the Regulations of the Port Authorities at the port of discharge.

Institute Classification Clause (01/01/2001)
QUALIFYING VESSELS
(i) This insurance and for marine transit rates as agreed in the policy or open cover apply only to cargoes carried by mechanically self-propelled vessels of steel construction classed with a Classification society which is
1.1 A Member or Associate Member of the International Association of Classification Societies (IACS), or
1.2 A National Flag Society as defined in Clause 4 below, but only where the vessel is engaged exclusively in the coastal trading of that nation (including trading on an inter-island route within an archipelago of which that nation forms part).
Cargoes under interests carried by vessels not classed as above must be notified promptly to underwriters for rates and conditions to be agreed. Should a loss occur prior to such agreement being obtained cover may be provided but only if cover would have been available at a reasonable commercial market rate on reasonable commercial market terms.
AGE LIMITATION
(2) Cargoes and/or interests carried by Qualifying Vessels (as defined above) which exceed the following age limits to will be insured on the policy or open cover conditions subject to an additional premium to be agreed.
Bulk or combination carriers over 10 years of age or
Other vessels over 15 years of age unless they:
2.1. Have been used for the carriage of general cargo on an established and regular pattern of trading between a range of specified ports, and do not exceed 25 years of age, or
2.2. Were constructed as containerships, vehicle carriers or double-skin open-hatch gantry crane vessels (OHGCs) and have been continuously used as such on an established and regular pattern of trading between a range of specified ports, and do not exceed 30 years of age.
CRAFT CLAUSE
(3) The requirements of this Clause do not apply to any craft used to load or unload the vessel within the port area.
NATIONAL FLAG SOCIETY
(4) A National Flag Society is a Classification Society which is domiciled in the same country as the owner of the vessel in question which must also operate under the flag of that country.
PROMPT NOTICE
(5) Where this insurance requires the assured to give prompt notice to the Underwriters, the right to cover is dependent upon compliance with that obligation.
LAW AND PRACTICE
(6) This insurance is subject to English law and practice.

*For a current list of IACS Members and Associate Members please refer to the IACS website at
www.iacs.or.g.uk

Warranted free from any liability for loss or damage which occurred
before issue of this Policy.
For TAIAN INSURANCE CO., LTD.

The Assured is requested to read this policy and if it is
incorrect, return it immediately for alternation.
Notwithstanding anything contained herein or attached hereto to the
contrary, this insurance is understood and agreed to be subject to English
law and practice only as to liability for and settlement of any all claims.
This insurance does not cover any loss or damage to the property
which at the time of the happening of such loss or damage is insured by or
would but for the existence of this policy be insured by any fire or other
insurance policy or policies except in respect of any excess beyond the
amount which would have been payable under the fire or other insurance
policy or policies except in respect of any excess beyond the amount which
would have been payable under the fire or other insurance policy or
policies had this insurance not been effected.
We, TAIAN INSURANCE CO., LTD., hereby agree, in consideration of the payment to us by or on behalf of the Assured of the premium
as arranged, to insure against loss damage liability or expense to the extent
and in the manner herein provided.
In witness whereof, I the Undersigned of TAIAN INSURANCE CO.,
LTD. on behalf of the said Company have subscribed My Name in the
place specified as above to the policies, the issued numbers thereof being
specified as above, of the same tenor and date, one of which being
accomplished, the others to be void, as of the date specified as above.

DOCUMENTATION OF CLAIMS
To enable claims to be dealt with promptly, the Assured or their Agents are
advised to submit all available supporting documents without delay, including
when applicable:
1. Original policy or certificate of insurance.
2. Original or certified copy of shipping invoices, together with shiping
specification and/or weight notes.
3. Original or certified copy of Bill of Lading and/or other contract of carriage.
4. Survey report or other documentary evidence to show the extent of the loss or
damage.
5. Landing account and weight notes at port of discharge and final destination.
6. Correspondence exchanged with the Carriers and other parties regarding their
liability for the loss or damage.
In the event of loss or damage which may involve a claim under the insurance,
no claim shall be paid unless immediate notice of such loss or damage has been
given to and a Survey Report obtained from this Company 's Office or Agents
specified in this policy.
No claim for loss by theft &/or pilferage shall be paid hereunder unless notice of survey has
been given to this Company 's agents within 10 days of the expiry of this insurance
Documents presenting claims, all of the concerned documents should be written in or translated into
English.

Not valid unless Countersigned by

72P 0922555

MARINE INSURANCE DEPT.

Calin Chen

PRESIDENT

MODERN
BUSINESS
ENGLISH

Part 4 （十、出貨通知及
抽檢文件）
商用英文信函寫法 367

附件十一　原產地證明書樣本(Certificate of Origin for ECFA)

海峽兩岸經濟合作架構協議原產地證明書
正本

如有任何塗改、損毀或填寫不清均將導致本原產地證明書失效

1.　出口商(名稱、地址)： 電話：　　　　傳真： 電子郵件：	編號： 簽發日期： 有效期至：
2.　生產商（名稱、地址）： 電話：　　　　傳真： 電子郵件：	5.受惠情況 ☐ 依據海峽兩岸經濟合作架構協議給予優惠關稅待遇； ☐ 拒絕給予優惠關稅待遇（請註明原因） ——————————— 進口方海關已獲授權簽字人簽字
3.　進口商（名稱、地址）： 電話：　　　　傳真： 電子郵件：	
4.運輸工具及路線： 離港日期： 船舶/飛機編號等： 裝貨口岸： 到貨口岸：	6.備註：

7.項目 編號	8.HS 編碼	9.貨品名稱、包裝件數及種類	10.毛重或其他 計量單位	11.包裝嘜頭或 編號	12.原產 地標準	13.發票價格、編號 及日期

| 14.出口商聲明
--本人對於所填報原產地證明書內容之真實性與正確性負責；
--本原產地證明書所載貨物，係原產自本協議一方或雙方，且
　貨物屬符合海峽兩岸經濟合作架構協議之原產貨物。

———————————
出口商或已獲授權人簽字

———————————
地點和日期 | 15.證明
依據「海峽兩岸經濟合作架構協議」臨時原產地規則規定，茲證明
出口商所做申報正確無訛。

———————————
地點和日期，簽字和簽證機構印章

電話：　　　　傳真：
地址： |

第　　頁，共　　頁

**MODERN
BUSINESS
ENGLISH**

現代商用英文

附件十二　煙燻證明書樣本(Fumigation Certificate)

彼得害蟲驅除有限公司
PETE PEST CONTROL CO., LTD.

TEL: (02)2799-2000. 2799-5000
FAX: (02)2799-7000

1947 創業
台北市內湖區環山路二段53巷6弄4號1
Http://www.pete.com.tw
北市建　公司(59)字第035043號

ALLEY 6, LANE 53, SEC 2,
-SHAN, ROAD,
I, TAIWAN

SINCE 1947

FUMIGATION CERTIFICATE

KT2010/8/27-1155

This is to certify that the merchandise described below have been fumigated as stated in the following declaration.

Name & address of exporter: KING-TECH VALVE PRECISION INDUSTRY INC.
1F. NO. 47, LANE 127M SECT 2, NORTH SINSHENG
ROAD TAIPEI TAIWAN

Name & address of consignee: BELLS MARKETING SDN BED
NO. 1 JALAN PERUNOING U1/17 GLENMARIE
INDUSTRIAL PARK 40150 SHAH ALAM, SELANGOR
DARUL EHSAN, P.O.BOX 42, 46700 PETALING JAYA,
SELANGOR, MALAYSIA.

NO. & description of merchandise: BRONZE HYDRAULIC BALANCING VALVE
INVOICE NO.: SB-100829
TOTAL: ONE (1) WOODEN CASE ONLY.

Means of conveyance: WAN HAI 305 V-S096

Fumigation or disinfection :
Date: April 06 2012   Treatment: FUMIGATION
Duration of exposure: 24 hours at   23 ℃
Chemical: Methyl Bromide   Concentration: 48g/M3
Distinguishing marks:

BMSB
PORT KLANG
O/NO. BM-S/2727/08/10
C/NO. 1

Yueh-Lan Shih

附件十三　原廠證明書樣本(Certificate of Original Manufacturer)

Smart Balancing

Vallentuna 2012-01-27

Certificate of Original Manufacturer

Smart Balancing Instrument SBI AB certifies that the Balancing Instrument Flex3 with serial number as below was manufactured by us and has passed the tests requirements.

Serial number: 1044-0090

Signed by:

Hellström, Marketing Manager

Vallentuna 2012-01-27

Smart Balancing Instrument SBI AB	Telephone:	Corporate identity no:	Bank: Handelsbanken
Cedersdalsvägen 11	+46 8 514 306 76	556629-6801	BIC (SWIFT): HANDSESS
SE-186 40 Vallentuna	Fax:	VAT-nr:	IBAN:
Sweden	+46 8 514 306 77	SE556629680101	SE37 6000 0000 0005 8264 1918

附件十四 校正證明書樣本(Certificate of Calibration)

FlowPlus

CERTIFICATE OF CONFORMITY

LOGIPLUS guarantees that this FLOWPLUS INSTRUMENT has successfully passed the series tests requirements following LOGIPLUS procedure LPP-99025/D, including the calibration of sensors and instruments. The calibration is valid for 12 months after the first use of the instrument.

FlowPlus Serial Number :	FP-0010163
Sensor range :	20 Bars
Sensor 1 serial Number :	75771
Sensor 2 serial Number :	75792
Controlled by :	PYT
Software version :	1.04
Kernel software version :	4
Hardware revision :	1.3
Date of calibration :	24/06/2012

CEBEC

ISO9001

logiplus

MODERN
BUSINESS
ENGLISH

Part 4

（十、出貨通知及
抽匯文件）
商用英文信函寫法

371

附件十五　稽核報告樣本(Audit Report)

AUDIT REPORT

Name & Address of Factory :

Audit Date & Place:

Nature of Audit:

1. According to the check list　　2. Random documents review

I. General information of the factory:

  1. Contact details: Contact person: _____

    Tel: _____　　Fax: _____

    E-mail: _____　　Web page: _____

  2. Found:

  3. Registration capital:

  4. Owner ship:

  5. Plant Area:

  6. Employees: Total _____ persons

         Engineering: _____ Production: _____

         Quality control: _____ Purchasing: _____

         Sales: _____ Others: _____

  7. Equipment:

  8. Major products:

  9. Major raw material suppliers:

  10. Major customers:

  11. Quality system registered:

  12. Sales in last year:

  13. Capacity utilization:

II. Design and Development

  1. Does the supplier have design ability?

  2. What is a new product developing process?

III. Documentation and Records

練習 21　出貨通知及押匯文件練習

1. 將下列翻譯成英文

(1)發票

(2)包裝表

(3)船位

(4)海運提單

(5)產地證明

(6)保單

(7)小提單

(8)出貨文件

(9)出貨通知

(10)寫出下列縮寫字的英文全名及中文意思：

S/O, D/O, B/L, AWB, ETD, ETA, P/L, RCV, UR, TTL, GW, SHPD, TDY, S.S., ABT, SHPG, BST, RGDS

2. 根據下列中文大意寫一封出貨通知函

主題：有關 L/C No. 3325，我們很高興通知貴公司我們已經 於今日 裝出 100 箱 的鞋
　　　子，經由長榮海運航線，船名 "EVA V-123"，預計離開基隆日期在 08/13，
　　　預計到達倫敦日期為 09/15。

說明：我們已 於今日 寄出一套副本出貨文件供參考。

結論：我們相信貨將 完好 抵達，謝謝惠顧。

生字／片語

　1. 經由長榮海運航線　through/via Ever Green Line

　2. 倫敦　London

　3. 完好　in good condition（放句尾）

　4. 惠顧　patronage

十一、催收貨款及要求寬鬆付款函
(Collection & Asking for Deferred Payment Letters)

1. 如果付款條件不是信用狀，而是託收(Bill for Collection)：

   D/P, D/A 及 O/A，出貨後，就要記得去追蹤貨款，打應收帳單(Debit Note)或對帳單(Statement)給買方。近來因L/C付款費用高且對出貨文件要求嚴謹，所以許多交易亦多改為分期付款方式(Instrallment)。（參三、報價及回覆函）。

2. 按照 D/P 付款條件，客人應於收到出貨文件後，馬上付款，但是有些長期客戶，固定每個月出數批貨，也拖延成每個月月底結算一次，就變成 O/A 方式了。

3. D/A 付款條件，則視訂單合約規定，付款時間看是出貨後 30, 60, 90, 120, 180 天不等，如為 D/A 60Days，那麼就於提單上出貨日算起 60 天後，客戶再到銀行付款即可。

4. 如有退貨，不良品扣款等狀況發生，賣方可打應付帳單(Credit Note)給買方，由新訂單的金額或買方帳上的欠款中扣抵。

5. 如果每個月有許多託收貨款，或未付款，或佣金結算等帳目，賣方也可逐項列出，打一張對帳單(Statement)給買方。

6. 以上付款條件的付款方式，可用下列幾種：
   (1)電匯 T/T　(Telegram Transfer / Cable Transfer)
   (2)信匯 M/T　(Mail Transfer/Letter Transfer)
   (3)銀行本票　(Bank check)
   (4)銀行匯票　(Bank Draft)
   (5)郵政匯票　(Money Order)
   一般較喜歡 T/T 付款，因為速度快。但要注意買賣雙方銀行如無直接通匯關係時，會增加轉匯款銀行的手續費。要事先查清楚，加入報價成本中，以免造成損失。

7. 附件一　應收帳單樣本(Debit Note)
   附件二　應付帳單樣本(Credit Note)
   附件三　對帳單樣本(Statement)

8. 催收貨款及要求寬鬆付款函常用單字／片語

(1)Debit Note (D/N) 應收帳單

(2)Credit Note (C/N) 應付帳單

(3)Statement 對帳單

(4)outstanding payment 未付款／欠款

(5)settle the payment 付清欠款

(6)receipt 收據

(7)due in this month 本月到期

(8)overdue 過期／逾期

(9)total amount 總金額

(10)remit (v) /remittance (v) 匯款；upon remittance 匯款同時

(11)wire ＝ T/T 電匯

(12)commission 佣金

(13)handling charge 處理費

(14)arrange the payment 安排付款

(15)to the following account 到下列帳戶

(16)deduct the amount from the next order 由下張訂單扣款

(17)D/A 承兌交單，D/P 付款交單，O/A(Open Account)月結

(18)Bill for Collection 託收

(19)installment 分期付款

(20)debit your account 記入貴方帳上（由貴方帳上扣款）

(21)credit your account 欠貴方款

(22)deferred payment 延期的付款

(23)催款公司 collection agency

(24)銀行匯票 bank draft

(25)銀行匯款 bank transfer；銀行本票 bank check

9. 催款函常用主題句／說明句／結論句

主題句(Main Idea)

(1)Attached please find our D/N No.001 for the shipment of your P/O No.123.

(2)We attach our Statement no.1234 showing total amount US$10,000 due in this month.

(3)With regard to your P/O No.100, we agree to return you the payment for the 500pcs of defective parts.

(4)With regard to the shipment of O/No. 810, we have remitted US$5,000 to you today through Bank of America.

(5)We confirm the receipt of your Statement and are checking with our accountant now.

(6)As to the outstanding payment due this month, we attach herewith our D/N no.125 for your settlement.

(7)We have sent you the samples for approval on June 10, please advise if the samples are approved. If so, please arrange your payment for the tooling cost.

(8)We regret that we have not received your payment for your outstanding payment due last month.

說明句(Explanation)

(1)Since the shipment has been made on June 1, please kindly settle this outstanding payment by T/T before end of June against our payment terms of D/A 30 days.

(2)Attached please find our Credit Note No. 810 and please deduct the amount from your next order as agreed.

(3)If everything is correct, we will wire the money to your account next week.

(4)Please kindly accept our deferred payment for two months due to the slow business.

(5)If you can not settle your outstanding payment by the end of this month, we will change the payment terms back to L/C instead of D/P.

結論句(Conclusion)

(1)Thanks for your order and look forward to receiving your payment soon.

(2)Please confirm when you receive our T/T payment.

(3)We will keep you informed upon remittance.

(4)Please do settle it by the end of this month, or we will charge you all interest and exchange rate loss.

範例 1　賣方催收貨款

From: VickyChen@merrybest.com

To: rosanobrave@admpx.com

Subject: Outstanding Payment D/N No.MB-2189

Dear Mr. Brave,

Attached please find our D/N No. MB-2189 for the shipment of your P/O No. ADM-86501.

Since the shipment has been made on Jan. 15, please kindly settle this outstanding payment

by T/T before Feb. 15 against our payment terms of D/A 30 days.

Thanks for your order and look forward to receiving your new order soon.

Best regards

Vicky Chen/Accounting Dept.

文章結構

主題：如附請發現我們的應收帳單編號 MB-2189 針對貴公司訂單 ADM-86501 的出貨。

說明：由於貨已於 1/15 裝出，請按 D/A 30 天的付款條件於 2/15 前以電匯方式付清
　　　欠款。

結論：謝謝您的訂單，並期盼很快收到貴公司的新訂單。

生字／片語

1. Outstanding payment　未付款／欠款；settle　付清／結清

2. D/N = Debit Note　應收帳單（參附件一）

3. by T/T (telegram transfer/telegraphic transfer/cable transfer)　用電匯

4. D/A 30 days　出貨後 30 天付款

　　（Document against Acceptance　承兌交單）

5. Accounting Dept. (Department)　會計部

範例 2　賣方因退貨，送出應付帳單，退還貨款

From: VickyChen@merrybest.com

To: richardyouth@flowchart.com

Subject: Rejection of 500pcs Defective Parts

Dear Richard,

With regard to your P/O No. 8011, we agree to return you the payment for the 500pcs of defective parts.

Attached please find our Credit Note No. 81221 and please deduct the amount of US$5,000 from your next order as agreed.

We are really sorry for the trouble caused to you and please consider this matter is settled.

Best regards
Vicky Chen

文章結構

主題：有關訂單 8011，我們同意退還 500 個不良品的貨款。

說明：如附請發現我們的應付帳單 81221，如所同意，請於下張新訂單中扣除金額US $5,000。

結論：我們為所造成的麻煩道歉，並請視此案已了結。

生字／片語

1. rejection　退貨；defective parts　不良品；as agreed　如所同意

2. Credit Note　應付帳單（參附件二）；deduct　扣除

3. Please consider this matter is settled.　請視此案已了結

範例 3　賣方寄對帳單催貨款

From: DanHuang@merrybest.com

To: mikealan@swanelec.com

Subject: Outstanding Payment

Dear Mike,

We attach herewith our Statement no. 81231 showing total amount US$15,500 due in this month.

Please kindly settle the above outstanding payment by T/T as soon as possible.

Thanks for your patronage and are awaiting your payment.

Best regards

Dan Huang

文章結構

主題：我們在此附上對帳單號碼 81231，顯示這個月到期的欠款總額　US$15,500。

說明：請儘快用電匯付清欠款。

結論：謝謝惠顧並等候您的付款。

生字 / 片語

1. Statement　對帳單（參附件三）
2. due in this month　這月到期的
3. patronage　惠顧；total amount　總金額

範例 4　買方回覆範例 3，告知正在核對帳款，並會於下週付款

From: mikealan@swanelec.com

To: DanHuang@merrybest.com

Subject: Your Statement No.81231

Dear Dan,

We confirm the receipt of your above Statement and are checking with our accountant to see if it is correct.

If everything is correct, we will wire US$15,500 to your account in International Commercial Bank next week.

We will keep you informed upon remittance.

Best regards

Alan

文章結構

主題：我們確認收到你們上述對帳單，正與我們的會計核對看是否正確。

說明：如果無誤，我們將於下週電匯 US$15,500 到你們在中商銀的帳戶內。

結論：匯款時將再通知貴公司。

生字／片語

1. accountant　會計；wire ＝ T/T　電匯
2. remittance (N)　匯款；upon remittance　匯款同時
3. International Commercial Bank　中商銀

範例 5　賣方催收貨款及買方要求延期付款

Re: Invoice No. KT-1201

We notice with regret that we have not received your payment US$6,800 for the shipment of our invoice no. KT-1201 till now. Since this invoice is one month overdue, please arrange your remittance in full amount without any further delay.

Re: Asking for deferred payment

Please accept our apology for our delay in paying your Invoice No. KT-1201. Due to the slow business here this year, we still have a lot of goods in stock. Would you please kindly allow us to have the deferred payment for further one month? We promise to settle this outstanding payment by the end of next month. Your great help and understanding to the above would be highly appreciated.

文章結構

主題：Invoice No. KT-1201

我們很遺憾注意到，我們至今尚未收到我方出貨發票 KT-1201，金額 US$6,800 貴公司的付款。因此發票已超欠一個月，請盡快安排全額匯款不得再延誤。

主題：要求延期付款

對於延期付款給貴公司發票KT-1201，請接受我方道歉。由於今年生意清淡，我們仍有許多貨在庫存中。可否允許再讓我們延期一個月付款？我們保證會在下個月底前結清欠款。感謝貴公司的大力協助及諒解。

生字 / 片語

1. notice with regret 很遺憾注意到；one month overdue 過期一個月
2. arrange your remittance in full amount 安排全額匯款
3. deferred payment 延期的付款
4. slow business 生意清淡; a lot of goods in stock 許多貨在庫存中
5. settle this outstanding payment 結清此欠款
6. Your great help and understanding would be highly appreciated.
   = We would highly appreciate your great help and understanding.
   感謝貴公司的大力協助及諒解

範例 6　代理通知工廠客戶更改付款爲D/A 120天，查詢是否屬實，並催其佣金

From: rosebrain@adamagency.com

To: williamwu@merrybest.com

Subject: Italtel Payment Terms

Dear William,

We are glad to inform you that the new payment terms for this customer is D/A 120 days instead of previous 180 days from now on. So, please check and inform us as soon as you receive the payment for your Invoices 52B-5050 and 52B-5093, so that we can see if it is true or not.

Also, please remit our commission to our account at the same time when you get the money from the customer.

Best regards

Rose Brain

文章結構

主題：很高興通知你們從現在起，此客戶的付款爲出貨後 120 天，取代以前的 180 天。

說明：因此，請查看並於收到發票 52B-5050 及 52B-5093 的貨款時儘快告知，以便證實客戶所說。

結論：當你們收到客戶錢的同時，也請匯出我方佣金。

生字／片語

1. D/A 120 days　出貨後 120 天付款；instead of　代替

2. previous　以前的；from now on　從現在起

3. at the same time　同時

MODERN
BUSINESS
現代 商用英文 ENGLISH

範例 7　買方通知匯款但賣方收到匯款卻短少部分金額

Payment notice for your tender 01-2012

We have remitted Euro 2,800.- to your account last week,. Please refer the attached remittance notice for your repair tender 01-2012.

Please arrange the shipment throught TNT (normal way) with "Freight Collect" under our TNT account number: 534921.

Re: Shortage of your T/T payment

We just received your remittance today, but the amount is Euro 2,580, not Euro 2,800. Please make up the shortage amount of Euro 220.- soon.

We will release the shipment after receiving the shortage amount.

文章結構

標單 01-2012 付款通知

有關貴公司的修理標單 01-2012，我們上週已匯款 Euro 2,800 到貴公司帳上，請參看如附的匯款通知。請盡快安排以運費到收方式經由 TNT 出貨。我們的 TNT 帳號為 534921。

有關電匯金額短少

我們今天剛收到你們的匯款，但金額為 Euro 2,580，而不是 Euro 2,800。
請盡快補足短少的金額 Euro220。我們將於收到短缺的金額後再放貨。

生字／片語

1. remittance notice 匯款通知
2. repair tender 修理標單
3. TNT 快遞公司
4. Freight Collect 運費待收／運費到付
5. make up the shortage amount 補足短少的金額
6. release the shipment 放貨

附件一　應收帳單樣本(Debit Note)

MERRYBEST INT'L CO.

E-mail Address: merrybest@msa.hinet.net

ADM PX IMPORT CO.　　　　　　　　　　　　Date: January 16, 2012

P.O. Box 21065 Rome

Italy　　　　　　　　　　　　　　　　　　D/N No. MB-2189

DEBIT NOTE

Dear Sirs,

We debit your account with the following amount:

Item	Quantity	Unit Price	Total Amount
Invoice No. 5201			FOB Taiwan
P/O No. ADM-86501			
Silicone Rubber Pads	50,000 pcs	US$0.30/pc	US$15,000.-

Total Amount: U.S. Dollars fifteen thousand only.

Please arrange your payment by T/T to our following account:

　　Overseas Chinese Bank

　　A/C No. 60123561

　　A/C Name: Merrybest Int'l Co.

We are looking forward to your payment soon.

Very truly yours,

Merrybest Int'l Co.

Vicky Chen / Accounting Dept.

附件二　應付帳單樣本(Credit Note)

MERRYBEST INT'L CO.

E-mail Address: merrybest@msa.hinet.net

To: Flow Chart Co.　　　　　　　　　　　　　　Date: January 31, 2012

　　P.O. Box 38096

　　Dollars, USA　　　　　　　　　　　　　　　C/No. 81221

CREDIT NOTE

Dear Sirs,

We hereby credit your account with the following amount:

Item	Unite Price	Total Amount
P/O. No. 8011		
500pcs defective parts	US$10/pc	US$5,000

Total amount: U.S. Dollars five thousand only.

Please deduct the above amount from your next order.

Very truly yours,

Merrybest Int'l Co.

Vicky Chen

Accounting Dept.

附件三　對帳單樣本(Statement)

MERRYBEST INT'L CO.

E-mail Address: merrybest@msa.hinet.net

To: SWAN Electronics Co.   Date: January 31, 2012

STATEMENT
(Ref. No. 81231)

Invoice No.	Date	Item	Amount
MB-803	1/3	Battery	US#2,500
MB-805	1/5	Coil	US#3,000
MB-815	1/15	Capacitor	US#2,250
MB-823	1/23	Resistor	US#3,250
MB-829	1/28	Adapter	US#4,500
		Total Amount:	US$15,500

Total amount: U.S. Dollars fifteen thousand and five hundred only.

Please settle the above amount by T/T to our bank account as soon as possible. Thanks.

Very truly yours,

Merrybest Int'l Co.

Dan Huang

Director

MODERN
BUSINESS
ENGLISH
現代 商用英文

練習 22 催收貨款及佣金函練習

1. 將下列句子翻譯成中文

(1)We regret that we have not received your payment for your outstanding payment due last month. Please do settle it by the fifth of this month, or we will charge you all interest and exchange rate loss.

(2)If you can not settle your outstanding payment by the end of this month, we will change the payment terms back to L/C instead of D/P.

(3)We have sent you the samples for approval on July 20. Please advise if the samples are approved. If so, please arrange your payment for the balance amount of 1/2 tooling cost soon.

2. 根據下列中文大意寫一封催款函

> 主題：有關貴公司的 這個月 的 欠款，我們在此附上應收帳單 40-122，金額 US$ 9,350。
>
> 說明：由於貨已於 上個月底 裝出，我們希望能在這個月底前收到貨款。
>
> 結論：請 於匯款同時 發 e-mail 告知。

3. Write a letter to reply the above letter and tell the seller that US$ 9,350 has been remitted by bank transfer few days ago.

4. 填入適當的英文單字

(1)您的貨款已逾期兩個月。

Your payment is two months _____.

(2)除非這個月底前收到付款，否則我們不得不將貴公司的帳交由催款公司。

Unless payment is received by the end of this month, we will have no choice but to turn your account over to a _____ agency.

(3)付款將用銀行匯票。

Payment will be made by _____ _____.

十二、抱怨退貨處理函(Complaints & Adjustment Letters)

1. 所有信函中，最難處理的就是抱怨函(Complaint Letter)，因為大都會牽扯到賠償或有所損失。

2. 一般客人抱怨的原因有下列幾項
   (1)裝錯貨(wrong shipment)：賣方過失、疏忽或忙季弄錯。

   (2)訂錯貨(wrong ordering)：買方過失。

   (3)卸貨卸錯港口(wrong destination)：運輸公司過失。

   (4)短裝(shortage)：出貨前即短裝或路上遺失。

   (5)包裝破損(package damage)：包裝材料有問題或路上摔壞。

   (6)品質不良(quality problem)：功能／外觀／尺寸／顏色等和確認樣品不符。

   (7)延遲出貨(delay in shipment)：原因有 16 點，參照第「九、包裝／生產／出貨／驗貨討論函」的說明。

   (8)市場不景氣，客人故意退貨(market claim)

   (9)此外，在各個流程中，也可能有抱怨事項發生，例如買方抱怨賣方漲價／樣品無法通過確認、賣方抱怨買方付款慢／索樣太多、沒下訂單等等。

3. 抱怨的責任歸屬有四
   (1)賣方的責任——例如上述原因(1)(4)(5)(6)(7)

   (2)買方的責任——例如上述原因(2)(8)(9)

   (3)運輸公司的責任——例如上述原因(3)

   (4)保險公司的責任——例如上述原因(4)(5)

4. 抱怨處理的方法
   (1)先請客人退回不良樣品(ask the buyer to return the defective samples)

   (2)收到樣品後，分析不良原因(analyze the reasons of the problems)

   (3)判定責任歸屬(judge whose responsibility)

   (4)如為賣方責任，賣方就必須提出更正報告(corrective action)告知以後如何改進，並負責賠償(make compensation)，賠償的方法有三：
   ①賠貨 make a replacement

②還錢 return money (refund)

③給折扣 give some discount

5. 不良品的處理方法

(1)退回工廠(return to the factory)

(2)轉賣當地其他客戶(re-sell to other customers)

(3)請客戶整修或挑用，付其整修挑用費

(ask the buyer to re-work or sort, then pay the re-working or sorting charge to the buyer.)

(4)自己派人去修(send people to re-work)

(5)丟棄(throw away)

6. 如果不是賣方過錯，而是運輸當中造成的損壞時，賣方可直接拒絕接受賠償，例如：

"We are sorry we can not accept your claim because it is not our fault. The damage should have happened during transit. Please ask for the compensation from the insurance company. = Please claim to the insurance company."

（很抱歉我們無法接受此抱怨，因為這不是我們的過失，損壞應是在運輸途中所造成，請向保險公司索賠）

7. 買方抱怨函寫法（參範例 1）

主題：收到訂單×××的出貨，但發現有下列問題：

(We have received the goods of O/No.×××, but find following problems:)

說明：告知不良原因，例如：

(1)品質比確認樣品差(The quality is worse.)

(2)顏色尺寸不對(The color and size are not correct.)

(3)東西裝錯(Some items are not what we ordered.)

(4)包裝破損(The cartons are broken.)

(5)數量與包裝表上所列不符

(The quantity is different from that shown on P/L)

結論：請儘快告知處理方法

(Please advise your disposal soon)

8. 賣方回覆，大都按上述第 4 點所述處理方法（參範例 2）

> 主題：先表示道歉，並訝異聽到此抱怨
>
> （We are sorry to hear from your letter of ×××complaining about the quality of our shipment.）
>
> 說明：請買方退回不良樣品，以便分析不良原因
>
> （Please return some defective samples for checking the problems.）
>
> 結論：如為賣方過錯，定當負責賠貨或還款
>
> （If it proves it is our fault, we will take the responsibility to make the replacement or return you money for the defective parts.）

9. **注意**：按照國際條約規定，買方如有抱怨：

(1)國際出貨：「必須於貨到港口 21 天內提出，並附上證明文件」

（Any claim by Buyer shall reach Seller within 21 days after arrival of the goods at the destination stated in B/L accompanied with satisfactory evidence thereof.）

(2)國內出貨：「必須於貨到用戶手上 30 天內提出」

（Any claim by Buyer shall be posted within 30 days after arrival of the goods at the place of the end user.）

(3)但如有負責維修保養的保固期限的產品例外，在保固期間有任何品質問題，賣方仍須負責。

10.此外，還要注意下列兩點國際條約上的規定：

(1)買方使用不當造成的損壞，賣方不予賠償

（Seller shall not be responsible for damages that may result from the use of goods.）

(2)賣方最大賠償金額＝不良品的金額

（Seller shall not be responsible for any amount in excess of the invoice value of the defective goods.）

除非訂單上，買賣雙方另有別的賠償規定，則例外。

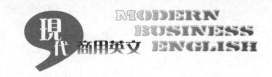

11. Claims: In the event of any claim arising in respect of any shipment, notice of intention to claim should be given in writing to the Seller promptly after arrival of the goods at the port of discharge and opportunity must be given to the Seller for investigation. Failing to give such prior written notification and opportunity of investigation within 21 days after the arrival of carrying vessel at the port of discharge, no claim shall be entertained.

In any event, the Seller shall not be responsible for damages that may result from the use of goods or for consequential or special damages, or for any amount in excess of the invoice value of the defective goods.

（索賠：任何出貨如有索賠事件產生，欲索賠的通知必須於出貨抵達卸貨港口後書面提出。如未能於承載的船抵達卸貨港 21 天內事先提出書面通知以及給予調查機會，索賠將不被接受。此外，賣方對於使用不當，或隨之而產生的損壞，或任何超過不良品的發票金額都不需要負責。）

12. 更正報告(Corrective Action)主要內容分（問題／造成原因／解決方法）三部分：

    I.   Problems:

    II.   Causes

    III. Solution

MODERN BUSINESS ENGLISH

Part 4

（十二、抱怨退貨
處理函）

商用英文信函寫法

391

13.抱怨退貨處理函常用單字／片語

(1)shortage　短裝；package damage　包裝破損；quality problem　品質問題

(2)delay in shipment　慢交貨；market claim　市場索賠

(3)ask the buyer to return the defective samples　要求買方退回不良樣品

(4)analyze the defective reasons　分析不良原因

(5)take the responsibility　負責

(6)corrective action　更正報告

(7)ask for compensation from ＝ claim to…　要求索賠

(8)make a replacement　賠貨

(9)return money = refund　還款

(10)give some discount　給折扣

(11)return to the factory　退回工廠

(12)re-sell　轉賣；re-work　整修；re-working charge　整修費用

(13)sort　挑用；sorting charge　挑用費用

(14)during transit ＝ during transportation　運輸途中

(15)cartons are broken　外箱破損

(16)disposal ＝ how to solve the problem　處理方法／如何解決問題

(17)defective parts　不良貨品；defective samples　不良樣品

(18)inspection report　檢驗報告

(19)inland freight　內陸運費；customs entry　海關費用

(20)brokerage charge ＝ broker charge　報關費用

(21)poor design and workmanship　不良設計和手工

(22)handling charge　處理費；freight collect　運費待收；freight prepaid　運費已付

(23)rejected material　退貨

(24)cancel the order　取消訂單；cancellation (n)　取消

(25)negotiate (v) ; negotiation (n)　交涉

(26)fulfill the order　履行訂單

(27)out of specification　超出規格（和規格不符）

(28)exchange rate loss　匯差；foreign currency fluctuation　外幣浮動

(29)hard position　困境

(30)subcontractor ＝ sub-factory　外包商；local vendor　國內廠商

(31)suffer cash flow problem　承受現金周轉問題

14.抱怨退貨處理函常用的主題句／説明句／結論句

主題句(Main Idea)

(1)We have just received the shipment of O/No.001, but we found there was one oil tank damaged during transportation.

(2)There were five cartons broken during transit.

(3)We inspected the shipment and found out of a total of 465pcs, there are 343pcs out of specification.

(4)We have received the shipment for our order no. 990. After testing, we found following problems:

(5)We received the shipment of your order number 001 and found quantity shortage.

説明句(Explanation).

(1)We have sent back the defective samples for evaluation.

(2)After receiving the defective samples, we will analyze the problems.

(3)If it proves that it is our fault, we will make the replacement

(4)We are sorry for that we can not accept your claim as it is not our fault.

(5)Please ask for the compensation from the insurance company. (Please claim to the insurance company.)

(6)We just returned you 500pcs defective parts via Fed. Express with freight collect.

(7)Please correct the deviation for the pending orders.

(8)Please send us free parts for replacement as this machine is still within the warranty period.

結論句(Conclusion)

(1)Please advise if you could make a replacement before we send back the damaged one for exchange.

(2)We are waiting for your disposal by return.

(3)As the machine is still within the one-year guarantee period, please send us the new meter for replacement as soon as possible.

(4)Please pay the rework charge and sorting charge, total US$ 3,000.-

(5)Please settle the outstanding payment soon, or we will charge you the exchange rate loss.

(6)We are sorry for the inconvenience (trouble) caused to you.

範例 1 買方抱怨一個油槽破損，要求退換

From: HankWhite@ctatank.com

To: BillyYen@muffytank.com

Subject: Damaged Oil Tank

Dear Billy,

We have just received the shipment of O/No. 010, but we found there was one oil tank damaged during transportaion. The relevant picture and inspection report were sent for your evaluation the day before. Please advise if you could help us make a replacement for this tank first, and we will then send back this damaged one for exchange. We are waiting for your disposal by return.

Best regards

Hank White

文章結構

主題：我們剛收到訂單 O/No. 010 的出貨，但是我們發現有一個油槽在運輸途中損壞。

說明：相關照片及檢驗報告前天已寄去供評估，請告知是否可協助先行更換，我們將隨後退回此損壞品做交換。

結論：我們等候你們儘快的處理。

生字 / 片語

1. oil tank　油槽；the day before　前天

2. for your evaluation　供評估

3. replacement　賠貨；during transportaion　運輸途中

4. for exchange　做交換

5. disposal　處理的方法

範例 2　回覆範例 1，賣方告知處理退貨方式

From: BillyYen@muffytank.com

To: HankWhite@ctatank.com

Subject: Damaged Oil Tank

Dear Hank,

We are sorry to hear from your e-mail dated 2/1 complaining about the damage of one oil tank. We will advise you our disposal as soon as we receive the picture and the inspection report you sent to us. If it proves that the oil tank was damaged during transit, we will then claim to the insurance company.

However, if it is our fault, we will take the responsibility to make the replacement as soon as we can. Sorry for the inconvenience caused to you.

Best regards

Billy Yen

文章結構

主題：很抱歉收到貴公司 2/1 來函抱怨一個油槽損壞。

說明：當我們收到貴方寄來的相關照片及檢驗報告時，將即刻告知我們的處理方式。如果此油槽證明為運輸途中損壞，我們將向保險公司索賠。然而如果是我方過失，我們將負責盡快賠貨。

結論：抱歉造成你們的不方便。

生字／片語

1. complaining　抱怨
2. inspection report　檢驗報告
3. during transit ＝ during transportation　運輸途中
4. claim to ＝ ask for the compensation from　索賠／要求賠償
5. insurance company　保險公司
6. fault　過失；take the responsibility　負責
7. make the replacement　賠貨；inconvenience　不方便

範例 3　買方告知退回不良品所需費用較高，建議在當地修理

From: Jeff Hoover@spotimport.com

To: TonyYeh@sunbrain.com

Subject: P.O.16800 Defective Parts

Dear Tony,

We have reviewed your letter, and estimated the costs as below based on the return of 625pcs defective parts.

Air freight L.A. to Taiwan	US$　392
Inland Freight Woodland Hills to LAX	$　22
UPS Charge San Jose to L.A.	$　158.25
Air freight Taiwan to L.A.	$　392
Inland freight	$　22
Customs Entry/Brokerage Charge	$　85
UPS Charge L.A. to San Jose	$　158.25
Total:	US$1,229.50

The above does not include:

(1)Your customs entry and brokerage charge.

(2)Inland transportation from airport to your factory.

(3)Rework cost.

(4)Inland transportation from your factory to airport.

It would seem that the rework charge of $1,562.50 charged here is cheaper. So, please review and confirm that we can rework locally. Thanks for your consideration.

Best regards

Jeff Hoover

文章結構

> **主題**：我們重審貴公司來函，並根據 625 個退回的不良品，預估費用如下：
>
> **說明**：洛杉磯到台灣的空運費   US$ 392
>
>   Woodland Hill 到 LAX 的內陸運費 $ 22
>
>   San Jose 到洛杉磯的快遞費 $ 158.25
>
>   台灣到洛杉磯的空運費 $ 392
>
>   內陸運費 $ 22
>
>   海關費／報關費 $ 85
>
>   洛杉磯到 San Jose 的快遞費 $ 158.25
>
>   總共   US$1,229.50
>
> 上述費用仍未包括：
>
> (1)你們的海關費及報關費
>
> (2)從機場到你們工廠的內陸運費
>
> (3)整修費
>
> (4)從你們工廠到機場的內陸運費
>
> **結論**：比較起來，似乎在此地的整修費 US$1,562.50 較便宜。因此，請重查看並告知是否我們可在本地整修，謝謝你們的考慮。

生字／片語

1. review   重審／重查；based on   根據
2. estimated the cost   預估成本費用
3. defective parts   不良品；air freight   空運費
4. Inland Freight   內陸運費；UPS charge   快遞費
5. Customs Entny   海關費；Brokerage Charge ＝ Broker Charge   報關費
6. rework cost   整修費；locally   在當地

範例 4　買方抱怨機器零件故障，請賣方速寄更換零件

From: merrybest@msa.hinet.net

To: JimRadio@flow.com

Subject: Components for Replacement

Dear Jim,

Please be advised that the drawings you sent to us are not what we want. For your better understanding, we enclose herewith the relevant drawings showing you the exact components which we need for replacement.

Our customer complained a lot about the poor design and workmanship of the panel, please correct the problems mentioned and provide all components we need before 1/24 our time, or send someone to bring them here to make the replacement and correction right away. Please advise soon.

文章結構

主題：請被告知你們傳來的圖面不是我們要的，為了便於你們了解，我們在此附上相關圖面，註明我們要求更換的正確零件。

說明：我們客戶抱怨面板的設計與手工極差，請於我方時間 1/24 前更正所提的問題，並提供我們所需的零件，或派人帶來這裡，馬上更換。

結論：請速告知。

生字／片語

1. drawings　圖面；components　零件；replacement　更換品／賠貨
2. For your better understanding　為了便於你們了解
3. complained　抱怨；poor design and workmanship　差的設計與手工
4. panel　面板；correct　更正

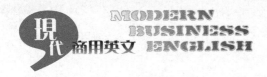

範例 5　賣方回覆範例 4 告知處理的方法

From: JimRadio@flow.com

To: merrybest@msa.hinet.net

Subject: Components for Replacement

Dear Ms. Huang,

After further conversations regarding the problems you have addressed in your mail of Jan. 22, we have decided as below:

(1)To save time, we will ship the components you need through our Singapore Rep. tomorrow.

(2)Our Representative in Singapore tonight will contact you to make arrangement to go to the job site and correct all problems you have addressed.

We sincerely apologize for the inconvenience we caused to you and hope all problems will be resolved very soon.

Best regards

Jim Radio

文章結構

主題：有關你們 1/22 來函上所說的問題，進一步談過後，我們決定如下：

説明：(1)為了省時，我們明天經由我們新加坡代表寄所需零件給你們。

　　　(2)今晚我們新加坡代表會與你們聯絡安排到現場更正所有你們提出的問題。

結論：我們誠心為給你們造成的不便致歉，並希望所有問題能很快解決。

生字／片語

1. addressed ＝ mentioned　所說／所提及；components　零件

2. Singapore Representative　新加坡代表；inconvenience　不方便

3. apologize (V)，apology (N)　致歉；job site　工作現場

範例 6　客戶告知喇叭出貨品質比確認樣品差，要退貨

LAUBE TECHNOLOGY

To: Samwell Int'l Inc.

Re: P/O 16688

Dear Sam,

We are attaching acceptable and unacceptable 36mm speaker samples for P/O 16688.

The customer claims there is a variation in sound output between the samples approved and the actual production they received. The approval samples were very consistent-only 1-2 db variation in output. They sampled 10 pieces from a 500 pieces shipment and they showed a variation of 3-5 db lower output in 50% of the sampling (see attached). This is unacceptable.

The customer has sent us two acceptable samples as reference for minimum specification and to make sure the shipment must meet this acceptable specification.

We have shipped 500 pieces so far to the customer. They will test all 500 pieces in the hope of getting enough parts to keep their production going.

The customer will return all unacceptable parts to us once they have completed their inspection. Also, we have 2,000pcs here which we need to return. Please advise your disposal ASAP.

Regards
Claude

文章結構

主題：我們附上訂單 16688 可接受及不可接受的 36mm 喇叭樣品。

說明：客戶抱怨他們收到的實際生產品和確認樣品在輸出聲音上有差異。確認樣品的輸出變化很穩定在 1-2db 範圍。他們從 500 個出貨中抽樣 10 個，有 50%（一半）的樣品其變化顯示為 3-5db 較低的輸出（參如附），這是不可接受的。

客戶寄給我們 2 個可接受樣品當做最低限規格參考，並請確定出貨要符合此可接受的規格。

目前，我們已經裝了 500 個給客戶，他們將全檢此 500 個以便能有足夠零件維持生產不斷線。

結論：客戶一完成檢驗，馬上會退回所有不能接受的零件，同時，我們這裡還有 2,000 個要退回，請儘快告知處理方法。

生字／片語

1. acceptable　可接受的；unacceptable　不可接受的
2. speaker　喇叭；claim　　抱怨
3. variation　變化／差異；sound output　輸出聲音
4. actual production　實際生產；approval sample　確認樣品
5. sampled(V)　抽樣；minimun specification　最低限規格
6. to keep their production going ＝ to avoid their production shut down
   避免生產斷線＝維持生產
7. lower output　較低的輸出
8. in 50% of the sampling　在一半的抽樣中
9. once they have completed their inspection　當他們一完成檢驗

範例 7　客戶通知所有供應商注意出貨品質，否則處理費高

DRUM TELECOM CO.

To Whom It May Concern:

In the time of world competition, it is extremely important that product supplied by you must be of the highest quality to avoid excess rework cost to you and to our company. Normally rejected material will be returned to you, but under certain circumstances it will be necessary for us to rework the product in order that our manufacturing lines are maintained.

Effective immediately, any rework or sorting performed by us will be charged to the Supplier at a rate of US$30.00 per hour. Also, there will be a US$50.00 handling charge for each lot rejected to cover costs of additional time and paperwork incurred by our Quality and Purchasing Departments.

It is imperative that products shipped by you to us meet our quality standards. Conforming product is an advantage both to you and our company. To achieve this goal, we must work closely together to become partners in quality. If we can be of assistance to you, please contact our Purchasing Department at any time.

Sincerely Yours,

Jim Boy
Director of Operations

MODERN BUSINESS ENGLISH
現代 商用英文

文章結構

主題：在此世界競爭之際，最重要的是貴方要提供高品質產品以免產生貴我雙方過
多的修理費。通常拒收的貨將被退回給你們，但是在某些情況下，為了維持
我們的生產線，我們必須整修產品。

說明：我方將向供應商索取每小時 30 元美金的修理或挑選費，即時生效。同時每批
退貨的處理費是美金 50 元，支付我方品管及採購部門所產生的額外作業及時
間。

結論：貴方裝出的貨絕對要符合我方品質標準，正確產品的出貨對貴我雙方都有利，
為達此目標，我們必須一起密切合作並在品質上成為合夥者。如果需要我方
協助，請隨時和我們的採購部聯絡。

生字 / 片語

1. In the time of world competition　在此世界競爭之際

2. extremely important　極重要

3. to avoid excess rework cost　避免過多整修費

4. Normally　通常；rejected material　退貨

5. under certain circumstances　在某些情況下

6. necessary　必須；Purchase Department　採購部

7. manufacturing lines are maintained　維持生產／持續生產

8. Effective immediately　即刻生效

9. per hour　每小時；handling charge　處理費

10. each lot rejected　被拒收的每一批出貨

11. additional time and paperwork　額外時間及作業

12. imperative　絕對的；to meet our quality standard　符合我方品質標準

13. conforming product　符合／正確的產品；advantage　有利

14. To achieve this goal　為達此目標；partners　合夥人

範例 8　買方告知機器零件故障，要求更換新零件

From: JackLee@textilemachine.com

To: Odani@hurb.com

Subject: Replacement for the meter

Dear Odani,

The textile machine we imported from you can't work today.

After checking, we found that the parts of meters were clogged and need to be replaced.

As this machine is still within the one year guarantee period, please send us the new meters for replacement as soon as possible.

Best regards

Jack Lee

文章結構

主題：我們從貴公司進口的紡織機器今日無法運轉。

說明：經檢查後，我們發現儀表零件故障，需更換。

結論：由於此機器仍在一年的保固期內，請儘快寄新的儀表來更換。

生字／片語

1. textile machine　紡織機器

2. can't work　故障／無法運轉

3. meters　儀表；clogged　故障／卡住

4. replaced (V)；replacement (N)　更換

5. within the one year guarantee period　一年保固期內

　　＝ under warranty

範例 9　買方挑選後仍不能用，只好將貨退回工廠

From: dickmiller@hardware.com

To: joelin@scfasteners.com

Subject: P/O 8898 Rejection

Dear Joe,

We had all parts sorted twice by warehouse staff. Since they had no experience with either sorting or understanding of the Fasteners. Therefore, US$1,600 for sorting charge was not properly done.

We could not but now return 5,600 pcs via Fed. Ex. Freight Collect. Enclosed please find the copy of AWB for reference. Please send back to us after re-working and correction.

Best regards

Dick Miller

文章結構

> 主題：我們叫倉庫員工將所有零件挑過兩次，由於他們對小五金的挑選或了解沒有
> 　　　經驗，因此雖然花了美金 1600 元挑選費，卻沒有做好。
>
> 說明：我們現在不得不用運費對方付方式，以飛遞退回 5,600 個，如附請發現空運提
> 　　　單供參考。
>
> 結論：請於整修更正後，寄回給我們。

生字 / 片語

1. sorting twice　挑選兩次；warehouse staff　倉庫員工
2. no experience　沒經驗；Fasteners　小五金（螺絲／螺帽）
3. sorting charge　挑選費；was not properly done　沒有做好
4. could not but　不得不；via Fed. Ex.　經由飛遞；correction　更正
5. Freight Collect　運費對方付；AWB＝air way bill　空運提單

範例 10　進口商告知工廠其客戶決定不外包生產其電源供應器的原因

BBC ELECTRONICS IMPORTER

To: Mr. Hugo Wang/Power Electronic Mfg.

Dear Mr. Wang,

<u>Re: Power Supply</u>

We are sorry to inform you that the customer has decided that they would produce this item by their own factory and no products would be built by any subcontractor, either in Taiwan or North America for the following reasons:

(1)Fear of the design being copied.

(2)Need a manufacturing cost of approx. 40% less than all the quoted prices they received.

(3)A recent slowdown of activity at buyer's factory that they have more space to produce by themselves.

We express a great disappointment to the buyer for all the work that you and your co-workers have contributed, as well as all the expenses that our two companies have incurred.

However, we have no way but close this deal with this customer due to slow business now. If you have any question, please write us.

Best regards

Johnson Black

文章結構

> 主題：很抱歉告知客戶已決定自己生產此項產品，而不到台灣或北美洲外包生產此產品，由於下列原因：
>
> 説明：(1)擔心設計被仿冒。
>
> (2)需有比所有他們收到的報價低 40%的製造成本。
>
> (3)最近買方工廠生意清淡，以致他們有空檔可以自行生產。
>
> 結論：(1)我們向買主表達極大的失望，由於你們和你們之共事者曾貢獻全力在此工作上，而且貴我雙方也發生了許多費用。
>
> (2)然而，由於目前的不景氣，我們也沒辦法，只好結束此客戶的這筆交易。貴方如有任何疑問，請寫信給我們。

生字／片語

1. Power Supply　電源供應器

2. be built ＝ be produced　被生產

3. Subcontractor　外包商

4. North America　北美洲

5. Fear　擔心；the design being copied　設計被仿冒

6. manufacturing cost　生產成本；approx.　大約

7. slowdown of activity ＝ bad business ＝ slow business　生意清淡

8. space　空檔；express　表達；disappointment　失望

9. co-workers　共事者；contributed　貢獻

10. expenses　費用；incurred　發生；close this deal　結束此案

11. we have no way　我們也沒辦法

12. as well as ＝ and　而且

範例 11　客戶抱怨不良率太高，賣方如不處理，將丟棄

From: jimmyloise@standing.com

To: dicklin@dve.com

Subject: P/O 3089 Rejection

Dear Dick,

We inspected the shipment and found out of a total of 465 pcs, there are 343 pcs did not meet specs. We assure the 2,000 pcs in our stock will have the same failure rate.

Have you corrected the deviation in length and width for the pending orders?

Please give an answer by the end of this week, or, we will scrap the parts and debit your account accordingly.

文章結構

主題：我們檢驗這批出貨，發現 465 個中有 343 個不符合規格。我們確定我們庫存的 2000 個當中，也會有相同的不良率。

說明：你們是否已更正未出貨訂單中產品的尺寸之誤差？

結論：請於本週底前給予回覆，否則，我們將廢棄這些零件並因此記入貴公司帳上。

生字／片語

1. inspected　檢驗；accordingly　因此；debit　記入帳上
2. out of a total of 465 pcs ＝ from a total of 465 pcs　從整個 465 個之中
3. did not meet specs (＝ specifications)　不符合規格
4. same failure rate　相同的不良率
5. corrected　更正；deviation　差異；in length and width　尺寸
6. pending orders　尚未出貨的訂單；scrap　廢棄／報廢

範例 12　客戶抱怨小出貨用海運，造成費用過高

RE: Extra Shipping Expenses

Dear Steve,

We think if you check your records you will find that we never make ocean freight shipments for such small shipment—we always have you consolidate shipments, because the broker fee is the same for either one carton or 48 cartons shipment.

Since this extra charge was caused by you, you should have to bear it. We have sent you a bill showing the actual extra charge caused, please send us your credit note to cover total US$111.25 by return.

文章結構

主題：我們認為如果你們查看你們的記錄，將會發現我們從未出過這麼小批的海運
　　　出貨──我們總是要你們併貨出口。因為 1 箱或 48 箱出貨的報關費相同。
說明：由於這些額外費用是貴公司造成，你們應該負擔。
結論：我們已經寄給你們一張實際額外費用的帳單，請寄給我們一張美金 111.25 元
　　　的應付帳單。

生字 / 片語

1. ocean freight shipment　海運出貨；broker fee　報關費
2. consolidate shipments　併貨出口；extra charge　額外費用
3. to bear it　負擔它；bill　帳單；actual　實際；credit note　應付帳單

範例 13　賣方抱怨買方付款太慢

Dear sirs,

Re: Overdue Outstanding Payments

As you know, to save your cost and to facilitate your paying procedure, we have especially allowed you to settle your monthly outstanding payment at the end of every month by one time.

However, we regret to find that we have suffered a serious cash flow problem and big exchange rate loss due to your delay in settling our outstanding payments.

As you have been clearly aware that the exchange rate between N.T. Dollar and Foreign Currency has been fluctuating quite seriously since early of this year, if we cannot receive your payment in time, we have to suffer a big loss not only on the discrepancy of exchange rate, but also on the interest for the advanced money to the local vendors who always request cash on delivery. Moreover, lots of time and effort had to be spent for tracing & following up the outstanding payments overdue, for which really increased lots of our extra work.

Therefore, we could not but would like to stress again that you must clear up all of your outstanding payments at the end of every month and send us the copy of bank evidence or reference number at the same time you make the remittance to us for easier tracing at our side.

If you could not settle your payment in time, we would then go back to the regular terms of payment by L/C or T/T in advance for all of your future deals.

Please understand our hard situation and confirm the above.

文章結構

> **主題：過期欠款**
>
> 　　如你們所知，為了節省貴方費用及方便貴方付款手續，我們已經特別允許讓貴公司在每個月月底一次付清當月欠款。然而，我們很遺憾發現由於貴公司延遲付清欠款造成我方蒙受嚴重資金周轉問題及很多匯差損失。
>
> **說明：**如你們清楚了解，自從今年年初以來，台幣對外幣的匯率浮動得很厲害，如果我們無法即時收到貴方付款，我們會蒙受不只匯差損失，還有預付款給國內廠商的利息，國內廠商總是要求交貨付現金。此外，還需花費許多時間及工作來追蹤欠款，這些真的增加我們許多額外的工作。
>
> **結論：**因此，我們不得不再次強調貴公司必須在每個月月底結清欠款，並於匯款同時，將銀行證明或參考號碼傳送給我們。以利追蹤。
>
> 　　如果貴方無法及時付清欠款，在未來的生意中，我們將改回一般信用狀或事先電匯的付款條件。請了解我方困境，並確認以上。

生字／片語

1. Overdue　逾期未付；Outstanding Payment　欠款／未付款
2. facilitate　方便於；procedure　手續；especially　特別地
3. settle　付清／結清；by one time　一次
4. suffered　蒙受；cash flow problem　資金周轉問題
5. exchange rate loss　匯差損失；delay　延遲
6. exchange rate　交換匯率；Foreign Currency　外幣
7. fluctuating　浮動；discrepancy　差額；interest　利息
8. advanced money　預支款；local vendors　國內工廠
9. clear up　結清；bank evidence　銀行證明；hard position　困境

MODERN BUSINESS ENGLISH

Part 4

十二、抱怨退貨
處理函

商用英文信函寫法

411

練習 23　抱怨退貨處理函練習

1. 將下列翻譯成英文

　　(1)請退回不良的樣品

　　(2)收到樣品後，我們將分析不良的原因

　　(3)如果是我方的過失，我們將負責賠貨

　　(4)我們很抱歉我們無法接受此抱怨，因為這不是我們的過錯，請向保險公司索賠

　　(5)我們收到訂單 5690 的出貨，我們發現數量短少 (shipment of O/No. 5690)

　　(6)請告知貴公司的處理方法(disposal)

　　(7)在運輸途中，有五個箱子破損(during transit)

　　(8)請付整修及挑選費總共 US$1,500

　　(9)請寄免費更換零件給我們，因為此機器仍在保固期內(still within the guarantee period)

　　(10)我們不接受訂單取消，請履行此合約

　　(11)我們　　於今日　　以運費對方付方式　　退回 5,000pcs. (with freight collect)

　　　　　　　　1　　　　4　　　　　　5　　　　　　2　　　3

　　(12)我們檢查 500 個，發現 200 個不符規格(didn't meet the specs)

　　(13)請儘快付清欠款，否則你們要付匯差(exchange rate loss)

　　(14)我們 為所造成的 不便 道歉

　　　　（＝我們 道歉 for 不便 which 我們 造成 給你們的）

2. 請依下列大意寫一封抱怨函

　　主題：我們剛收到訂單 1100 的出貨，但是發現下列問題：

　　說明：(1)尺寸和顏色不對。

　　　　　(2)有些箱子破損。

　　結論：請儘快告知處理方法。

---

3. 請自由發揮，寫一封賣方回覆上述抱怨的信函

　　（提示：可先請買方寄回不良樣品，以便分析原因，再告知解決的方法）

4. 將下列句子翻譯成中文

　　(1)We received your shipment for our O/No. 110 yesterday. However, upon unpacking (opening) the cartons, we found 5 boxes were broken and some parts were damaged.

(2)Due to the inferior quality, we returned all goods to you yesterday and please send us the replacement with the quality we requested as soon as possible.

(3)We found that the shortage was caused during transit. So, please ask the compensation from (or claim to) the insurance company.

(4)There are some defects on the right corner of the products, which will cause the problem in use.

(5)We enclose herewith a check for US$1,000 to compensate your loss.

5. 克漏（填入適當的英文單字）

(1)我們收到訂單 001 的出貨，發現數量短少。

We received the shipment of O/No.001 and found _____ _____

(2)我們在此附上檢驗報告，請盡快賠貨。

We enclose herewith the _____ report. Please make the _____ soon.

(3)請退回不良樣品供評估。 如為我方過失， 我們將退還貨款。

Please return the _____ _____ for evaluation.

If it is our _____, we will _____ you money.

(4)請按照我方所要求的品質生產產品並履行合約 。

Please _____ the products with the quality we required and _____ the contract.

(5)請付挑選費和整修費用。

Please pay the _____ charge and _____ charge.

6. Write a letter to reply the following complaint letter telling the buyer that you will make up the shortage and change the supplier of PDA to solve the problem.

We received your new replacement last week. However, after unpacking, we found there were two cables shortage. Besides, our engineer found that the connection was unstable. Sometimes there was no connection between PDA and the Sensor. Although after resetting, it started working. But this situation would be very inconvenient in using it.

Is this the problem of PDA or Sensor? or the problem of software design?

Is there any solution to solve this problem?

十三、代理及合約(Agencies & Agreements)

1. 代理種類有三種：

   (1)賣方銷售代理(Seller's Sales Agent)

   在某些國家或地區，賣方因不熟悉當地的法令，語言或生活習慣，為了便於推銷，會在當地找一個可靠的代理商，協助推廣銷售其產品，此代理即稱為賣方的銷售代理(Seller's Sales Agent)或稱業務代表(Sales Representative)。

   (2)進口銷售代理(Import Exclusive Agent) / 經銷代理(Distributorship)

   某些國外廠商製造的產品欲打入台灣市場，或進口商覺得某些產品在國內極有銷售潛能，為了避免競爭或獲取較好利潤，就會爭取獨家代理權(Exclusive Agency or Sole Agent)或總經銷權(Distributorship)。

   (3)買方採購代理(Buyer Agent or Purchase Agent)

   有些買主在某些地區購買許多東西，如果每一項產品都要親自找廠商問價比價，耗時費力又花錢，因此買主就會在當地找一個可靠的採購代理，協助其詢價、索樣，下訂單甚至驗貨等工作。

2. 合約（Agreement 或 Contract）的定義，即兩人以上的當事人所締結的雙方欲遵守的條約，因此重點是一定要有雙方的簽字，一式兩份，各留一份。

   (1)合約／契約格式沒有一定，繁簡隨意；內容由當事人雙方自行決定，因此，不論什麼條款，只要雙方同意，簽了字，就有法律效用，可受法律約束。

   (2)合約的種類很多，只要當事人雙方覺得有必要寫出來，明文規定，雙方共同協議來遵守的任何事件，都可簽定成合約或契約。

3. 商業上常見的合約種類有下列幾種：

   (1)銷售合約(Sales Agreement)和訂單(Purchase Order)內容相同，參附件一。

   (2)代理合約(Agent Agreement)，參附件二。

   (3)佣金合約(Commission Agreement)，參附件三。

   (4)經銷合約(Distributorship Agreement)，參附件四。

   (5)付款合約(Payment Agreement)，參附件五。

   (6)授權書 (Authorized Letter)，參附件六。

   (7)專利權使用合約(Patent/Royalty Agreement)，參附件七。

4. 不論什麼合約，雖然格式內容不拘，但下列的基本條件在合約內要寫出：

(1)甲乙雙方(Party A, Party B)：公司全名及詳細地址。

(2)見證人(Witness)：可找律師，第三者或甲乙雙方互為見證，一般基於互信原則，都以甲乙雙方互為見證。

(3)生效日(Effective Date)：大都以雙方簽字後即生效，也可在合約上訂定生效日。

(4)有效期(Expiry Date)：1，3，5 年不等，當事人雙方自行約定。

(5)終止日(Termination)：如為長期的合約，多半會註明此點，以防萬一中途互有不滿，可隨時取消合約。

終止日算法：任何一方以掛號書面提出後，60-90 天後，此合約才終止。

(6)抱怨(Claim)：合約上有時會註明，有糾紛時，要如何解決，如果沒特別註明，就按國際條約執行。

(7)仲裁(Arbitration)：在每個國家都有國際仲裁協會(International Arbitration Association)，附設在商會或工會中，當事人雙方如有無法解決的糾紛發生時，在合約上也可規定由第三國的仲裁協會來裁定。但因此法費時又花錢，所以當事人雙方大都會私下解決。

(8)條款(Terms and Conditions)：雙方自行約定，逐項列出。

(9)簽字(Signature)：合約上一定要有雙方的簽字才有效。

5. 在代理合約(Agent Agreement)中，較重要的條款如下：

(1)代理的地區(Exclusive Area)

(2)代理的產品(Exclusive Products)

(3)最低銷售額(Minimum Sales Amount)

(4)佣金比率(Commission Rate)及付佣金時間

(5)其他條件則視需要，雙方同意，再逐項列出，否則都按國際條款執行即可。

6. 代理佣金的付法有兩種：

(1)抽佣式(Commission Basis)，也稱開放式(Open Way)

即代理協助工廠找到買主後，買主(Buyer)直接付款給工廠／賣方(Seller)；而工廠／賣方直接出貨給買主，代理不負責任何金錢事宜，等賣方拿到錢，按合約上規定的佣金比率及期限內，付佣金給代理商。

(2)賺差額，又稱總經銷(Distributor)

此種代理方式則是買方下訂單及付款給經銷代理商，經銷代理再下訂單付款給工廠，工廠出貨給代理商，代理商再交貨給買主，賺取差額。此種方式的代理，買方知道工廠名字，但不可直接接觸，工廠則不一定知道買主名字，但如有代理地區的客戶直接找工廠，工廠必須將客戶名單資料轉知代理人處理。

7. 申請當銷售／進口代理信函的寫法：（參範例 1 及範例 7）

主題：我們從×××得知貴公司想在×××地區找一個代理商，我們有興趣當貴方此地的銷售／進口代理。

說明：(1)自我介紹經驗／能力：例如，有多年經驗，了解市場並與許多買主關係不錯，有很好的銷售管道及很強的推銷員，可以推銷得很好。

(2)可提供什麼服務

(3)佣金比率的要求多少

(4)抱怨將會如何處理

結論：希望貴方考慮我們，並告知代理條件。

---

8. 申請當買方採購代理信函的寫法：（參範例 2）

主題：我們從×××得知貴公司想在×××地區找一個採購代理，我們有興趣應徵此工作。

說明：(1)自我能力介紹

(2)強調對代理地區的工廠非常瞭解且關係不錯，可爭取到最好的價錢及支持。

(3)可提供最迅速的服務，例如：蒐集資料、尋找合格供應商、確保準時出貨、協助買主驗貨、解決糾紛及處理退貨等。

結論：請考慮以上，並告知採購代理合約的條款。

9. 代理及合約函常用單字／片語

(1)Seller's Agent 賣方代理；Buyer's Agent 買方代理；Import Agent 進口代理

(2)Exclusive Agent = Sole Agent 獨家代理

(3)Agent Agreement 代理合約；Distribution Ageement 經銷商合約

(4)Sales Agreement 銷售合約

(5)Commission Agreement 佣金合約

(6)Payment Agreement 付款合約

(7)Non-Disclosure and Confidentiality Agreement 保密合約

(8)Patent/Royalty Agreement 專利權使用合約

(9)Party A 甲方；Party B 乙方；Witness 見證人

(10)Terms and conditions 條款

(11)Effective date 生效日；Expiry date 有效期；Termination 終止日

(12)Abritration 仲裁

(13)Exclusive Area 代理地區；Exclusive Products 代理產品

(14)Minimum Sales Amount 最低銷售額

(15)Commission Rate 佣金比率

(16)Penalty 罰則

(17)Force Majeure 不可抗力事件

(18)Letter of Authorization / Authorization Letter 授權書

(19)after sales service 售後服務

(20)sales channel 銷售管道；sales network 銷售網

(21)collect information 蒐集資料

(22)source the best qualified supplier 尋找最合格的供應商

(23)inspect the goods before shipment 出貨前驗貨

(24)sales program 銷售計劃；sales method 銷售方法

(25)company profile 公司簡介；company organization 公司組織

(26)forecast 預測；market situation 市場狀況

MODERN
BUSINESS
ENGLISH
Part 4 (十三、代理及合約)
商用英文信函寫法
417

10.代理及合約函常用的主題句／說明句／結論句

主題句(Main Idea)

(1)We know from your advertisement that you are looking for a sales agent in UK.

(2)We are very interested in applying for your Exclusive Agent in Asia.

(3)We are glad to grant you as our exclusive agent in Asia.

說明句(Explanation).

(1)With many years' experience in international trade, we know very well the market situation here.

(2)We have very good sales channel and many experienced sales personnel.

(3)Please provide us your company profile showing your organization for consideration.

(4)We have very good relationship with local suppliers and can render very good and quick service to you.

(5)We know very well the manufacturers' situation here.

(6)We have very good relationship with most of local suppliers.

(7)We have confidence in getting very good prices and supports for you.

(8)We have confidence in getting good market share for you.

(9)We can render very good and quick service to you, such as: to collect all information you want, source the best qualified suppliers for you, make sure shipments on time, inspect the goods before shipment, etc.

結論句(Conclusion)

(1)We have confidence to promote very well your products in this area.

(2)Please let us know your terms and conditions for a sole agent.

(3)Please consider the above and advise us your terms and conditions for further discussion

(4)Please confirm your interest in cooperating with us and e-mail us your relevant terms and conditions for an exclusive agent soon.

(5)We agree to grant you as our sole agent for a trial period of one year in order to get business moving.

MODERN
BUSINESS
現代 商用英文 ENGLISH

附件一　銷售合約樣本(Sales Agreement)

SALES AGREEMENT

Article 1: Shipment

Time of shipment   : Within 45 days after L/C opened

Partial shipment   : not allowed

Transshipment   : allowed

Port of loading   : Any Taiwan Port

Port of discharge   : Saigon

Shipping mark   : SUN/VT

Notice of shipment   : Within two days after the sailing

Article 2: Payment

2.1 By an irrevocable L/C at sight of 100% total amount of the contract.

2.2 Beneficiary: ABC Co.

2.3 Advising Bank: Bank of Taiwan

2.4 Bank of Opening L/C: Bank of Saigon

2.5 Time of opening L/C: Within 7 days after signing the contract.

2.6 Payment documents:

Payment shall be made upon receipt of the following documents:

+ 2/3 original clean on board bill of lading marked "Freight Prepaid" made out to the order and notify the buyer.

+ 1/3 original clean on board bill of lading sent by DHL to the buyer.

+ Insurance policy/certificate in duplicate, covering "all risk", 110% of invoice value.

+ Signed commercial invoice duly signed

+ Certificate of Quality, Quantity issued by Manufacturer

+ Certificate of Origin issued by Chamber of Commerce

+ Detailed packing list

+ Telex or fax Shipping Advice sent to the buyer showing B/L no, ETD, ETA, etc.

（續下頁）

（續上頁）

Article 3: Inspection

The buyer will ask Vinecontrol or SGS to inspect the cargo at discharge port. If the delivered cargo matches the specification, the inspection fee shall be borne by the buyer. Vice versa, the seller shall bear the inspection fee.

Article 4: Force Majeure

Neither party will be held responsible if either of both parties can not perform to the contract due to the international accepted force majeure such as Act of God, Strike, War, etc. Should the effect of the force majeure continue for more than 120 days, the buyer and seller shall discuss through friendly negotiation as soon as possible their obligation to continue the performance under the terms and conditions of this contract.

Article 5: Arbitration

All disputes or differences which can not be settled amiably will be arbitrated by Vietnam International Arbitration Center at the Chamber of Commerce & Industry of Vietnam. The arbitration will be final and binding upon both parties. All of the charges will be borne by the losing party.

Article 6: Penalty

In case delay shipment happens, the penalty for delay interest will be based on annual rate 15 percent of the total contract amount, but not over one month. If the seller or buyer wants to cancel the contract, 5% of total contract value would be charged as penalty to that party.

Article 7: General condition

Any amendment should be made in written form and confirmed by both parties.

Signed by the Seller   Signed by the Buyer

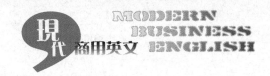

銷售合約中文大意

銷售合約

第 1 條款：出貨

出貨時間：開狀後 45 天內

分批出貨：不准

轉運：准許

裝運港：任何台灣港口

卸貨港：西貢

出貨嘜頭：SUN/VT

出貨通知：開航後兩天內

第 2 條款：付款

2.1 不可撤銷全額即期信用狀

2.2 受益人：ABC Co.

2.3 通知銀行：台灣銀行

2.4 開狀銀行：西貢銀行

2.5 開狀時間：簽約後 7 天內

2.6 付款文件：

需於收到下列文件後付款：

＋三份正本中的兩份提單、註明「運費已付」收貨人：to the order；通知人：買方，送入銀行

＋三份正本提單中的一份，用 DHL 直接寄給買主

＋保單 2 份，含「全險」，保額＝發票金額的 110%

＋簽名的商業發票

＋工廠出具的品質，數量證明書

＋商會出具的產地證明

＋詳細的包裝明細表

＋電報或傳真的出貨通知發給買方，註明提單號碼、預計離港日、預計到港日等

（續下頁）

（續上頁）

第 3 條款：檢驗

買方將要求越南品管或商檢局到御貨港口驗貨，如果交送的貨符合規格，檢驗費由買方負擔；相反的，如沒通過檢驗，賣方要負擔檢驗費。

第 4 條款：不可抗力事件

如果由於國際上可接受的不可抗力事件例如天災、罷工、戰爭等造成任何一方無法履行此合約，任何一方都無需負責。

如果此不可抗力事件持續超過 120 天，買賣雙方將儘快友善討論是否繼續依此合約條款履行。

第 5 條款：仲裁

如有任何無法和平解決的爭論或不同意見時，將由越南工商會中的越南國際仲裁中心來做判決，其判決可約束雙方，所有費用將由敗方負擔。

第 6 條款：罰款

如有任何延遲出貨情形發生，延遲出貨利息的罰款將根據此合約總金額的年利率15%計算利息，但不得超過一個月。

如果買賣任何一方要解除合約，需給另外一方此合約總金額 5%的解約金。

第 7 條款：一般條件

如有任何修改，需以書面寫出並由雙方確認同意。

賣方簽字　　　　　　　　　　買方簽字

附件二　代理合約樣本(Agent Agreement)

AGENT AGREEMENT

On this DD of MM, YY M/S _____ address: _____

(herein after called Principal) and M/S _____ address: _____

(herein after called Agent) agree and hereby certify to follow the terms/conditions set below:

1. The Principal appoints the Agent as the Turkey Distributor for its computer systems and other products produced by Principal. The principal agrees to give the Agent 10% commission upon receiving the payment for each shipment.

2. The Agent will be responsible for the marketing, promoting and distributing the computer systems in Turkey to a satisfactory level, say minimum annual sales amount up to US$500,000. If the Agent could not reach this target amount, this Agreement will be terminated automatically.

3. The Principal will provide a two years warranty on the systems supplied by the Principal to the Agent, and would take the responsibility for any defective items.

4. The Principal will provide all necessary manuals and brochures for the publicity and support for its products.

5. The Principal will be responsible for making timely shipment to the Agent and will follow the Agent's shipping instruction in order to avoid any discrepancy during customs clearance.

6. The trial period of this Agreement is one year. During the trial period, the Agent must try best to obtain a satisfactory ratio of the market share.

7. The Agent will make Sales forecast every month and the Principal will provide updated prices every month including 10% commission for the Agent.

This Agreement will be effected from the date signed by both parties and is automatically extended if no objection.

The Principal   The Agent

代理合約中文大意

代理合約

在××年××月××日甲公司＿＿＿＿＿地址＿＿＿＿＿（隨後稱為被代理人）與乙公司＿＿＿＿＿地址＿＿＿＿＿（隨後稱為代理）同意並在此保證按照如下條款執行此合約：

1. 被代理人指派代理當其土耳其的總經銷，銷售其所生產的電腦系統與其他產品。並同意於收到每批出貨的貨款同時給予代理 10%的佣金。

2. 代理將負責市場開發、推銷及經銷電腦系統到土耳其市場，並達到令人滿意的最低年銷售額美金五十萬元。如果代理無法達到此目標額度，此合約將自動中止。

3. 被代理人將對其供應給代理的電腦系統提供兩年的保證期，並對任何不良品負全責。

4. 被代理人將提供所有必須的手冊及型錄以為其產品的宣傳及支援。

5. 被代理人將負責定期出貨給代理，並按照代理的出貨指示以避免其清關時造成任何差額。

6. 此代理合約的試驗期為一年，在此試驗期間，代理必需盡力爭取一個令人滿意的市場占有率。

7. 代理將於每個月做銷售預測，而被代理人則於每個月提供一次最新報價，內含 10%佣金給代理。

此合約將於雙方簽字後生效，如果雙方無異議，將自動延長。

　　被代理人　　　　　　　　　　　　代理

附件三　佣金合約樣本(Commission Agreement)

COMMISSION AGREEMENT
(MARKETING PARTNERSHIP AGREEMENT)

This agreement is made between ABC Co. located at _____ ,
acting as the Representative and XYZ Co. located at _____ ,
acting as the Exporter under following terms and conditions:

1. The commission is calculated by FOB selling price.

   The Representative and the Exporter share the profit.

   (i. e. FOB price deducts EX-works price then divided into two).

2. The Representative and the Exporter shall pay their own local expenses respectively.

3. The territory is covering USA and North America.

4. Payment should be made by the L/C from the buyer directly to the Exporter. However, in case the buyer sends the payment by T/T or Bank Check to the Representative, the Representative shall transfer the T/T or Bank Check to the Exporter upon receipt.

5. This agreement is effective from Jan. 1, _____ and valid to Dec. 31, _____ .

The Representative   The Exporter

_____   _____

(Signature)   (Signature)

佣金合約中文大意

佣金合約

（銷售合夥關係合約）

此合約由 ABC 公司，地址_____扮演當 XYZ 公司的代理和 XYZ 公司，地址_____
扮演當出口商，按下列條款簽定：

1.佣金按照 FOB 賣價計算，代理和出口商分享利潤。
　（換言之，FOB 價扣除離廠價，除以二）

2.代理和出口商各自負擔自己的當地費用。

3.版圖包括美國和北美。

4.付款將由買主直接開信用狀給出口商，然而，如果買主用電匯或銀行本票付款給代
　理，代理將於收到同時，馬上轉交出口商。

5.此合約於_____年 1 月 1 日開始生效，有效至_____年 12 月 31 日。

代理　　　　　　　　　　　　出口商

_____　　　　　_____
（簽名）　　　　　　　　　　（簽名）

附件四　經銷合約樣本 (Distributorship Agreement)

DISTRIBUTORSHIP AGREEMENT

This agreement is made between KUNSHAN KING-TECH VALVE PRECISION INDUSTRY INC. located in Kunshan, China (hereinafter called MANUFACTURER) and HQI ENGINEERING CORP. located at No.3, Hsing-Yi Road, Taipei City , Taiwan (hereinafter called DISTRIBUTOR). Both parties have agreed to follow the terms and conditions set below:

1. Manufacturer, KUNSHAN KING-TECH, appoints HQI as its sole distributor for marketing and selling its "SB" brand valve products in Taiwan.
2. DISTRIBUTOR shall actively promote the sale of the above products in Taiwan, and take the responsibility for the after-sales service.
3. MANUFACTURER shall take the responsibility for the quality shipments and any defective items.
4. This agreement is effective from the date both parties sign their signatures and valid for a period of five years.
5. Based on mutual trust, other terms and conditions are followed by the International terms and conditions.

MANUFACTURER:
KUNSHAN KING-TECH VALVE
PRECISION INDUSTRY INC.

DISTRIBUTOR:
HQI ENGINEERING CORP.

_____

_____

Signed by:

Signed by:

Date:

Date:

佣金合約中文大意

經銷合約

此合約為位於大陸崑山的 KUNSHAN KING-TECH，隨後稱為製造商，和台灣HQI，隨後稱為經銷商，之間所簽訂，雙方同意下列條款：

1. 製造商 KUNGSHAN KING-TECH 指派 HQI 作為其獨家經銷商，在台灣行銷其"SB"品牌的閥類產品。

2. 經銷商將主動在台灣宣傳銷售上述產品並負責售後服務。

3. 製造商將負責出貨品質及任何不良品。

4. 此合約於雙方簽字後生效並有效 5 年。

5. 基於雙方互信原則，其他條款依照國際條款執行。

製造商：　　　　　　　　　　經銷商：
KUNSHAN KING-TECH VALVE 　　HQI ENGINEERING CORP.

_____　　_____

簽字：　　　　　　　　　　　簽字：
日期：　　　　　　　　　　　日期：

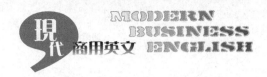

附件五 付款合約樣本(Payment Agreement)

PAYMENT AGREEMENT

This agreement is made between _____ (herein after called the Seller) and _____ __ (herein after called the Buyer) for the mutual benefit cooperation on long term business basis. Both parties agree and hereby certify to follow the terms and conditions set below:

1. Payment is to be made by irrevocable L/C at sight from the Buyer or the customers of the Buyer, and to be reached the Seller at least three weeks prior to shipment. Or, the Seller will not respond to bear the loss of shipment delay.

2. The Seller will reserve the commission of the Buyer into the unit price if the payment is made directly from the customer of the Buyer, and the commission will be settled soon after the shipment is made and the payment is received from the customer. The rate of commission will be finalized case by case.

3. Regarding the payment for the small amount shipment less than US$2,000 or any special favorable payment terms such as Open Account, or D/P , the Buyer agrees to settle the payment by T/T as soon as the shipment is made.

4. If any claim happens to the shipment which has not been paid, the Buyer can not refuse to pay the money before getting the Credit Note from the Seller.

This agreement will be effected from the date that the Seller and the Buyer sign their signatures.

The Seller   The Buyer

付款合約中文大意

付款合約

此合約是由＿＿＿＿（隨後稱賣方）和＿＿＿＿（隨後稱買方）根據雙方長期有利合作生意關係而簽訂。雙方同意並在此保證根據下列所設定的條款執行：

1. 付款將由買方或買方客戶以不可撤銷即期信用狀付款，並於出貨前三週到達賣方手上，否則，賣方將不負責任何延遲出貨的損失。

2. 如果付款由買方客戶直接付給賣方，賣方將在單價中保留買方佣金，佣金將於出貨及收到貨款後，馬上結清，佣金比率將以個案決定。

3. 有關低於美金 2,000 元以下的小金額出貨的貨款，或任何特別有利的付款條件，例如記帳／月結或付款交單，買方同意於出貨後，馬上電匯付款。

4. 如果抱怨發生在尚未付清款項的出貨，買方在得到賣方的應付帳單前，不得拒付貨款。

此合約將於買賣雙方簽字後生效。

賣方 　　　　　　　　　　　　　　　　買方

＿＿＿＿＿＿＿＿＿＿　　　　　　　　＿＿＿＿＿＿＿＿＿＿

附件六　授權書樣本(Letter of Authorization/Authorized Letter)

To: HBS Industrial Inc.

LETTER OF AUTHORIZATION

Dear Sirs,

BEST VALVES LTD , located at International Business Park, TAIWAN Centre, hereby appoints HBS Industrial Inc., No. 10, Yun-Ho , Taipei, Taiwan, R.O.C as our exclusive Authorized Representative in Taiwan effective from the dated signed on this Letter of Authorization and valid for one year,

HBS Industrial Inc. shall be responsible for the marketing of full range of BEST VALVES and provide installation, services, spare parts and after sales maintenance services in the territory of Taiwan.

These activities will be fully supported by BEST VALVES Ltd through HBS own workshops and technical personnel.

Their responsibilities will cover all BEST VALVES installations in Taiwan, both in the past and in the future.

Yours truly,
BEST VALVES LTD

Joe Nill
Director
Date:

授權書中文大意

授權書

To: HBS

位於台灣國際商業園區的 BEST VALVES LTD 在此指派 HBS Industrial Inc.當作本公司在台灣的授權代表，由此授權書簽字起為期一年。

HBS 要負責行銷整個 BEST 的閥類產品，並提供在台灣地區的所有安裝、零件、維修、保養及售後服務。

BEST VALVES 會全力支持由 HBS 本廠及其技術人員所提供的上述工作事項。

其所負責包括在過去及未來在台灣所安裝的所有 BEST 閥類產品。

Yours truly,
BEST VALVES LTD.

Joe Nill
董事長
日期：

附件七　專利權使用合約(Patent/Royalty Agreement)

PATENT/ROYALTY AGREEMENT

This agreement is made between ABC Co. located at _____ (herein after called Party A) and XYZ Co. lacated at _____ (herein after called Party B) for the Patent of Party A used by Party B under following terms and conditions:

1. Party A hereby agrees to authorize Party B to use its Patented Product Model No. AB052 in Taiwan.

2. Party B agrees to pay for the Royalty to the Party A for every piece sold in Taiwan based on 5% of the selling price.

3. Party B will remit the Royalty to the Party A by T/T at the end of every month according to the quantity sold.

4. This agreement is effective from the date both parties sign their signatures for a period of one year.

5. Based on mutual trust, other terms and conditions are followed by the International terms and conditions.

Party A　　　　　　　　　　　Party B
ABC Co.　　　　　　　　　　　XYZ Co.

_____　　　　　_____
(Signature)　　　　　　　　　(Signature)

專利權使用合約中文大意

專利權使用合約

此合約由 ABC 公司，住址＿＿＿＿＿（隨後稱甲方）和 XYZ 公司，住址＿＿＿＿＿＿＿（隨後稱乙方）按下列條款簽定甲方讓乙方使用其專利：

1.甲方在此同意授權乙方在台灣使用其專利產品編號 AB052。

2.乙方同意按照在台灣的賣價付給甲方 5%的權利金。

3.乙方同意於每月月底按照出貨數量用電匯將權利金付給甲方。

4.此合約於雙方簽字日算起，開始生效，有效一年。

5.根據共同信任原則，其他條款則按照國際條約。

甲方　　　　　　　　　乙方
ABC 公司　　　　　　　XYZ 公司

＿＿＿＿＿＿＿＿＿　　　＿＿＿＿＿＿＿＿＿

（簽名）　　　　　　　（簽名）

範例 1 申請當銷售代理(Sales Agent Application)

Dear Sirs,

Re: Application for a Sales Agent

We know from a friend (your advertisement) that you are looking for an Agent in Europe.

With more than 10 years' experience in international trade, we know very well the market situation here and have very good connections with many local buyers. Moreover, we have very good sales channels and many experienced sales personnel. So, we have confidence to promote very well your products in this area.

With regard to the commission, we only request 5% on each deal based on FOB price, which can be settled in the end of every month after shipment.

Please consider the above and advise your interest and terms and conditions for your Agent.

文章結構

主題：我們從朋友（廣告）處得知貴方正在尋找歐洲代理。

說明：(1)具有十多年國貿經驗，我們對此地市場非常了解且和許多國內買主關係良好，此外，我們有很好的銷售管道及經驗豐富的業務員，因此我們有信心可在此將貴方產品推銷得很好。

(2)有關佣金，我們只要求 FOB 價的 5%，可於出貨後月底結算。

結論：請考慮以上，並告知貴方的興趣及代理條款。

生字／片語

1. Europe 歐洲；experience 經驗；market situation 市場情況

2. good connection 關係良好；sales channel 銷售管道

範例 2　申請當買方代理（Buyer Agent）

Dear Sirs,

<u>Re: Application for a Buyer Agent</u>

We know from your advertisement that you are looking for a Buyer Agent to handle your business here in Taiwan, and feel very interested in applying for it.

Since 1986, we have had very long experience in handling the same business for many foreign buyers. As we know very well the manufacturers' situation here, and have very very good relationship with most of local suppliers, so we have confidence in getting very good prices and supports.

Besides, we can render very good and quick service to you, such as : to collect all information you want, source the best qualified suppliers for you, make sure shipments on time, inspect the goods before shipment, etc. In case any claim happens, we can try hard to get the fair compensation from the suppliers within the shortest time on your behalf.

Please take the above into consideration and don't hesitate to let us know your terms and conditions for further discussion.

Thanks for your attention and look forward to hearing from you soon.

Very truly yours,

MODERN BUSINESS ENGLISH
現代商用英文

文章結構

> 主題：我們從廣告上得知貴公司正在尋找一個買方代理來處理你們在台灣的生意，
> 我們很有興趣應徵此工作。
>
> 説明：(1)自從 1986 年以來，我們已有長期為國外買主處理相同業務的經驗。由於我
> 們非常了解國內工廠情形，且和許多當地的供應商關係良好，因此我們有
> 信心可拿到最好的價錢及支持。
>
> (2)此外，我們可提供非常好又快速的服務給貴公司，例如蒐集貴方想要的資
> 料、為貴方尋找最合格的供應商、確保準時出貨、出貨前去驗貨等。如有
> 任何抱怨發生，我們可以代表貴方於最短時間內爭取到合理的賠償。
>
> 結論：請考慮以上，並不要遲疑告知貴方條款，以便進一步討論。謝謝貴方的注意
> 並盼望很快聽到貴方的回音。

生字／片語

1. Application (N)，Apply (V)　申請
2. Buyer Agent　買方代理；Sales Agent　銷售代理
3. advertisement　廣告
4. manufacturers' situation　工廠情況
5. good relationship　良好的關係
6. local suppliers　國內供應商；have confidence in getting　有信心拿到
7. render　提供；service　服務
8. collect all information　蒐集所有資料
9. source the best qualified suppliers　尋找最合格的供應商
10. inspect the goods　驗貨；before shipment　出貨前
11. get the fair compensation　得到合理的賠償
12. on your behalf　代表貴方；further discussion　進一步討論

範例 3　進口代理要求目錄及訂購樣品供推銷展覽用

From: WoodChang@brightstar.com

To: orestangel@angelscientifica.com

Subject: Distribution Agreement

Dear Mr. Angel,

Thank you very much for your e-mail of Feb. 15, and the attached Distribution Agreement. Attached please find a copy with our signature for your file.

For the exhibition, we decided to order one set each of your newest model NFE-370 and Mini LG1 Blood Bank. Please send us your Proforma Invoice with your best prices and delivery time immediately.

We appreciate your support with the literature and posters which are to be put inside the above shipments. For complete promotion, we need at least 500 catalogs, please prepare them for us as early as possible. The freight cost to ship them can be at our account.

We are looking forward to hearing from you soon and expecting to increase our mutual business under your best support.

Yours Sincerely

Bright Star Enterprise Co.

Wood Chang/General Manager

文章結構

主題：感謝貴公司 2/15 來信及如附經銷代理合約，如附請發現我們簽回的一份供留檔。

說明：(1)為了展覽，我們決定訂購各一組貴公司最新型號NFE-370及迷你血庫LG1，請儘快寄來最好價錢、交期的預付發票。

(2)我們感謝你們支持送予說明書及海報，請和上述出貨一起裝出。為了全面推銷，我們需要至少 500 份目錄，請儘快準備，裝運運費由我方負擔。

結論：盼望儘快聽到貴方回音並期盼在貴方大力支持下能增加雙方的生意。

生字／片語

1. Distribution Agreement　經銷代理合約
2. signature (N)　簽名；for your file　供留檔
3. For the exhibition　為了展覽
4. Blood Bank　血庫
5. literature　文字說明書；posters　海報
6. For complete promotion　為了全面推銷
7. freight cost　運費；at our account　由我方負擔
8. mutual business　雙方共同的生意
9. under your best support　在貴方大力支持下

範例 4　工廠回覆範例 3

From: orestangel@angelscientifica.com

To: WoodChang@brightstar.com

Subject: Distribution Agreement

Dear Mr. Chang,

We received your mail of Feb. 16 and the signed Distribution Agreement with tnanks.

As to the samples of Chest Freezer and Blood Bank you ordered for the Taipei International Instruments Show, we enclose our Proforma Invoice showing you our best prices and delivery time for the Agent. The literature, catalogs and posters will be shipped with the above sample order.

We do hope under your strong promotion and our complete support, you will get a good business in Taiwan this year.

Best regards

Orest Angel

文章結構

主題：貴方 2/16 來信及簽回的代理合約收到，謝謝。

說明：有關台北國際儀器展用的大型冷凍器及血庫樣品訂購，我們附上給代理的最好價錢交期的預付發票。文字說明書、目錄與海報將與上述樣品一起裝出。

結論：希望在貴方大力推銷及我方全力支持下，你們今年可在台灣爭取到好的生意。

生字 / 片語

1. Chest Freezer　大型冷凍器；Blood Bank　血庫
2. Taipei International Instruments Show　台北國際儀器展

範例 5　工廠回覆應徵者其代理條件要求

To: Mr. Bob Patrice

From: Min-Ray Valve Co.

Dear Bob,

Thank you for your letter of Feb. 1 with interest in applying for being our Agent in Australia. Please be advised that so far we still have no any exclusive Agent in Australia except have 5-6 regular customers who place us orders regularly. The reason why we still have not granted any of them as our exclusive Agent, is because none of them can reach the min. sales amount we request.

We hope the exclusive Agent can place regular orders for at least US$ 50,000~US$60,000 per month, which is same as what our Agent in S. Africa does now.

So, the basic terms we would like to stipulate in the contract are as follows:

1. The trial period of the exclusive Agreement is one year.
2. During the trial period, you have to try your best to promote our Valves into your market, and we will not contact any other new customers except we have the right to keep our original old customers whose names will be specified on the contract.
3. After this trial period, if both feel satisfied, we can make a formal exclusive Agency Contract, and release all of our regular customers to you.
4. The min. Sales Amount is US$50,000 per month.
5. The commission or profit is to be calculated by yourself.

Please consider the above and let us know of your interest soon.

MODERN
BUSINESS
ENGLISH

Part 4 (十三、代理及合約)
商用英文信函寫法

441

文章結構

主題：謝謝貴方 2/1 來信，有興趣申請當我方澳洲代理。

說明：(1)請被告知目前我們除了有 5～6 個定期下訂單給我們的固定客戶外，在澳洲尚無獨家代理。我們沒有授權他們其中一家當我方代理的原因是他們都無法達到我們最低銷售額的要求。

(2)我們希望獨家代理每個月能下給我們至少美金 50,000～60,000 元的長期訂單，與我們目前南非代理相同的額度。

(3)因此，在此合約內，我們希望規定的基本條件如下：

①獨家代理合約的試驗期為一年。

②在試驗期間，貴方在貴國市場要盡力推銷我們的閥，而我們除了有權保留合約中規定的原來老客戶的聯絡權外，不再與任何其他新客戶聯絡。

③在試驗期後，如果雙方都滿意，我們可簽定正式代理合約，並將所有固定客戶轉讓給貴方。

④最低銷售額是每個月美金 50,000 元。

⑤佣金及利潤由貴方自行計算。

結論：請考慮以上並儘快讓我們知道貴方的興趣。

生字／片語

1. Australia　澳洲；exclusive Agent　獨家代理

2. regular customer　固定／長期客戶；regularly　定期地

3. grant　同意／授予；min. sales amount　最低銷售量

4. S. Africa　南非；basic terms　基本條件；formal　正式的

5. stipulate　規定；trial period　試驗期

6. Valves　閥；original old customers　原來的老客戶

7. release　釋放；profit　利潤；calculate　計算

範例 6　進口商要求產品上掛其品牌代理銷售

BUSINESS ASSOCIATES CO.

To: Samtech Computer Co.

Dear Sir / Madam:

Our company, Business Associates Co., has begun a new venture along with the distribution of main components, peripherals, and related products through its offices in the United States, and its affiliates in England, Saudi Arabia, Egypt and South America.

Our policy in this regard is to incorporate the most recent technological developments into our systems and the above products distributed on a continuing basis.

Kindly mail us your catalogues, price lists and sales terms. We would also appreciate your advising on the possibility of selling your products under our company's logo and name with a statement indicating that the product is manufactured by your company for Business Associates Co.

Your prompt response and mailing of your catalogues via DHL will be highly appreciated.

Looking forward to receiving your catalogues and reply.

Yours Faithfully,
Business Associates Co.

Andrew Moham
President

文章結構

主題：本公司，商業聯合公司，已經開始一個新的投資生意，在美國及其在英國，
沙烏地阿拉伯、埃及和南美的各分公司經銷主要的零組件，電腦周邊和相關
產品。

說明：(1)有關此方面，我們的政策是根據持續性合作，併入最新技術發展的產品到
我們的系統中並經銷上述產品。

(2)因此，請寄給我們貴公司的目錄，價目表及銷售條件，我們亦將感謝貴公
司能告知是否可用本公司商標及名義銷售貴方產品的可能性，在產品上將
註明製造商為貴方。

結論：貴方儘快的回覆及用快遞寄來貴方目錄將被十分感謝，盼望收到貴公司的目
錄及回覆。

生字／片語

1. venture　投資／冒險；joint venture　合資

2. distribution　經銷；components　零組件

3. peripherals　電腦周邊；related products　相關產品

4. the United States　美國；England　英國；Egypt　埃及；Saudi Arabia　沙烏地阿
拉伯；South America　南美

5. affiliates ＝ branch offices　分公司；system　系統

6. policy　政策；incorporate　併入

7. recent technological developments　最新技術發展產品

8. on a continuing basis　根據持續性合作

9. sales terms　銷售條件；possibility　可能性

10. under our company's logo and name　用我方商標及名義

11. statement　聲明；via DHL　用快遞

範例 7 進口商寫信給國外廠商爭取單項產品代理權

BULL MEDICAL IMPORTER

To: Shell Medical Co. Ltd.

Dear Sirs,

As informed, our company specializes in selling medical instruments and equipments here in Taiwan for more than four years and have very good sales network and relationship with local hospitals, universities and government institutions.

Since we have been selling your Blood Bank in Taiwan market during last few years and we find there is still a potentiality in this product. That's why we would like to ask you to grant us as your exclusive Agent here for only this Blood Bank.

Through the Exhibition held in Taipei this July, we would like to fully promote this product and have confidence to get more business for you. So, we would like to get your support in supplying us some samples and catalogs, and sign a Distribution Contract with us.

Please confirm your decision and e-mail your relevant terms and conditions for an exclusive Agent to us soon.

Best regards
Felix Wang

文章結構

主題：先自我介紹

　　　　如所告知，本公司專門在台銷售醫學器材，已有 4 年多經驗，並有很好的銷售網，同時和國內各醫院、大學及政府機構關係良好。

說明：(1)由於在過去幾年我們一直在台銷售貴公司的血庫，也發現此產品仍具潛力，那即是為何我們希望要求貴公司同意授予本公司此項產品的在台獨家代理權。

　　　　(2)經由今年七月在台北舉行的展覽，我們將全力推銷此產品，並有信心為貴公司爭取更多生意。因此，我們希望獲得貴公司提供樣品及目錄的支持並和我們簽訂一份經銷合約書。

結論：請確認你們的決定並儘快 e-mail 相關代理條款給我們。

生字／片語

1. As informed　如所告知
2. specializes in　專門於
3. medical instruments and equipments　醫學器材
4. sales network　銷售網
5. government institutions　政府機構
6. potentiality　潛能；grant　授予
7. fully promote　全力推銷

範例 8　工廠回覆範例 7，索取其銷售計畫書

From: ShellTatcho@shellmedical.com

To: FelixWang@bullmedical.com

Subject: Exclusive Agent for Blook Bank

Dear Mr. Wang,

Thank you for your e-mail of Jan. 29. We are interested in your proposal and would like to know more details about the followings:

1. What is the quantity you have already sold?
2. How large is the market you forecast in the future?
3. What are your sales program, sales method and sales channel?
4. Are you doing any other business besides Blood Bank?
   What are your other lines?
5. Please submit us your Company Profile showing your company organization and financial status.

When we receive your reply, we will be in a better position to make a decision about an Exclusive Agency for Taiwan.

We look forward to your reply.

Best regards

Shell Tatcho

President

文章結構

> 主題：謝謝貴公司 1/29 的 e-mail，我們很有興趣貴公司的提案，並希望詳細了解下
> 　　　列事項：
> 說明：1.貴方已銷售多少數量？
> 　　　2.貴方預測未來市場量有多大？
> 　　　3.貴方的銷售計畫、銷售方法及管道為何？
> 　　　4.除了血庫以外，貴方是否也做別項生意？其他項目是什麼？
> 　　　5.請提供貴公司簡介註明公司組織及財務狀況。
> 結論：當我們收到貴方回覆，我們才有較好立場決定台灣的獨家代理權，盼覆。

生字／片語

1. proposal　提案
2. forecast　預測
3. sales program　銷售計畫
4. sales method　銷售方法
5. sales channel　銷售管道
6. other lines　其他項目
7. submit = send　提供
8. Company Profile　公司簡介
9. company organization　公司組織
10. financial status　財務狀況
11. in a better position　有較好的立場

範例 9　工廠告知在當地已有代理，無法授權或銷售

From: Hiroshi@japanmedical.com

To: bertyen@lincon.com

Subject: Exclusive Agency

Dear Mr. Yen,

We are very sorry to inform you that we have just signed an Agreement with a Japanese domestic wholesaler, and in the Agreement, it states that territory includes Far East Market.

Therefore, it is impossible for us to grant you our exclusive Agency and to offer you the product from now on. The name and address of this Japanese domestic wholesaler is as follows:

      Susume Medical Inc.

      No. 2-4, Hongo Bldg.

      Tokyo, Japan

      Fax: 3-888-2626

We regret that we cannot be of assistance to you in this instance, but trust that you will get the best service from the above company.

With best regards

Hiroshi / Manager

文章結構

主題：很抱歉告知我們剛和一家日本國內大盤商簽約，在合約中，敘述版圖包括遠
　　　東市場。

說明：因此，我們不可能授予貴公司獨家代理權，也不能再報此產品的價錢給你們。

　　　此日本國內大盤商的名字及住址如下：

　　　　　　　　Susume Medical Inc.

　　　　　　　　No. 2-4, Hongo Bldg.

　　　　　　　　Tokyo, Japan

　　　　　　　　Fax:3-888-2626

結論：我們很抱歉在此事件上無法協助，但是相信上述公司定會提供貴公司最好的
　　　服務。

生字／片語

1. domestic　國內的；wholesaler　大盤商

2. states　敘述；territory　版圖

3. Far East Market　遠東市場

4. exclusive Agency　獨家代理權

5. from now on　從現在開始

6. assistance ＝ help　協助

7. instance　事例／事件

8. in this instance ＝ in this case ＝ in this matter　在此事件上

練習 24　代理及合約練習

1. 將下列句子翻譯成中文

(1) We are very interested in your Dynamo Radio MG-500 and would like to be your exclusive agent in Asia.

(2) We are an experienced company and we know very well the market situation. Besides, we have very good connections with many buyers and also have very good sales channels. So, we have confidence in promoting your product very well in this area.

(3) Please kindly advise us your interest in cooperating with us and let us know your terms and conditions for a sole agent.

(4) Regarding the sole agent agreement, it is our policy to sign a sole agent agreement only until the company can sell our products successfully over a period of time.

(5) We agree to grant you as our sole agent for a trial period of one year in order to get business moving.

2. 請依下列大意寫一封應徵當代理的信函

> 主題：我們從廣告上得知貴公司正在尋求代理，我們很有興趣想知道貴方的條件。
> 說明：具有多年銷售經驗，我們非常了解此地市場，且有很好的銷售網，因此我們有信心可將貴方產品在此地推銷得很好。
> 結論：先感謝貴方的考慮，並等候儘快回覆。

3. Write a reply letter to the above letter asking the applicant to provide more information about their sales programs, sales method and sales channel, etc. together with their Company Profile showing company organization and financial status for consideration.

Part 5

MODERN BUSINESS ENGLISH MODERN BUSINESS ENGLISH
MODERN BUSINESS ENGLISH MODERN BUSINESS ENGLISH

其他雜務聯絡
接洽信函

Miscellaneous Letters

除了商業流程中討論的信函之外，買賣雙方也會有一些雜項事務須討論，例如：

一、通知拜訪及回覆歡迎函(Notice of Visit)

二、邀請函(Invitation Letters)

三、搬遷新址通知函(Notice of Removal)

四、設立新分公司通知函(Notice of New Branch Office)

五、人員異動及更換新人通知函(Notice of New Employee)

六、請求預訂飯店及確認協助函(Hotel Reservation)

一、通知拜訪及回覆歡迎函(Notice of Visit)

範例 1　買方通知拜訪時間及欲討論事項

From: LynnLorsch@texas.com

To: FredLin@shunyuan.com

Subject: Our Visit in China

Dear Fred,

Chuck Brain and I will visit your factory in ShangHai on Feb. 1. The topics we would like to discuss with you during our visit are as follows:

Quality issues, test procedure, on-time delivery, cost down, marking and packing. Please let me know whether you can pick us up at the Pu-Dong airport or mail us the directions to go to your factory.

Thanks

Lynn Lorsch

文章結構

主題：Chuck Brain 與我將於 2/1 參觀貴廠。

說明：在我們拜訪期間，希望與貴公司商討的主題如下：品質問題、測試步驟、準時交期、成本降低、嘜頭及包裝等。

結論：請告知是否可到浦東機場接我們或告知前往貴廠的路線圖。

生字／片語

1. Chuck Brain　人名；topics　話題／主題

2. Quality issues　品質問題；test procedure　測試步驟

3. on-time delivery　準時交期；cost down　成本降低

4. marking and packing　嘜頭及包裝；Pu-Dong airport　浦東機場

範例 2 回覆範例 1，歡迎並確認拜訪時間没問題

From: FredLin@shunyuan.com

To: LynnLorsch@texas.com

Subject: Your Visit in China

Dear Lynn,

We are glad to know from your e-mail of 1/18 that Mr. Brain and you have fixed your visiting schedule and will visit our factory on Feb. 1. We sincerely welcome your visit and confirm the time is OK.

We will pick you up at the Pu-Dong airport and then drive you to our factory. All details for the topics you listed will be well prepared for discussion during your visit in our factory.

If there is any information you still want us to prepare in advance, please don't hesitate to advise us. Look forward to seeing you.

Best regards

Fred Lin

文章結構

主題：很高興由您 1/18 來函得知 Chuck Brain 與您已確定拜訪行程，將於 2/1 參觀本廠。

說明：我們誠意歡迎你們的拜訪並確認時間可以。我們將到浦東機場接你們並開車到我們工廠。你們所列的主題細節將會準備好於參觀本廠時討論。

結論：如有任何資料需要我們事先準備的話，請盡快告知。期盼與您見面。

生字／片語

1. have fixed your visiting schedule 已確定拜訪行程

2. sincerely welcome 誠意歡迎；confirm 確認；discussion 討論

3. prepare in advance 事先準備；don't hesitate ＝ soon 不要遲疑／盡快

二、邀請函(Invitation Letters)

範例 1　邀請國外技師來討論品質問題

To: Mr. Jorge Rodriguez Leal/WDM

Date: Jan. 22, 2012

INVITATION LETTER

We would like to invite your service engineer Mr. Eudelio Vallejo to come over here to Taiwan to check the problems of the Pumps we bought from your factory last month. We hope he can arrive here before Jan. 28, 2012 and it may need to take 7-10 days for the repairing.

We are awaiting his coming soon.

Very truly yours,

Merrybest International Co.

Mary Huang

V. President

文章結構

主題：邀請函
　　　我們希望邀請你們的售後服務工程師，Vallejo 先生到台灣來檢查我們上個月向貴廠採購的幫浦之問題。
說明：我們希望他能在 1/28 之前到達，大約需要 7～10 個工作天做維修工作。
結論：我們等候他儘快到訪。

生字／片語

1. Invitation Letter ＝ Letter of Invitation　邀請函
2. service engineer　售後服務工程師
3. Pumps　幫浦
4. we bought from your factory last month　（形容詞子句）我們上個月向貴廠採購的
5. for the repairing　為了維修工作

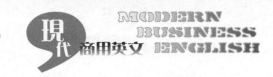

範例 2　邀請買主來參觀展覽攤位

Dear Sirs,

Invitation for Taipei Electronic Show

This is to inform you that our company will attend the Taipei Electronic Show held in Taipei World Trade Center during March 8-15. Our booth number is AB012.

You are mostly welcome to come over here to take a look. After the show, we can also have a further business discussion.

Please don't hesitate to let us know your flight details when you fix your visiting schedule, so that we can arrange the people to pick you up at the airport.

We are looking forward to seeing you soon.

Best regards

文章結構

主題：這是通知貴公司本公司將參加在 3/8～15 在台北世貿中心舉行的台北電子展，我們的攤位號碼是 AB012。

說明：我們熱烈歡迎貴公司來參觀，展出後，我們同時也可做進一步的生意討論。

結論：當您確定拜訪行程時，請不要遲疑讓我們知道您的班機明細，以便安排人員到機場接機。盼望很快見到您。

生字／片語

1. attend　參加；Taipei Electronic Show　台北電子展
2. World Trade Center (WTC)　世貿中心；booth=stand　攤位
3. Flight details　班機明細；fix　確定

三、搬遷新址通知函(Notice of Removal)

範例 1　泰國客戶通知遷新址

MESSAGE:

This is to inform you that effective from 1st of April. we are moving to the following new premise/building:

    New Address: Thailand Asia Representatives

                900/85 Phaholyothin Rd., Bangkok

                Phone no: 662-299-0202

                Fax no: 662-299-0203

                E-maill address: no change

Please update your file accordingly. Thank you.

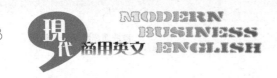

範例 2　通知客戶搬新址

NOTICE OF REMOVAL

Dear Sirs,

Please be informed that our company is going to move to the new and modern premises in the center of Taipei on and after Feb. 1. New address is as below:

> ABC Compamy
> Fl. 12, No. 5, Tun Hwa S. Road, Taipei, Taiwan, R.O.C.
> Phone number, Fax number and E-mail address are the same without change.

Please revise your record and send all your new correspondence to the above new address from that time.

Thanks and best regards

文章結構

主題：請被告知本公司將於 2 月 1 日起搬遷至台北市中心的一棟新的現代化建築物，
　　　新址如下：

説明：ABC 公司
　　　台北市敦化南路 5 號 12 樓
　　　電話、傳真號碼及 E-mail 不變

結論：請修正貴公司記錄，並於那時起將所有信件寄至上面新址，謝謝。

生字／片語

1. Notice of Removal　搬家通知
2. the new and modern premises　一棟新的現代化建築物
3. in the center of Taipei　台北市中心
4. Please revise your record ＝ Please update your file　請修正／更新貴方檔案

四、設立新分公司通知函(Notice of New Branch Office)

範例 1　通知在香港設立分公司

Dear Sirs,

We are glad to inform you that we have just set up a new Branch Office in Hong Kong on Jan. 1, this year to handle all of our shipments made in China.

Please kindly address your new inquiries or orders direct to our HK Branch Office for rendering you more efficient service.　The e-mail address and phone no are as below:

TACHA INTERNATIONAL TRADE

E-mail Address: tacha@yaboo.com

Phone No.: 852-1-7865432　Fax No.: 852-1-7865433

Attn: Mr. Ken Kin

文章結構

主題：我們很高興通知貴公司我們剛於今年元月 1 日在香港設立一個新分公司來處理大陸的出貨。

說明：請將新詢價及訂單直接發給我們香港分公司，以便提供更有效率的服務。

結論：新址及電話號碼如下：

生字／片語

1. set up　成立／設立
2. HK Branch Office　香港分公司
3. rendering　提供；efficient service　有效率的服務

五、人員異動及更換新人通知函(Notice of New Employee)

範例 1　以老闆名義發出的更換新人通知函

Dear Sirs,

Re: Notice of the New Employee

Please kindly be informed that Miss Betty Lee has just quit the job (has transferred to another Dept./has been assigned to other Dept.). Miss Jane Wang is assigned to take her place and will take care of your business from now on.

We will appreciate your best cooperation and support as usual.

Best regards
Hunter Lin/Manager

文章結構

主題：請被告知 Betty 李已離職（轉調其他部門）。

説明：Ms. Jane Wang 被指派接管她的工作並從現在起負責貴公司業務。

結論：我們感謝您和以往一樣的合作及支持。

生字／片語

1. quit the job ＝ resign　辭職
2. transfered to another Dept.　轉調其他部門
3. assigned　指派；take her place　代替她的職務

範例 2　接任新職者自行發函通知往來客戶

Re: Notice of the New Employee

Dear Sirs,

Please be informed that Miss Betty Lee has just quit her job and I am assigned to take over her job from now on.

It would be my great pleasure to handle all of your business and certainly I will do my best to serve you and meet your full satisfaction.

Your cooperation and support as usual will be highly appreciated.

Best regards
Jane Wang

文章結構

主題：請被告知 Betty 小姐已離職，而我被指派從現在開始接管她的工作。

說明：這將是我的榮幸來處理貴公司業務，當然我定將盡全力來服務貴公司並達到貴方完全的滿意。

結論：您和以前一樣的合作及支持將被極力感激。

生字 / 片語

1. meet your full satisfaction　達到貴方完全的滿意
2. take over her job　接管她的工作
3. from now on　從現在起
4. meet your full satisfaction　符合貴公司完全的滿意
5. as usual　如往常一樣

六、請求預訂飯店及確認協助函(Hotel Reservation)

範例 1　客人來電要求協助預訂飯店房間

Re: Hotel Reservation

Dear George,

We have just fixed our flight schedule to arrive at Taoyuan International Airport on Jan. 10, at 4:00 P.M.

Would you be kind enough to reserve two single rooms at the hotel near your office for us from Jan. 10 to 15.

Also, please ask the hotel to arrange the car to collect us at the airport.

Thank you for your assistance to the above and see you soon.

文章結構

主題：我們剛確定飛機行程，於 1/10 下午 4:00 抵達桃園國際機場。

說明：請協助在貴公司附近的飯店預訂兩間單人房，時間是 1/10～15。

結論：請要求飯店安排接機，謝謝協助並盼望很快與您見面。

生字 / 片語

1. Taoyuan International Airport　桃園國際機場
2. reserve　預訂；single rooms　單人房
3. collect us ＝ pick us up　接機

範例 2　回覆範例 1，告知客戶已訂好房間及安排接機

Re: Hotel Reservation

Dear Lynn,

We have reserved two single rooms for you and Mr. Brain at the Howard Plaza, where is just two blocks from our office, from Jan. 10-15 as requested.

We have also arranged the Hotel Bus to pick you up at the airport. We predict you may arrive Taipei at about 5:30 P.M., so we will await and meet you at the hotel for having dinner together.

Seeing you soon.
George

文章結構

主題：如所要求，我們已在離我們公司兩個路口的福華飯店，幫您與 Brain 先生預定了兩間單人房，時間從 1/10～15。

說明：我們也已安排飯店的車子去機場接機，我們預估你們大約於下午 5:30 到台北，因此我們將在飯店等候與你們見面並一起吃晚飯。

結論：盼很快見面。

生字 / 片語

1. Howard Plaza　福華飯店；two blocks　兩個路口
2. Hotel Bus　飯店的車子；predict　預估

Part 6

推薦／應徵函及
履歷表

Recommendation Letters/Application
Letters & Resumes

　　一般學校剛畢業的學生或想轉換工作的有經驗的工作者，都必須面臨求職應徵工作的程序。而大部分的求職者，都以從電腦網路及報紙上的徵人廣告優先尋找機會，也有從青輔會、學校或親朋好友各方面推薦而來。不論什麼方式得到的應徵機會，去各公司行號面試前，一定要先將中英文履歷表(Resume)及應徵函(Application Letter)準備好，如有推薦函(Recommendation Letter)及成績單(Transcript)也可一併附上供參考。

履歷表內容主要包括下列幾項：
(1)個人資料(Personal Information)：姓名／出生日／住址／電話
(2)家庭狀況(Family Situation)：簡單介紹家中成員
(3)教育背景(Educational Background)：以最後畢業的那個學校名稱及科系為主
(4)工作經驗(Working Experience)：可按年份順序列出公司名稱／職稱／工作性質；如為剛畢業的學生，如有暑期兼職工作經驗，也可列出；如果都沒經驗，則強調學校修過的類似課程，或自己的英語說寫能力
(5)其他語言能力(Other Language Ability)：例如日語、法語、德語等
(6)電腦軟體／打字／速記等能力(Software & other ability)
(7)興趣喜好(Hobbies)：游泳、登山、運動、旅遊、讀書、音樂等
(8)個性(Character)：強調自己優點部分
(9)證照(License)：例如 TOEIC 成績、英檢及國貿丙乙級技術士等證照

　　以上幾點可以列表方式或用書信分段寫出，重點是要強調自己的能力及給人自信，讓對方覺得你是最適當人選。

　　如為列表方式，前面最好附一封信函，告知是從哪裡得知此工作機會，並強調自己能力及表示對此職缺極感興趣，相信自己一定能盡力表現並勝任愉快。

範例 1　應徵函(Application Letter)——無工作經驗者

APPLICATION LETTER

ATTN: Personnel Dept.

Dear Sirs,

I am glad to know from the advertisement that you have a job vacancy for a Sales (an assistant/ secretary), and feel interested in applying for this position.

Enclosed please find my Resume for reference. As I am very interested in International Trade, I have studied very hard in all English courses such as English letter writing and English conversation, etc. Though I still have no any full-time working experience, I have confidence to take the job well.

I would highly appreciate it if you could grant me an opportunity of interview as well as a possibility to work in your esteemed company.

Thank you very much for your consideration and look forward to hearing from you soon.

Very truly yours,

July Lin

文章結構

> 主題：很高興從廣告上得知貴公司有一個業務職缺，很有興趣來應徵此職位。
>
> 說明：如附請發現我的履歷表供參考。由於本人很有興趣國際貿易，也很努力學習各種英文課程例如英文書信和會話等，雖然我尚無全職工作經驗，我仍有信心可勝任此職。
>
> 結論：我將非常感謝您給予我面談機會以及為貴公司服務的可能性。非常謝謝您的考慮，並盼望聽到您的回音。

生字 / 片語

1. advertisement　廣告
2. job vacancy　職缺
3. assistant　助理
4. Resume/Curriculum Vitae　履歷表
5. International Trade　國際貿易
6. English letter writing　英文書信
7. full-time working experience　全職工作經驗
8. opportunity　機會
9. interview　面談
10. possibility　可能性
11. esteemed company　令人尊敬的公司

範例 1 附件　履歷表（Resume）──剛畢業無經驗者

RESUME/CURRICULUM VITAE

Personal Information（個人資料）

English Name: July Lin

Birth Date: May 1, _____

Birth Place: Taipei, Taiwan

Address: No. 10, Sec. 2, Shin-Yi Road, Taipei, Taiwan

Phone No.: 02-27654321　Cell phone: 0988-222-333

E-mail address: july0800@yahoo.com.tw

License（證照）

TOEIC SCORE: 800

LICENSE OF GEPT（全民英檢）

LICENSE OF INTERNATIONAL TRADE ID（國貿業務丙級技術士技能檢定通過）

Educational Background（教育背景）

I just graduated from Fu-Jen University this year and my major was Business Management.

Language & Software Ability（語言及軟體能力）

Beside English, I can also speak and write a little Japanese.

I can operate Window XP, Word, and E-mail.

Working Experience（工作經驗）

I have no full-time working experience as I just graduated from school this year. However, I had part-time job for the bank during summer vacation.

Hobbies（興趣喜好）

I like swimming, mountain climbing, going to the movies, playing tennis, traveling and listening to the music.

Character（個性）

I am a responsible person with active, easy going and open minded personality.

MODERN
BUSINESS
現代 商用英文 ENGLISH

範例 2　應徵函（Application Letter）——有工作經驗者

APPLICATION LETTER

ATTN: Personnel Dept.

Dear Sirs,

It is my great pleasure to know from a friend that your company is looking for a Sales working for Export Dept. I am very interested in applying for this job vacancy.

As I have more than two years' experience working in a trading company, and I am also interested very much in English, I have confidence to take this job well.

Please take my enclosed Resume for reference and kindly grant me an opportunity of interview. If there is a possibility to work in your respective company, I will certainly work very hard to prove my ability.

Your consideration to the above is highly appreciated and looking forward to hearing from you very soon.

Sincerely yours,

David Wu

文章結構

主題：這是本人榮幸能由朋友處得知貴公司正在尋找外銷部門的業務，我很有興趣
　　　應徵此職缺。

說明：(1)由於本人在貿易公司工作有兩年多經驗，加上本身對英文非常喜好，因此
　　　　我很有信心將此工作做得很好。

　　　(2)請參考如附件的個人履歷表，並能給予面談機會。如有可能為貴公司服務，
　　　　定當努力表現以證明個人的能力。

結論：您對以上的考慮將被大力感謝，並期盼貴方回音。

生字／片語

1. Export Dept.　外銷部門；trading company　貿易公司

2. respective company＝esteemed company　令人尊敬的公司

3. to prove my ability　以證明我的能力

範例 2 附件　履歷表（Resume）──已畢業有經驗者

RESUME/CURRICULUM VITAE

Personal Information（個人資料）

English Name: David Wu

Birth Date: May 1, ____

Birth Place: Taipei, Taiwan

Address: No. 20, Sec. 2, Tun Hwa S. Road, Taipei

Phone No.: 27776666

E-mail Address: david@yahoo.com

License（證照）

TOEIC SCORE: 800

LICENSE OF GEPT（全民英檢）

LICENSE OF INTERNATIONAL TRADE ID（國貿業務丙級技術士技能檢定通過）

Educational Background（教育背景）

2006～2008　University of Taiwan, MBA

2002～2006　National Taipei College of Business, International Trade

Language & Software Ability（語言及軟體能力）

Beside English, I can also speak and write Spanish(German, French). I can operate Window XP, Word, and E-mail.

Working Experience（工作經驗）

2008～2010　　ABC Company as a Sales Assistant

2010～2012　　XYZ Company as an Export Sales

Hobbies（興趣喜好）

I like swimming, mountain climbing, going to the movies, playing tennis(volleyball, badminton), traveling and listening to the music.

Character（個性）

I am an active, outgoing and understanding person with good sense of responsibility.

範例 3　學生應徵外國公司實習機會

Friedhelm ENSLIN　　GieBstraBe 3　　7200 Tuttlingen　　W-Germany

Tel: 0751-53584

07461-2282

Fax: 07461-78958

Application for a second internship,

Physical Techniques-Engineering

To whom it may concern,

As a student, in the fourth semester, at Weingarten Fachhochschule of Polytechnic (similar to an Institute of Technology), I must complete a practical semester of the second internship, over a six months period. To begin from this March.

Physical techniques-engineering combines Physics and Engineering. In the seventh semester I plan to concentrate my studies in the area of measurement techniques and scientific instruments. My other interests are aviation technology and optics.

I would like to have the opportunity to work in a foreign country, with its different economy, industry and culture, for my personal and professional development. Your company would give me a good foundation to complete my practical course successfully. I would be pleased to get the chance to prove my abilities in your company.

Please find my formal application form and Curriculum vitae enclosed. Thank you for taking the time to read my letter. I hope to hear from you soon.

Yours faithfully

Friedhelm Enslin

文章結構

主題：作為一個 Weingarten 技術學院第四學期的學生，我必須完成一個第二階段為期半年的實習課程，由今年 3 月開始。

說明：物理技術工程包括物理和工程兩方面，在第七學期，我計劃專攻測量技術和科學儀器方面，我其他興趣是飛行技術和光學。為了個人及專業發展原因，我希望有機會在一個不同經濟、工業和文化的外國公司工作。貴公司將可提供我一個好的環境，成功完成我的實習課程，我將很榮幸能有機會在貴公司證明我的能力。

結論：請發現我如附的履歷表及推薦函，以為正式申請。感謝您花時間閱讀我的信，並希望很快聽到您的消息。

生字 / 片語

1. semester　學期
2. Institute of Technology　技術學院
3. complete　完成
4. Practical semester　實習學期
5. internship　實習課
6. Physical techniques-engineering　物理技術工程
7. concerntrate　專心於
8. measurement techniques　測量技術
9. scientific instruments　科學儀器
10. different economy, industry and culture　不同的經濟、工業和文化
11. Curriculum Vitae ＝ Resume　個人履歷表
12. aviation technology and optics　飛行技術和光學

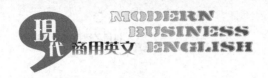

範例 3 附件　教授推薦函

FACHHOCHSCHULE RAVENSBURG-WEINGARTEN

Dipl.-Math. Roland Zurawka   March 13, _____

Recommendation for Friedhelm Enslin

student at the Ravensburg-College of polytechnic

Friedhelm Enslin is a student in the department of physics engineering at the polytechnic Ravensburg-Weingarten.

I know Mr. Enslin personally from lectures and exercises in Mathematics. He attended the events regularly and achieved a very good mark in the examination Mathematics 1.

Besides, he belongs to the small group of students, who collaborates actively the class of mathematics.

Mr. Enslin is open-minded and has good relationship to his fellow students and his professors. I think, that he will profit a lot if he works in a foreign country.

(Dipl.-Math. Roland Zurawka)

文章結構

主題：Friedhelm Enslin 是一個 Weingarten 技術學院物理工程系的學生。

說明：我是在數學演講實習課親自認識他，他定時出席上課，並且在數學考試達到
　　　很好的成績。此外他積極參與數學活動的一個小團體。
　　　Enslin 先生是一個接受意見的人，和同學朋友及教授相處良好。

結論：我認為如果他在外國公司工作將極有益。

生字／片語

1. personally　親自地；lectures　演講；exercises　實習；Mathematics　數學
2. achieved a very good mark　達到良好成績
3. collaborates　共同合作
4. open-minded　接受意見的人

範例 4　公司推薦函（國外研究所進修用）

To:UCLA　　　　　　　　　　　　Date:

Attn:Chairman of Computer Graduate Institute

Dear Sirs,

　　It is my real pleasure to recommend Mr. Wen-Chieh Jack Lee, a very able young man with a strong sense of responsibility and great enthusiasm for helping others, for graduate study at your esteemed institution.

　　Mr. Lee has worked as an Electronic Engineer at Computer department of Samwell International Inc. (Samwell Group), one of the leading companies here in Taiwan doing trading and manufacturing in Electronic Components, Electronic Assemblies, Computers and Microprocessing Devices, for more than two years. During the period of his service with our company, he has exhibited an outstanding capacity for hardwork and whole-hearted devotion to his job such as setting up the automatic production line, testing instrument system, and computer management system, etc.. As his superior, I appreciate very much his initiative and aggressiveness in duty performance and great help to our company.

　　In view of his previous achievements in his working and good command of the English language, I believe that he will be able to make an outstanding graduate student at your institution.

Although his resignation will be a great loss of our company, I still would strongly recommend his admission to your institution for further study, and will highly appreciate your favorable consideration of his application.

　　　　　　　　　　　　　　　　　　Very truly yours
　　　　　　　　　　　　　　　　　　Samwell International Inc.
　　　　　　　　　　　　　　　　　　Rita Min
　　　　　　　　　　　　　　　　　　Manager

文章結構　中文翻譯

To：加州大學　　　　　　　　　　　　日期：

收件人：電腦研究所所長

本人很榮幸能推薦李傑克先生，一位極有責任感及樂於助人的年青人，到貴所進修。

李先生在台灣主要的電子電腦零組件裝配廠之一的Samwell集團任職兩年多的電子工程師，在他任職期間，他表現傑出能力，努力工作並全心致力於他的工作例如建立自動化生產線、測試儀器系統及電腦管理系統等。做為他的上司，我很感謝他在工作表現上的進取心及對本公司的幫忙。

考慮到他之前的工作上的成就及良好的英文能力，我相信他在貴所將會是一位傑出的研究生。

雖然他的離職是本公司的一個大損失，我仍極力推薦他能進入貴所做進一步研習，並感謝您考慮他的申請。

範例 5　教授推薦函（國外研究所進修用）

TO: UCLA

Attention: Chairman of Computer Research Institute

Date: _____

RE: Letter of Recommendation

Dear Sirs,

I am very glad to know Mr. Wen Chieh Jack Lee, one of my favorite students, is going to apply for admission to your institution for further education.

During the period of Mr. Lee's studying in my department, I was deeply impressed with his deep thinking, good reaction, great English ability both in speaking and writing, and intellectual ability especially in computer field for which he had won the Award for his creative Intelligence Encirclement Chess in a computer software competition.

Moreover, he is a hard-working, sincere and ambitious young man with an eager desire for knowledge. During his study, he always had his work well-prepared before class, and occasionally asked some questions which were always stimulating and sensible.

In view of his performance in the school, I am firmly convinced that he will make a successful graduate student in the field of Computer Department.

Sincerely Yours,

××××（name)
Chairman of Chemical Engineering Dept.

文章結構　中文翻譯

To：加州大學

收件人：電腦研究所所長

日期：＿＿＿＿＿＿

主題：推薦函

我很高興我最喜愛的學生之一，李傑克正要申請貴所的入學許可做進一步研習。

在李先生在我系上研讀期間，我對他的深思、反應靈敏、良好的英文說寫能力以及他曾以智慧圍棋獲得電腦軟體比賽獎之傑出表現印象深刻。

此外，他是一位工作努力，有誠意和野心的年青人。在他讀書期間，他總是於課前預習，並於課堂中問些敏感激勵的問題。

參照他在學校的表現，我深信他在電腦研究所將會是一位成功的研究生。

化工系系主任

範例 6　教授推薦函（國外研究所進修用）

To:╳╳╳╳╳　University　　　　　　　Date: Jan. 10, _____

Attn: International Office (Department)

Subject: Letter of Recommendation

Dear Madams and Sirs,

　　It is my great pleasure to strongly recommend one of my favorite students, Lu Chi-Hung, to apply for the admission to your institute for further education.

　　Mr. Lu, one of the students I directed, was an excellent student when studying in my class of Japanese Political Research. He not only was the best student who got the highest grade in my class, but also always kept his average grade in the rank of top one in every semester during two years' study in his master programs in our institute. It proves that he possesses deep foundation in all respects of the Japanese researches.

　　I was deeply impressed with his deep thinking, good analytic ability, great speaking ability and the good potentiality in academic researches as he always actively participated the discussion by asking questions and bringing up his idea during his study in my class.

　　With excellent language ability in both Japanese and English, he often assisted his classmates to settle all kinds of problems in study. He is also a very independent person with very good performance in coordination. So, I believe he would be able to learn the knowledge efficiently by his good language ability when studying abroad.

　　Moreover, Mr. Lu also possesses more life experience and working experience than any other regular students as he had lots of experiences in communicating with Japanese during his previous works. Since he has abundant knowledge of Japanese culture, I believe that he will be able to be an outstanding student and have more outstanding performance and achievement in academic research than any others.

　　In view of his good performance mentioned above, I am firmly convinced that Mr. Lu will be a successful student if he has the chance to continue his further study in doctor's programs under a good research environment in your institute. Your favorable consideration will be highly appreciated.

Very truly yours.

Chinese Culture University

Dr. Chen Peng-Jen (Doctor of Tokyo University, Japan)

Dean of Japanese Graduate Institute

文章結構　中文翻譯

能強力推薦我最喜愛的學生之一，呂先生，來申請貴所的進修，這是我極大的榮幸。

呂先生是我指導的學生之一，當其在我日文政治研究的課程學習時，是一位極為優秀的學生。他不但是我班上成績最好的學生，而且在其兩年的研究所碩士班課程中，平均每學期都是保持第一名。這證明他在各方面的日本研究皆具有很深的基礎。

我對他的深度思考，良好的分析能力，極佳的語言及溝通能力及學術研究的潛力印象深刻，因為他在我課堂上總是積極參與討論，並提出問題及其看法。

因其具有極佳英語和日語的語言能力，他也經常協助他的同學解決各種學習上的問題。他也是一位非常獨立且具有良好協調能力的人，所以我相信以其語言能力，當其在國外進修時，必能迅速學習新的知識。

此外，呂先生也具有比其他一般學生更豐富的生活及工作經驗，因為在他過去的工作中，他曾有許多和日本人溝通的經驗。因其具有豐富的日本文化知識，我相信他在學術研究方面，將會比其他研究生有更為傑出的表現和成就。

考慮以上所述的傑出表現，我堅信呂先生如有機會能進入貴所良好的研究環境中，繼續其博士課程的進修，他將會是一位成功的研究生。感謝您的贊同考慮。

範例 7　申請研究所的讀書計劃

To:×××　　　　　　　　　　University, Japanese Dept.

Research Proposal

I was born in 1959 in Taiwan and immigrated to Japan with my family when I was three years old as my father went to study in the Tokyo University at that time. I had lived in Japan for eight years that made me possess good language ability in communicating with Japanese. I chose to major in German in the university not only because I was interested in the western culture but also because I wanted to possess one more kind of foreign language ability. Although my school report in German Department was not as good as that in my Japanese graduate school, I constantly keep on studying German after graduation.

After finishing my military service, I went to the Tokyo University in Japan to study Music in the Aesthetics(Artistic) Department. However, due to the unforeseen happenings in my family, I came back to Taiwan to start my career and have been working in the companies that all are related to Japan for more than ten years. Two years ago, I decided to transfer to the academia and then entered the Japanese Research Institute in Chinese Culture University for my master's degree.

The title of my master's thesis is "The Study of the Reception of Western Aesthetic (Artistic) Thought by Modern Japanese." -- I mainly studied how the Japanese accepted modern western aesthetic thought, and also involved that how the simultaneous Chinese accepted the modern western aesthetic thought through Japanese. The purpose that I chose this subject was trying to connect the Japanese aesthetics that I learned before in Japan with the western aesthetics. I tried to study how the modern aesthetic thought was transferred into Japan and how to influence Japanese aesthetics to be modernized and westernized.

We know that many Japanese scholars studied how the Japan got westernized during Meiji period. In addition, there were also some Chinese scholars in China studied that how the China contacted with the western culture through Japan, especially how the Chinese obtained the modern technology, for which Japan did play an important role.

With regard to the process that Japanese received western aesthetic thought during Meiji period, there are a large number of academic works studied and published only in the recent years in Japan. However, there are not so many scholars studying in the field that how the Chinese received the western aesthetic thought through Japan, "the passage" In my master's thesis, I tried to compare the similarities and the differences between China and Japan in receiving the western aesthetic thought, and I discovered the whole process about how the Japanese received the western aesthetic thought during Meiji period. I hope to continue my study in this field and do further deep discussion in my doctor's thesis if I have the chance to get into your graduate institute for further study.

I hope to understand more about the western culture and to see the process that Japanese absorbed the western aesthetic thought in Meiji period from different view point from other scholars studying in Japan. That's also the reason why I hope to have the chance to continue my further study in your institute rather than to study in Japan. I will study more deeply about how the European culture was spread and absorbed through other Japanese information and British information in my doctor's degree, as I believe that this kind of academic research would be more valuable and different from the Japanese study done in Taiwan.

If possible, I also like to do more deep study in comparison between China's westernization and Japan's westernization, especially emphasize the reception of modern western aesthetics. I realize that your institute is a well-known and esteemed British university only accepting the excellent students from all over the world. I have confidence to be an outstanding and successful graduate student in the filed of Japanese study if I am granted the chance to study in your institute. Thank you for your favorable consideration for my application.

Very truly yours,

Lu Chi-Hung

文章結構　中文翻譯

　　1959 年我出生於台灣。當我三歲時，由於父親留學東京大學，全家移居日本八年，因而具有近乎於日本人的日語能力。大學時為了能夠多學習一種外語，並對西方文化產生興趣，因此主修德文。雖然當時沒有留下像在日本研究所那樣優異的成績，但是我在畢業後仍然不斷地進修德文。

　　退伍後，我前往東京大學美學系攻讀音樂美學，後因家庭變故，回國開始工作。這十多年都服務於與日本相關的公司。兩年前決定轉往學術界，因此進入中國文化大學日本研究所攻讀碩士學位。碩士論文為「近代日本接受西方藝術思想的研究」，主要是探討明治時代日本如何接受近代西洋藝術思想，並同時涉及當時中國如何透過日本間接接受西洋藝術思想。我之所以訂定這個題目，其目的是希望將以往在日本涉及的美學與跨文化研究加以結合，試圖探索近代西洋藝術思想如何傳入日本，並且影響當時日本藝術的近代化、西化。

　　我們知道很多日本學者研究過明治時代日本是如何西化。另外，也有一些中國大陸學者針對中國大陸透過日本接觸西洋文化，尤其在現代技術的獲取上，日本扮演了重要角色，在這方面他們也作了很多的研究。

　　然而，針對明治時代日本所受到的西洋藝術思想的過程，則是到了近幾年來日本學界才開始有了大量的專著，而中國經過日本這個「通道」吸收西洋藝術思想的這個研究範圍，則中國學者較少討論。我在碩士論文上，就試著比較中國與日本對西洋藝術上接受的異同做了一個比較，並且對於明治時代日本的西洋藝術思想過程做了整體的掌握。

　　由於碩士論文的題目設定上顯得過寬以及本身在能力以及時間受到限制，雖然在探討明治日本吸收西洋藝術思想上做了整體的探討，但是在內容的深度上尚有需要加強的部分。因而我必須將此留待博士論文上再予以加深以及更進一步的論述。

　　本人希望能有機會進入貴所，繼續研究我在碩士階段尚未周全的部分。我選擇申請貴所而不是日本的研究所，其原因是希望能強化對西方文化的了解，從不同於留學者角度來看明治時代日本在西洋藝術思想的吸收過程。我希望今後能夠透過其他日本資料和英國的資料，將歐洲文化的傳播和吸收方面做更為深入的研究，以使我有不同的見解，而使這方面的研究在台灣的日本研究界能有與眾不同的學術價值。

　　另外，如果可能，我也盼望能在中國與日本的西化之比較上再做更深入的研究，尤其著重於近代西洋藝術的接受部分。深知貴校是名望崇高的英國大學，只有接受全世界最為優異的學生，因此本人深信具有在貴所繼續從事日本研究的能力與決心，因而申請了貴所。

國際貿易實用參考附表

Appendices

為了便於參考查閱，在此附上下列幾項附表：

一、國際貿易常用專有縮寫名詞
　　(Noun Phrases Abbreviations for International Trade)

二、國際貿易電報常用縮寫字
　　(Abbreviations)

三、國際貿易流程及主要進出口流程表
　　(Flow Chart of International Trade)

四、國際及國內航空公司英文代號及訂位電話

五、美國各州州名及縮寫

六、世界主要貿易國家幣別

七、中國大陸及全球主要城市英文名稱及城市代碼
　　(English Names and Codes of the Main Cities in China and Global Countries)

八、中文拼音及羅馬拼音對照表

九、何謂國際貿易
　　(International Trade Business)

十、國際貿易銷售管道及供應鏈中的相關行業
　　(Supply Chain in Trade Business)

十一、國際性的貿易規約及模式
　　　(International Trade Conventions)

十二、貨櫃材積表／國際尺碼換算表／溫度換算表

十三、品質條件、數量單位、包裝種類及出貨嘜頭補充說明
　　　(Additional Remarks of Quality Terms, Quantity Units, Packing Methods & Shipping
　　　Mark)

十四、進出口保險運作流程

十五、進口貨物通關流程簡介

十六、出口貨物通關流程簡介

十七、國貿業務丙級學科基礎貿易英文題庫

十八、ECFA 整體經濟效益／ECFA 項目及進口稅率

一、國際貿易常用專有縮寫名詞
(Noun Phrases Abbreviations for International Trade)

1. 保險常用縮寫名詞（Insurance Terms）

A.R.　All Risk　全險(= ICC A)－通常會外加兵險及罷工險(Strike)

F.P.A.　Free of Particular Average　平安險(= ICC C)

S.R.C.C.　Strike Riot Civil Commotion　罷工暴動內亂險

T.P.N.D.　Theft, Pilferage and Non-Delivery　竊盜未能交貨險

W.R.　War Risk　兵險／戰爭險

WPA　With Particular Average　水漬險(FPA + WPA = ICC B)

2. 價錢條件／國貿條規縮寫名詞（Price Terms）－INCOTERMS 2010

Rule I：任何或多種運送方式的規則（7 種）

EXW　Ex-Works／Ex-Factory 工廠交貨條件規則

FCA　Free Carrier 貨交運送人條件規則

CPT　Carriage Paid To 運費付訖條件規則

CIP　Carriage, Insurance Paid To 運保費付訖條件規則

DAT　Delivered At Terminal 終點站交貨條件規則

DAP　Delivered At Place 目的地交貨條件規則

DDP　Delivered Duty Paid 稅訖交貨條件規則

Rule II：海運及內陸水路運送的規則（4 種）

FAS　Free Alongside Ship 船邊交貨條件規則

FOB　Free on Board 船上交貨價條件規則

CFR　Cost and Freight 運費在內條件規則

CIF　Cost, Insurance and Freight 運保費在內條件規則

3. 付款條件縮寫字（Payment Terms）

L/C　Letter of Credit　信用狀

COD　Cash on Delivery　交貨付現

CWO　Cash with Order　下單付現

B/C　Bill for Collection　託收票據

D/P	Document against Payment	付款交單
D/A	Document against Acceptance	承兌交單
O/A	Open Account	記帳／月結
D/N	Debit Note	應收帳單
C/N	Credit Note	應付帳單

4. 付款方法縮寫字（Payment Method）

T/T	Telegraphic (Telegram/Cable)Transfer	電匯
M/T	Mail Transfer	信匯
D/D	Demand Draft	即期匯票

5. 裝船常用縮寫字

CY	Container Yard	整櫃
FCL	Full Container Load	整櫃
CFS	Container Freight Station	併櫃
LCL	Less than container load	併櫃
ETD	Estimated time of departure	預計離港日
ETA	Estimated times of arrival	預計到港日
NW	Net weight	淨重
GW	Gross weight	毛重
D/O	Delivery Order	交貨單（小提單）
S/O	Shipping Order	裝貨單
S. S.	Steam Ship	汽船

6. 合約常用縮寫字

P/O	Purchase Order	訂單
P/I	Proforma Invoice	預付發票／預約發票
S/C	Sales Contract / Sales Confirmation	銷售合約／銷售確認書

7. 其他常見縮寫字

I/P	Insurance Policy	保單
AWB	Air Way Bill	空運提單（MAWB 主提單／ HAWB 子提單）
B/L	Bill of Lading	海運提單

L/I　　Letter of Indemnity　賠償保證書

L/G　　Letter of Guarantee　保證書

L/A　　Letter of Authority　授權書

I/L　　Import License　輸入許可證(= I/P Import Permit)

E/L　　Export License　輸出許可證(= E/P Export Permit)

AQL　　Acceptable Quality Level　可接受品質標準

OEM　　Original Equipment Manufacturing 原廠委託製造

ODM　　Original Designed Manufacturing 原廠委託設計

3C　　　Capacity, Character, Capital 能力／特性／資本額

C1, C2, C3　　Channel 1 免審免驗；C2 文件審核；C3 貨物審核

T/V　　關貿網路簡稱

8. 船運

CY/CY 整裝／整拆

CY/CFS 整裝／併拆

CFS/CY 併裝／整拆

CFS/CFS 併裝／併拆

F.I. (free in)裝貨船方免責（裝貨費用由貨方負擔）

F.O./ F.D.(free out/ free discharge) 卸貨船方免責（卸貨費用由貨方負擔）

F.I.O.(free in and out) 裝卸貨船方免責（裝卸貨費用由貨方負擔）

F.I.O.S.T.(free in／out／stowed／trimming)裝卸、堆積及平艙船方免責（以上費用由貨方負擔）

Berth Term 裝卸貨船方負責條件（裝卸貨費用由船方負擔）

Multimodal Transport 複合運送

Shipping conference 航運同盟

Demurrage and dispatch 延滯費與快速費

C.Q.D. (customary quick dispatch)習慣速度裝卸

CFT Cubic Feet 立方英呎（才）（1CFT = 1728 立方吋）

CBM Cubic Meter 立方公尺 (1CBM = 35.315 CFT)

TEC 20 呎貨櫃簡稱

二、國際貿易電報常用縮寫字(Abbreviations)

國際貿易電報上常用的縮寫字的原則如下：

1. 去掉字母中的母音字(a, e, i, o, u)
   但是如果母音字在第一個字母時，則不可省
   例：PLEASE = PLS, ORDER = ORDR

2. 取前三個字母：星期／月份
   例：MONDAY = MON, JANUARY = JAN

3. 字尾 ED 用 D 代替；ING = G；MENT = MT
   例：RECEIVED = RCVD, SHIPPING = SHPG
   PAYMENT = PAYT, SHIPMENT = SHPT

4. 保留前面三／四個字母或前後重要字母
   例：DEPARTMENT = DEPT, COMMISSION = COMM
   MANAGER = MGR, MINIMUM = MIN

5. 取同音字
   例：YOU = U, ARE = R, YOUR = UR, BEFORE = B4
   I OWE YOU = IOU（我欠你）

6. 航空公司代碼縮寫，參附表四

7. 美國州名縮寫，參附表五

8. 一般貿易常用縮寫字（照字母順序）
   ABOUT = ABT（大約）
   ABOVE MENTIONED = A.M.（上述）
   ACCEPT = ACPT（接受）
   ACCOUNT = ACCT (A/C)（帳戶）
   ADVISE = ADV（告知）
   AFTER RECEIVING THE ORDER = ARO（收到訂單後）
   ALREADY = ALDY（已經）

AMOUNT = AMT（金額）

ANSWER = ANS（回答）

ARE = R（是）

ARRIVED = ARVD（到達）

ARTICLE = ART（東西）

APPROXIMATE = APPROX（大約）

AS FOLLOWS = A/F (AS FLW)（如下）

AS SOON AS POSSIBLE = ASAP（儘快）

ATTENTION = ATTN（收件人）

BALANCE = BAL（剩餘）

BEST REGARDS = BST RDGS（致意）

BETWEEN = BTWN（之間）

BETTER = BTR（較好）

CONFIRM = CFM（確認＝ V）

CONFIRMATION = CFMN（確認＝ N）

COMPANY = CO（公司）

CARE OF = C/O（轉交）

CARTON = CTN（紙箱）

COMMISSION = COMM（佣金）

CONTAINER = CTNR（貨櫃）

CARBON COPY = CC（副本抄送）

CATALOG = CAT（目錄）

CENTIMETER = CM（公分）

DELIVERY = DEL（交期）

DEPARTMENT = DEPT（部門）

DESTINATION = DESTN（目的地）

DIFFERENT = DIFF（不同）

DOCUMENTS = DOCU（文件）

DOLLARS = DLRS（元）

DOZEN ＝ DZ（打）

DRAWING ＝ DWG（圖面）

EACH ＝ EA（每個）

EXPORT ＝ EXP（出口）

FLIGHT ＝ FLT（班機）

FOLLOWING ＝ FLWG（下列的）

FOOT/FEET ＝ FT（英呎）

FOR EXAMPLE ＝ E.G.（例如）

FREIGHT ＝ FRT（運費）

FROM ＝ FM（從）

GOVERNMENT ＝ GOVMT（政府）

HAVE ＝ HV（有）

HIGH TECHNOLOGY ＝ HI-TECH（高科技）

HORSE POWER ＝ HP（馬力）

HOWEVER ＝ HWVR（然而）

INCLUDE ＝ INCL（包括）

INCORPORATION ＝ INC（公司）

INFORM ＝ INF（通知）

IMPORT ＝ IMP（進口）

IMMEDIATELY ＝ IMMD（馬上）

INQUIRY ＝ INQ（詢問）

INVOICE ＝ INV（發票）

INSPECTION ＝ INSPTN（檢驗）

KILOMETER ＝ KM（公里）

KILO ＝ KG（公斤）

LEAVE ＝ LV（離開）

LETTER ＝ LTR（信）

MANUFACTURER = MFR（工廠）

MAJOR = MAJ（主要的）

MEMORANDUM = MEMO（留言條）

MESSAGE = MSG（留言）

MINOR = MIN（次要的）MINIMUM = MIN（最小／最少）

METRIC TON = M/T（公噸）

MAXIMUM = MAX（最大）

NEGOTIATE = NEGO（交涉）

NEW TAIWAN DOLLAR = NTD（新台幣）

ORDER = ORD（訂單）

ORDER NUMBER = O/N（訂單號碼）

PACKING = PKG（包裝）

PACKING LIST = P/L（包裝表）

PAIR = PR（雙）

PAYMENT = PAYT（付款）

PIECE = PC（個－單數）

PIECES = PCS（個－複數）

PLEASE = PLS（請）

POST SCRIPT = PS（附註）

POUND = LB（磅）

PRICE LIST = P/L, PRC LST（價目表）

QUALITY = QLTY（品質）

QUANTITY = QTY（數量）

QUOTATION = QTN（報價單）

REFERRING = RE（有關）

RECEIVED = RCVD（收到）

REPRESENTATIVE = REP（代表）

SAMPLE = SMPL（樣品）

SCHEDULE = SCHDL（交期）

SHIPPING = SHPG（出貨的）

SHIPPED = SHPD（裝出）

SHIPMENT = SHPT（出貨）

SHOULD BE = S/B（應該）

SPECIFICATION = SPEC（規格）

SQUARE = SQ（平方）

THANKS = TKS（謝謝）

THROUGH = THRU（通過）

TODAY = TDY（今天）

TOMORROW = TOMO (TMRW)（明天）

TOTAL = TTL（總共）

URGENT = UGT（緊急）

UNIT PRICE = U/P（單價）

VERY IMPORTANT PERSON = VIP（重要人物）

WILL = WL（將）

WITH = WZ（和…）

WITHIN = W/I（在…內）

WITHOUT = W/O（沒有）

三、國際貿易流程及主要進出口流程表
(Flow Chart of International Trade)

1. 市場開發／推銷(Promotion/Marketing)

   (1) 選定產品：不同市場，不同產品需求，不同設計樣式顏色喜好。

   (2) 市場調查

   ①法令限制調查：此產品有沒有被管制禁止進口、有沒有被告傾銷(dumping)、有沒有配額(quota)問題。

   ②進口商付款能力調查：有沒有外匯管制問題。

   ③地理環境調查：交通運輸港口情形、氣候（溫度溼度雨量）、人口、語言、教育、宗教風俗習慣（對顏色、材料、設計有無特殊禁忌）。

   ④個人所得和生活水準等調查。

   (3) 準備資料

   ①公司目錄(catalogue)

   ②公司簡介(company profile)

   ③報價單(quotation/price list)

   ④樣品(sample)

   (4) 尋找客戶

   ①到貿協圖書室查閱。

   ②電腦網站查詢。

   ③向商會、公會索取會員名單。

   ④到各國在台辦事處查詢。

   ⑤透過各國大使館、領事館協助。

   ⑥去函各國貿協，要求介紹或免費刊登廣告。

   ⑦從各相關報章雜誌廣告媒體查詢。

   ⑧從國內外各相關展覽中蒐集。

   (5) 選擇推銷方式→寄開發信(Sales Letter)及目錄(Catalog)、刊登廣告、參加展覽、拜訪客戶、設分公司、找代理、郵購、寄售、租保稅倉庫。

2. 買方主動尋找國外廠商或接到賣方推銷函後發出詢問函(Inquiry letter)給工廠或出口商詢價或詢問索取資料。

3. 收到客戶詢問函(Inquiry letter)或來函後，處理詢價函或回覆客戶疑問。

4. 和工廠聯絡，蒐集工廠資料：價錢、目錄、規格、樣品等。

5. 報價(Make the Quotation)：計算出成本(Cost)，加上利潤，做出報價單(Quotation)，寄給買方。

6. 討價還價(Counter Offer)：買方收到報價後，會和賣方討價、還價。

7. 索送樣品或規格供確認(Send Samples or Specs for Approval)：買方接受賣方價錢後，會要求賣方送樣品或規格圖面供買方確認。索送樣品時要考慮樣品費用及成本：

   (1) 完全免費。

   (2) 需收樣品費和郵費／運費。

   (3) 樣品免費，買方要付郵費／運費。

   (4) 先收樣品費，下訂單後扣回。

8. 下訂單(Place the Order)：買方確認接受報價單上條件及樣品後，下訂單(Purchase Order)給賣方。訂單條件與所接受的報價單上的條件完全相同。

9. 確認訂單(Confirm the Order)：賣方接到買方訂單後，核對確認後，簽回一份；或寄出銷售確認書(Sales Conformation or Pro-forma Invoice)給買方簽回一份。 此 S/C 或 P/I 是賣方提供給買方申請輸入許可證及開信用狀用。

10. 付款(Payment)：收到賣方確認訂單或賣方銷售確認書(S/C or P/I)後，買方於簽約後15天內（按照國際條約規定）到銀行申請輸入許可證，並開出信用狀(Letter of Credit = L/C)或按照訂單約定安排付款事宜。台灣開狀費用計算：(1)開狀手續費：三個月期，按 L/C 金額的 0.25%；六個月期則收 0.4%；(2)開狀保證金：由銀行自行決定，最低為信用狀金額的 10%；(3)郵電費：依不同寄送信用狀方式收費不同，台灣的銀行收取費用：全電約新台幣 2,000-3,000 元；郵寄信用狀約新台幣 1,000 元。外國銀行收費則更高。
   信用狀寄送方式有三種：(1)Full Cable L/C（全電信用狀）；(2)Mailed L/C（郵寄信用狀）；(3)Brief Cable(Short Cable)+Mailed L/C（短電＋郵寄信用狀）。

11. 安排生產(Arrange Production)：賣方收到買方信用狀(L/C)或訂金後，下訂單給工廠安排購料、生產、檢驗、包裝等事宜。

12. 驗貨(Inspection)：買主親自前往工廠或指定檢驗公司派員到工廠驗貨。

13. 催信用狀或催買方付款(Push Payment)：如果付款為出貨前電匯(T/T)或開信用

狀，出貨前如尚未收到，賣方要催促買方，收到貨款或信用狀再出貨。

14. 安排出貨(Arrange the Shipment)：當貨品於預定時間內生產完成，並收到信用狀或貨款後：

(1) 開始製作發票(Invoice)和包裝表(Packing List)等出貨文件。

(2) 請報關行到船公司訂船位，簽 Shipping Order(S/O)。

(3) 如為 CIF 出貨，同時向保險公司投保。

(4) 申請輸出許可，請報關行辦理出口簽證，申請出口報單。

(5) 於結關日前將貨送至指定的貨櫃場或機場貨運站，交由報關行安排報關通關登機上船出貨事宜。

15. 押匯(Negotiation with the Bank)

(1) 貨裝出後，請報關行持 S/O 到船公司或航空公司領取提單；如為 CFR 或 CIF 出貨，需繳清運費後，才能領到提單，提單上會蓋上 Freight Prepaid 字樣。（如為 FOB 出貨，提單上則會蓋上 Freight Collect）。

(2) 按照信用狀上規定，做出整套押匯文件。基本文件為：發票、包裝表、提單、保單、產地證明等。（其他文件則按信用狀要求做出）

(3) 將押匯文件送入銀行押匯。銀行會扣押匯手續費（銀行會按押匯金額的 0.1%或 0.2%計算）、14 天利息（7、12、14、18 天不等）、郵電費等。另寄一套副本押匯文件給買方。

(4) 如為 D/A、D/P 付款方式，賣方則將正本出貨文件直接寄給買主或透過銀行轉去亦可。

16. 贖單(Reimbursement/Settle the Payment)：如為 L/C 付款方式，買方於收到銀行轉來的正本出貨文件後，需到銀行繳清信用狀上貨款，贖回正本押匯文件所有單據。如為其他付款方式如 D/A、D/P，賣方會將正本文件直接寄給買方，買方於約定時間內付款給賣方。

17. 清關提貨(Clean the Goods from the Customs)：

(1) 貨到買方港口時，買方／進口商請進口報關行拿正本提單(B/L)到船公司換小提單(Delivery Order)，如為 FOB 出貨，買方要繳清運費才能領取 D/O。

(2) 將 Invoice、Packing List、D/O 和進口產品型錄交給報關行，進行清關報關手續。

(3) 待海關至倉庫或貨櫃場驗過貨後，估定應繳納的進口關稅及其他稅捐。例如：進口關稅、商港建設捐、推廣貿易服務費、貨物稅等。

注意：進口稅的科稅標準是按完稅價格（即 CIF 價格）來核算。

(4) 繳清關稅後，海關會簽發貨物放行單交給報關行，即可清關提貨。報關行會和買
主／進口商聯絡將貨送至指定地點或倉庫。

18.買方／進口商抱怨／索賠(Complaint/Claim)：從貨到達買方港口開始，就有可能
產生抱怨，例如：海關驗貨時發現短少破損或買主收到貨時，發現品質不符、短少、
包裝破損等。買方／進口商會持照片或相關證明文件向賣方進行索賠。

注意：國際條約上規定，任何索賠需於貨到港口 21 天內提出。

19.賣方處理抱怨／索賠(Handle the Complaint/Claim)：收到買方抱怨／索賠要
求時，要先查明原因、判定責任、做出更正解決方法。買賣雙方如仍要繼續合作，
雙方大都不會訴諸漫長繁瑣的法律途徑，而盡量按國際慣例處理方式，私下和平解
決。抱怨糾紛產生的主要原因如下：

(1) 品質不符。

(2) 數量不符（重量、數量短少）。

(3) 包裝不良。

(4) 賣方裝錯貨。

(5) 買方訂錯貨。

(6) 運輸公司卸貨卸錯港口。

(7) 未投保或保險不足。

(8) 出貨文件與信用狀條款不符。

(9) 工程設計問題。

(10)溝通不良（工廠內部／出口商與工廠／買方與賣方）。

(11)不可抗力事件。

(12)延遲出貨

　　①生產不順。

　　②模具有問題。

　　③原料短缺。

　　④人工短缺。

　　⑤機器故障。

　　⑥樣品無法通過買方之確認。

　　⑦賣方報價錯誤偏低。

　　⑧買方信用狀慢開或貨款未付。

　　⑨政府法令變更。

　　⑩沒有船位或班機。

　　⑪假日太多。

　　⑫天災、罷工、戰爭、意外等不可抗力事件。

注意：其中以品質不符和延遲出貨造成的糾紛最頻繁。

主要進出口流程表

主要出口流程   主要進口流程

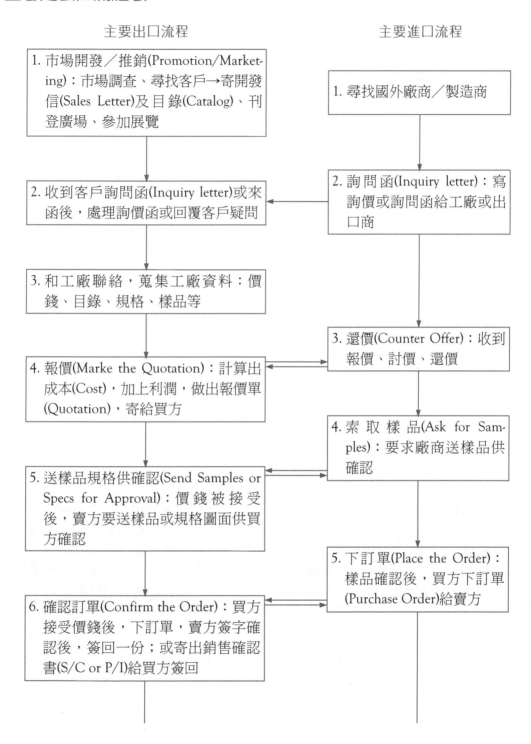

1. 市場開發／推銷(Promotion/Marketing)：市場調查、尋找客戶→寄開發信(Sales Letter)及目錄(Catalog)、刊登廣場、參加展覽

1. 尋找國外廠商／製造商

2. 收到客戶詢問函(Inquiry letter)或來函後，處理詢價函或回覆客戶疑問

2. 詢問函(Inquiry letter)：寫詢價或詢問函給工廠或出口商

3. 和工廠聯絡，蒐集工廠資料：價錢、目錄、規格、樣品等

3. 還價(Counter Offer)：收到報價、討價、還價

4. 報價(Marke the Quotation)：計算出成本(Cost)，加上利潤，做出報價單(Quotation)，寄給買方

4. 索取樣品(Ask for Samples)：要求廠商送樣品供確認

5. 送樣品規格供確認(Send Samples or Specs for Approval)：價錢被接受後，賣方要送樣品或規格圖面供買方確認

5. 下訂單(Place the Order)：樣品確認後，買方下訂單(Purchase Order)給賣方

6. 確認訂單(Confirm the Order)：買方接受價錢後，下訂單，賣方簽字確認後，簽回一份；或寄出銷售確認書(S/C or P/I)給買方簽回

（續上頁）

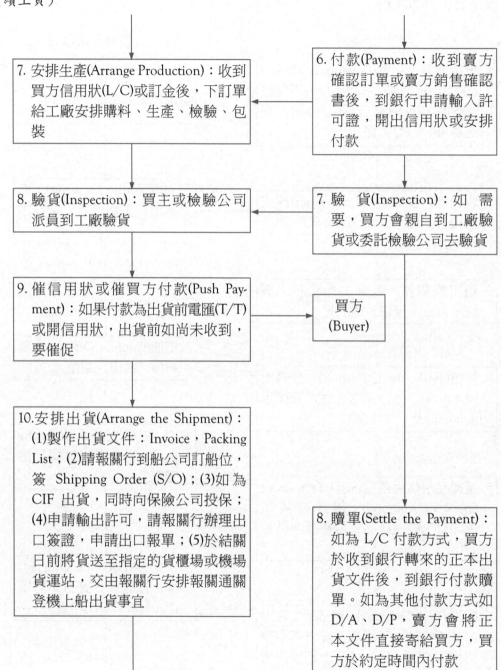

7. 安排生產(Arrange Production)：收到買方信用狀(L/C)或訂金後，下訂單給工廠安排購料、生產、檢驗、包裝

6. 付款(Payment)：收到賣方確認訂單或賣方銷售確認書後，到銀行申請輸入許可證，開出信用狀或安排付款

8. 驗貨(Inspection)：買主或檢驗公司派員到工廠驗貨

7. 驗 貨(Inspection)：如 需要，買方會親自到工廠驗貨或委託檢驗公司去驗貨

9. 催信用狀或催買方付款(Push Payment)：如果付款為出貨前電匯(T/T)或開信用狀，出貨前如尚未收到，要催促

買方 (Buyer)

10.安排出貨(Arrange the Shipment)：(1)製作出貨文件：Invoice，Packing List；(2)請報關行到船公司訂船位，簽 Shipping Order (S/O)；(3)如為 CIF 出貨，同時向保險公司投保；(4)申請輸出許可，請報關行辦理出口簽證，申請出口報單；(5)於結關日前將貨送至指定的貨櫃場或機場貨運站，交由報關行安排報關通關登機上船出貨事宜

8. 贖單(Settle the Payment)：如為 L/C 付款方式，買方於收到銀行轉來的正本出貨文件後，到銀行付款贖單。如為其他付款方式如 D/A、D/P，賣方會將正本文件直接寄給買方，買方於約定時間內付款

MODERN
BUSINESS
ENGLISH

Part 7

國際貿易實用參考附表

501

（續上頁）

1. 押匯(Negotiation with the Bank)：(1)貨裝出後，到船公司或航空公司領取提單；(2)按照信用狀上規定，做出整套押匯文件。基本文件為：發票、包裝表、提單、保單、產地證明等；(3)送入銀行押匯、銀行會扣押手續費、14 天利息、郵電費等；(4)如為 D/A、D/P 付款方式，則將正本文件直接寄給買主或透過銀行轉法亦可

9. 清關提貨(Clean the Goods from the Customs)：貨到買方港口時，請進口報關行拿正本提單到船公司換小提單(Delivery Order)，進行清關報關工作。待海關驗過貨，估定、繳清關稅後，即可放行清關提貨

12. 抱怨／索賠處理(Handle Complaint/Claim)：收到買方抱怨／索賠要求時，要先查明原因、判定責任、做出更正解決方法。買賣雙方如仍要繼續合作，雙方大都不會訴諸漫長繁瑣的法律途徑，而盡量按國際慣例處理方式，私下和平解決

10. 抱怨／索賠(Complaint/Claim)從貨到達買方港口開始，就有可能產生抱怨，例如：海關驗貨時發現短少破損或買主收到貨時，發現品質不符、短少、包裝破損等

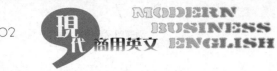

國際貿易作業流程補充資料

1. **國際貿易進出口主要流程為何？**

   市場開發→詢價→聯絡工廠計算報價→報價→討價還價→索送樣品→下訂單→確認訂單→付款→安排生產→驗貨→催信用狀或催買方付款→安排出貨→製作出貨文件→押匯→買方贖單→買方辦理清關提貨→買方抱怨索賠→賣方處理抱怨。

2. **何謂外匯管制？外匯管制造成的主要問題為何？**

   外匯管制即外幣或匯款只准進來，但是管制出去（限制匯出國外的金額）所有開發中國家（例如：中國大陸、菲律賓、印尼、馬來西亞、越南、泰國、中東國家、非洲、南美州、南歐、東歐等國家）都有實施外匯管制，所以和這些國家交易時要特別注意下列兩個問題：(1)付款可能有問題或付不出來；(2)付款緩慢。

3. **市場推銷的方法主要有哪些？**

   (1)寄開發信；(2)刊登廣告；(3)參加國內外展覽；(4)拜訪客戶；(5)貿協或朋友引介；(6)租保稅倉庫；(7)郵購；(8)設分公司；(9)找代理；(10)寄售。

4. **何謂傾銷(dumping)？一般國際上對傾銷國的處罰方式為何？**

   所謂傾銷即「大量低價銷售產品至別國」。一般國際上對傾銷國的處罰方式有四：(1)禁止進口；(2)課傾銷稅；(3)限制底價；(4)配額(quota)

5. **買方討價還價時，賣方有四種選擇：**

   (1)不接受；(2)接受；(3)條件式接受；(4)折衷接受。

6. **客戶索取樣品時，賣方要特別注意：**

   (1) 樣品品質＝出貨品質。

   (2) 樣品沒確認前，不要先行出貨。

   (3) 如剛好沒有客戶要求的樣品時，可先送類似樣品提供客戶參考，等正式訂購時，再送正確樣品供正式確認。

   (4) 樣品費(sample charge)及郵費(postage)是否要求，有四種狀況：

   　①樣品費和郵費全部免費。

②樣品免費，客人需付郵費。

③樣品費和郵費都要事先付，因成本高。

④請客人先付樣品費和郵費，客人正式下訂單時，再退還或從訂單中扣除。

7. 報價單(Quotation)上應列出的報價基本條件有哪些？

(1)編號(model No./style No./part No./item No.)

(2)品名規格(specification/description)。

(3)單價(unit price)：包括幣別(US$)／價錢／單位(pc, set, pair, dozen, kg)／價錢條件(FOB, CFR, CIF)／目的地港／裝運方式(by air, by sea, by air parcel, by courier) 例：US$ 10/pc at CIF New York by sea。

(4)交期(lead time ／ shipping date ／ shipment date ／ delivery time)。

(5)最低訂購量(minimum order quantity)。

(6)包裝方式(packing method)：Standard export packing。

(7)付款條件(terms of payment)。

(8)付款方法(payment methods)。

(9)有效期(validity)。

8. 價錢條件：EX-Works、FAS、FOB/FCA、CFR/CPT、CIF/CIP、DAT/DAP/DDP 之用法？

(1) EXW 工廠交貨價——離開工廠以外的費用皆由買方負擔。

(2) FAS 船邊交貨價——含報關通關費用；如不含可於訂單合約上註明（限海運及內陸水路運送）。

(3) FOB 船上交貨價——離開出口港後的運保費由買方負擔（限海運及內陸水路運送）。

(4) CFR 運費在內價——保費由買方負擔（限海運及內陸水路運送）。

(5) CIF 運保費在內價——賣方負責運費及保險費（限海運及內陸水路運送）。

(6) FCA 貨交運送人價——成本和 FOB 相同，但可用於各種運送方式。

(7) CPT 運費付訖價——成本和 CFR 相同，但可用於各種運送方式。

(8) CIP 運保費付訖價——成本和 CIF 相同，但可用於各種運送方式。

(9) DAT 終點站交貨價——貨交至買方指定的終點站（例如：碼頭／車站等），不含進口關稅及進口報關費用，賣方負責卸貨，取代舊版的 DAF/DEQ/DES。

(10)DAP 目的地交貨價——貨交至買方指定的目的地（註明詳細門牌地址），不含進口關稅及進口報關費用，賣方不負責卸貨，取代舊版的 DDU。

(11)DDP 稅訖交貨價——貨交至買方指定的地點，含進口關稅及進口報關費用，賣方成本最高，風險最大的條件。

　　注意：EXW 離廠價＋搬運費＋內陸運費＋報關費(broker charge)＋海關通關費(customs charge)＋銀行費用(banking charge)＝ FOB 離港價。

投保金額＝出貨發票金額＋ 10%（例：出貨金額為 US$10,000×1.1 ＝保額 US$11,000）

9. 主要付款條件有 9 種：

(1) 不可撤銷保兌及可轉讓即期信用狀(L/C)。

Payment: by irrevocable, confirmed and transferable Letter of Credit at sight.

(2) 遠期信用狀(Usance L/C)。

Payment: by irrevocable Letter of Credit at 30 days sight in our favor.

(3) 付款交單(Document against Payment)＝ D/P（託收的一種）。

(4) 承兌交單(Document against Acceptance)＝ D/A（託收的一種），後列天數例如：

Payment: by D/A 30 days。

(5) 記帳/月結(Open Account)＝ O/A。

(6) 分期付款(Installment)。

(7) 寄售(On Consignment)。

(8) 交貨付現(Cash on Delivery)＝ COD。

(9) 下訂單同時付現(Cash with Order)＝ CWO。

10.除了 L/C 外，D/P、D/A、O/A、CWO、COD 或 On Consignment 的付款方法有 5種：

(1) 電匯(Telegram Transfer/Cable Transfer)＝ T/T。

(2) 信匯(Mailed Transfer)＝ M/T。

(3) 銀行本票(Bank Check)。

(4) 銀行匯票(Bank Draft)。

(5) 郵政匯票(Money Order)。

注意：T/T 是付款方法，不是付款條件，所以不可打 Payment: by T/T；如要強調以 T/T 付款，一定要加註付款時間，例如：Payment: by T/T before shipment 或 Payment: by T/T within 30 days after shipment.。

11. 何謂信用狀(L/C)及其功能？

(1) 信用狀即買方先將貨款（全額或至少 10%）付給開狀銀行。

(2) 開狀銀行即按照 P/I 內容開出信用狀給買方的通知銀行＝其分行或往來銀行）。

(3) 通知銀行收到 L/C 後會通知賣方（受益人）來領取正本 L/C，賣方需付約 NT $1,000 的通知費。

(4) 賣方收到 L/C 後，即可安排訂船位／班機出貨事宜。

(5) 貨經由空運／海運／郵包／快遞裝出後，賣方會拿到提單（B/L，AWB 或 Receipt）。

(6) 賣方按照 L/C 上規定，將所有文件準備好（發票／包裝表／提單／產地證明／保單等）。

(7) 賣方（受益人）將準備好的出貨文件，和正本 L/C 於有效期內一起送入銀行押匯。

(8) 押匯銀行會核對出貨文件是否符合 L/C 上條件要求，如無誤，即扣除手續費／利息／郵電費等，然後將貨款先付給受益人。

(9) 押匯銀行將所有押匯文件郵寄去給買方的開狀銀行。

(10)買方的開狀銀行收到正本押匯文件後，會核對，如無誤，就付款給賣方銀行。

(11)然後買方的開狀銀行會通知買主（申請人）來銀行付清貨款將正本文件贖回（此即所謂贖單）。

(12)買方拿到正本文件後，請進口報關行持 B/L 去船公司換取 D/O（小提單），然後至海關進行通關報關手續，繳清進口稅及海關費用等即可將貨提出。

L/C 的優點：買賣雙方皆由銀行做擔保，買方不須擔心賣方拿錢不出貨；賣方亦不用擔心貨裝出後會收不到錢。

L/C 的缺點：銀行費用過高；如有任何錯誤即造成瑕疵，賣方會拿不到錢或較慢拿到。

注意：L/C 上的最後出貨日(Latest shipping date)＝提單上的裝運日期(On board date)＝海運的「開航日」（非「結關日」）／空運的「放行日」（非「進倉日」）。

P.S. D/A、D/P 付款大都用於小額出貨或信用良好的客戶，其出貨文件則不一定要透過銀行傳送，正本出貨文件（發票／包裝表／提單／產地證明等）大都由賣方直接寄給買方。

12.何謂押匯？押匯時應準備哪些單據文件？

出口商按信用狀(L/C)上規定備妥各項文件後，送入外匯押匯銀行，填具押匯申請書，連同信用狀正本，向銀行申請辦理押匯，經銀行審核單據無誤，扣除押匯手續費後，將貨款兌換成當地幣值付給送押匯的出口商，這整個過程即稱為押匯。

押匯必備文件為：(1)信用狀正本；(2)信用狀上所規定的所有文件——至少要提具提單／發票／包裝表／產地證明及其他文件（視 L/C 上規定準備）；(3)銀行押匯申請書。

13.何謂瑕疵押匯？

所有押匯文件必需完全按照 L/C 上規定做出，如缺任何文件或押匯文件上有任何一個字和信用狀(L/C)上所敘不同，或和 L/C 規定不符（特別是出貨時間超過最後出貨日），即為瑕疵押匯，買方可以拒付(unpaid)；賣方最好要求買方事先修改 L/C。

買方如接受瑕疵文件，押匯銀行會由受益人帳上扣除瑕疵費(discrepancy fee)。

14.驗貨的步驟如下：

(1) 準備好客人的訂單／信用狀及確認樣品或規格圖面。

(2) 請工廠按訂單規定，做出正確的發票(Invoice)／包裝表(Packing List)

(3) 驗貨的順序：

①先核對文件：核對發票／包裝表是否和訂單規定一致。

②再核對嘜頭：出貨文件（發票／包裝表）和出貨外箱上是否和訂單規定一致。

③清點箱數。

④按 AQL 標準，隨意開箱抽點數量。

⑤按 AQL 標準，抽驗出貨品質（比照確認樣品或確認的規格圖面來驗）。

客人如無指定驗貨標準，一般都按MIL-105D，AQL Level II，Major defect（主缺點）：0.65, Minor defect（次缺點）：1.5 來驗貨

⑥如發現有不良品，但在允收範圍內，請工廠更換不良品後出貨；如驗貨結果為退貨，請工廠 Sorting（挑選）或 Reworking（整修）後，再重驗。

15.台灣進出口通關手續在自動化作業系統下，有三種通關方式：

(1)C1(Channel 1)：免審免驗通關方式。

(2)C2(Channel 2)：文件審核通關方式。

(3)C3(Channel 3)：貨物查驗通關方式。

16.抱怨索賠處理方法：

(1)先請客人退回不良樣品；(2)賣方收到不良樣品後分析不良原因；(3)判定責任歸屬；

(4)如為賣方責任，賣方需提出更正報告(Corrective Action)並負責賠償，賠償方法有

三：賠貨／還錢／給折扣。

不良品的處理方法有五種：(1)退回工廠；(2)轉賣當地其他客戶；(3)請客人整修或挑

用，賣方付整修挑用費；(4)賣方自己派人至買方處去整修；(5)丟棄。

17.代理分三種：

(1)銷售代理 (Seller's Agent)

(2)經銷代理 (Distributorship)

(3)買方代理 (Buyer's Agent)

經濟國貿

編　號	作　者	書　名	年　份	版　次	定　價
52MET00104	楊雲明	個體經濟學	2007.10	四版	580 元
B2005	楊雲明	總體經濟學	2005.01	三版	580 元
52MET00202	劉亞秋	經濟學原理	2007.09	再版	550 元
52WET00401	楊雲明	經濟學	2010.08	初版	680 元
52MET00501	周賓凰	計量經濟學：理論、觀念與應用	2010.10	初版	600 元
52WET00601	周賓凰 徐耀南 王絹淑	綠色經濟學：理論、政策與實務	2011.03	初版	420 元
52MET00801	陳建良 譯	賽局理論	2006.02	初版	520 元
52WET00903	徐仁輝 何宗武	公共議題經濟學	2010.09	三版	380 元
52MET01301	謝登隆	個體經濟學：生活與個案	2010.07	初版	620 元
52MET01401	謝登隆 徐繼達	總體經濟學：理論、政策與個案	2007.03	初版	500 元
52MET01503	方博亮 林祖嘉	管理經濟學	2007.06	三版	600 元
52MET02305	蔡孟佳	國際貿易實務	2011.09	五版	620 元
52MET02504	黃瑪莉	現代商用英文	2012.09	四版	580 元
52MET02603	黃瑪莉	現代商用英文會話	2007.03	三版	480 元
B2073	黃瑪莉	現代商用英文會話（光碟）	2002.08	初版	480 元
52MET02702	蔡孟佳	國際貿易實務題庫	2009.04	再版	350 元
B4016	蔡孟佳	國際貿易經營專題（一）	2003.08	初版	400 元

❀本表所列書目定價如與書內版權頁不符，以版權頁為主

財務金融（一）

編　號	作　者	書　名	年　份	版　次	定　價
52MFF00111	李榮謙	貨幣銀行學	2012.01	十版增訂	680 元
52MFF00304	李榮謙	國際金融學	2009.10	四版	600 元
52MFF00401	林左裕	衍生性金融商品	2009.06	初版	480 元
52MFF00502	陳威光	衍生性金融商品： 選擇權、期貨、交換與風險管理	2010.03	再版	580 元
52MFF00602	陳威光	選擇權：理論、實務與風險管理	2010.09	再版	680 元
52MFF00701	李榮謙	新時代的貨幣銀行學概要	2011.09	初版	550 元
52WFF00806	蕭欽篤	國際金融	2009.09	六版	600 元
52WFF00905	謝劍平	財務管理：新觀念與本土化	2009.09	五版	720 元
52WFF01104	謝劍平	財務管理原理	2011.07	四版	500 元
52MFF01201	郭敏華	行為財務學： 當財務學遇上心理學	2008.03	初版	400 元
52WFF01304	謝劍平	投資學：基本原理與實務	2012.06	四版	550 元
52WFF01505	謝劍平	現代投資學：分析與管理	2011.02	五版	720 元
52MFF01604	謝劍平	期貨與選擇權： 財務工程的入門捷徑	2010.02	四版	580 元
52WFF01704	謝劍平	當代金融市場	2012.06	四版	580 元

✤本表所列書目定價如與書內版權頁不符，以版權頁為主

財務金融（二）

編　號	作　者	書　名	年　份	版　次	定　價
52WFF01803	謝劍平	固定收益證券： 債券市場與投資策略	三版即將發行		
52MFF01901	賴碧瑩	現代不動產估價：理論與實務	2009.01	初版	500 元
52MFF02001	沈中華	金融機構在中國的機會與挑戰： 對台灣銀行業的策略建議與提 醒	2010.01	初版	350 元
52MFF02101	朱浩民	中國金融制度與市場	2010.01	初版	400 元
52WFF02201	謝劍平	國際財務管理： 跨國企業之價值創造	2011.07	初版	450 元
52MFF02401	李春長	不動產經濟學	2012.01	初版	600 元
52MFF02502	謝劍平 林傑宸	證券市場與交易實務	2012.05	再版	580 元
52WFF02601	陳繼堯	海上保險	2012.05	初版	300 元
52MFF03103	謝劍平	現代投資銀行	2010.05	三版	650 元
52MFF03604	林傑宸	基金管理：資產管理的入門寶典	2011.07	四版	480 元
52MFF04102	吳啓銘	企業評價：個案實證分析	2010.03	再版	480 元
52MFF04304	林左裕	不動產投資管理	2010.09	四版	680 元
52MFF04502	陳繼堯	汽車保險：理論與實務	2006.03	再版	480 元
52MFF05302	謝劍平	金融創新：財務工程的實務奧秘	2010.06	再版	600 元
52MFF05602	陳奉瑤 章倩儀	不動產經營管理	2011.02	二版	420 元
52MFF05702	杜麗娟 許秉翔	休閒產業的財務管理	2009.03	再版	400 元
52MFF05801	陳彩稚	財產與責任保險	2006.09	初版	600 元

✦本表所列書目定價如與書內版權頁不符，以版權頁為主

企管行銷（一）

編　號	作　者	書　　名	年　份	版　次	定　價
52MMM00102	廖勇凱 楊湘怡	管理學：理論與應用	2009.05	再版	420 元
52MMM00302	于卓民	企業概論	2007.01	再版	700 元
52MMM00603	廖勇凱 楊湘怡	人力資源管理：理論與應用	2011.02	再版	420 元
52MMM00802	廖勇凱	國際人力資源管理	2009.09	再版	480 元
52WMM00904	吳青松	國際企業管理：理論與實務	2012.07	四版	500 元
52MMM01001	陳龍潭 廖勇凱	企業策略管理	2011.07	初版	400 元
52MMM01302	陳明哲	動態競爭策略微探	2010.06	再版	500 元
52MMM01403	許振邦 中華採購 管理協會	採購與供應管理	2011.03	三版	680 元
52WMM01502	李易諭	品質管理	2007.05	再版	650 元
52MMM01601	廖勇凱	企業倫理學：理論與應用	2008.09	初版	350 元
52MMM01703	邱志聖	策略行銷分析：架構與實務應用	2010.05	三版	550 元
52MMM01803	邱志聖	行銷研究：實務與理論應	2011.06	三版	520 元
52MMM01903	蕭富峰	行銷管理	2011.05	三版	620 元
52WMM00201	何明城 晏介中 審訂	彼得杜拉克管理個案全集	2012.06	初版	480 元
52WMM02001	邱順應	廣告文案	2008.01	初版	500 元
52MMM02102	蘇雄義	供應鏈管理	2012.02	再版	580 元
52MMM02202	蕭富峰	消費者行為	2012.01	再版	480 元

❖本表所列書目定價如與書內版權頁不符，以版權頁為主

企管行銷（二）

編　號	作　者	書　名	年　份	版　次	定　價
52MMM02301	黃鴻程 廖勇凱	商業自動化之連鎖事業經營	2008.07	初版	380 元
52WMM02401	吳筱玫	傳播科技與文明	2008.04	初版	300 元
52MMM02503	于卓民 巫立宇 蕭富峰	國際行銷學	2009.01	三版	480 元
52WMM02601	楊慕理	大學校院公關運作：理論與實踐	2009.08	初版	280 元
52MMM02701	陳明惠	創意管理	2009.08	初版	250 元
52WMM02801	丁明勇	組織行為學	2009.01	初版	480 元
52WMM02901	陳明惠	創意領導：技巧驅動變革	2010.05	初版	400 元
52MMM03001	清大科管學院	服務科學入門 10 講	2009.08	初版	280 元
52MMM03101	屠益民 張良政	系統動力學	2010.04	初版	350 元
52MMM03201	蕭富峰 張佩娟 卓峰志	廣告學	2010.07	初版	520 元
52MMM03303	于卓民 審訂	行銷個案分析	2010.07	三版	520 元
52MMM03403	羅應浮	門市服務：乙級檢定創意 Q&A	2012.06	三版	480 元
52MMM03501	楊慕理	地方文化產業：整合行銷傳播	2011.03	初版	380 元
52MMM03601	郭小喬 楊鳳美	門市服務：丙級檢定實戰秘笈	2011.07	初版	300 元
52MMM03701	劉常勇 審訂	創業管理：策略與資源	2006.03	初版	650 元
52MMM03901	林芬慧 主編	網路行銷：e 網打盡無限商機	2011.08	初版	450 元
52WMM04001	趙滿鈴	網路行銷特訓教材	2011.03	初版	450 元

✿本表所列書目定價如與書內版權頁不符，以版權頁為主

企管行銷（三）

編 號	作 者	書 名	年 份	版 次	定 價
52MMM04201	蕭富峰	行銷管理概論	2011.06	初版	450 元
52MMM04301	邱順應	廣告修辭新論	新版即將發行		
52MMM05101	吳秉恩 審訂	領導學：原理與實踐	2006.06	初版	480 元
52MMM05301	樓永堅	行銷管理	2007.01	初版	560 元
52MMM05501	于卓民 張力元 蘇瓜藤 主編	高科技產業個案集	2006.05	初版	350 元
52MMM05702	李正綱 陳基旭 張盛華	現代企業管理：理論與實務導向	2007.02	再版	480 元
B2013	吳美連	人力資源管理：理論與實務	2005.08	四版	550 元
W2062	吳筱玫	網路傳播概論	2003.01	初版	420 元
B2100	司徒達賢	策略管理新論： 觀念架構與分析方法	2005.08	再版	680 元
B2144	劉建順	現代廣告學	2004.01	初版	520 元
B2151	于卓民 審訂	國際企業管理	2004.08	初版	620 元
B2152	于卓民 審訂	國際企業管理個案	2005.01	初版	260 元
B2153	沈永正 審訂	消費者行為個案分析	2004.08	初版	320 元
B2158	何明城 審訂	管理溝通策略	2005.02	初版	380 元
B2169	劉建順	現代公共關係學： 整合傳播與公共報導導向	2005.01	初版	500 元

✢本表所列書目定價如與書內版權頁不符，以版權頁為主

資訊科技

編　號	作　者	書　　名	年　份	版　次	定　價
52MIT00104	林東清	資訊管理： e化企業的核心競爭能力	2010.09	四版	700元
52WIT00206	謝清佳 吳琮璠	資訊管理：理論與實務	2009.07	六版	600元
52MIT00304	吳仁和	資訊管理：企業創新與價值創造	2012.06	四版	650元
52MIT00503	林東清	知識管理	2009.07	三版	550元
52MIT00605	吳仁和 林信惠	系統分析與設計： 理論與實務應用	2010.03	五版	650元
52MIT00704	吳仁和	物件導向系統分析與設計： 結合 MDA 與 UML	2012.06	四版	700元
52MIT00801	鄭炳強	軟體工程：從實務出發	2007.08	初版	520元
52MIT01002	方文昌 汪志堅	電子商務	2010.1	再版	500元
52MIT01101	吳仁和 陳翰容 沈德村 洪誌隆 林麗敏	醫療資訊管理	2010.09	初版	580元
52MIT01202	梁定澎	決策支援系統與企業智慧	2006.01	再版	750元
52MIT02301	吳道文	專案規劃：資訊概念篇	2006.03	初版	600元
B2109	林信惠 黃明祥 王文良	軟體專案管理	2005.07	再版	600元

✚本表所列書目定價如與書內版權頁不符，以版權頁為主

會計統計

編　號	作　者	書　　名	年　份	版　次	定　價
B2021	洪清和	初級會計學（上）	2004.09	五版	350 元
52MAS00205	洪清和	初級會計學（下）	2006.03	五版	460 元
52MAS00309	洪清和	中級會計學（上）	2011.07	九版	680 元
52MAS00410	洪清和	中級會計學（下）	十版即將發行		
52MAS00903	郭敏華	財務報表分析	2009.07	三版	480 元
52MAS01003	田美蕙	會計報表實務	2010.07	三版	450 元
52MAS01204	陳志愷	稅務會計	2010.09	四版	620 元
52MAS01304	陳志愷	稅務法規	2011.07	四版	520 元
52MAS01403	蕭子誼	商業會計法	2010.09	三版	520 元
52WAS01503	曾炳霖	商法案例實證解析	2010.09	三版	500 元
52WAS01603	陳建勝 陳美菁 朱瑞淵 呂明哲	統計學：商業與管理的應用	2012.09	三版	650 元
52WAS01801	William W. S. Wei 繆震宇 審閱	時間序列分析： 單變量與多變量方法	2012.01	初版	680 元
52WAS01905	吳琮璠	審計學：新觀念與本土化	2012.08	增修版	620 元
52WAS02001	吳琮璠	新會計學： 實務應用與法律觀點	2012.02	初版	500 元
52MAS03102	林震岩	多變量分析： SPSS 的操作與應用	2007.01	再版	700 元
52WAS03203	謝劍平	財務報表分析： 洞察企業良窳的技能	三版即將發行		

✛本表所列書目定價如與書內版權頁不符，以版權頁為主

現代商用英文
Modern Business English

國家圖書館出版品預行編目資料

現代商用英文 / 黃瑪莉著. -- 四版. -- 臺北市： 智勝文化, 2012.09 　面；　公分 ISBN 978-957-729-898-0(平裝)   1.商業書信 2.商業英文 3.商業應用文 493.6　　　　　　　　101016468

作　　者/黃瑪莉
發 行 人/紀秋鳳
出　　版/智勝文化事業有限公司
　　　　地　　址/台北市 100 館前路 26 號 6 樓
　　　　電　　話/(02)2388-6368
　　　　傳　　真/(02)2388-0877
　　　　郵　　撥/16957009 智勝文化事業有限公司
　　　　登 記 證/局版臺業字第 5177 號
出版日期/2012 年 9 月四版
定　　價/580 元
ISBN　　978-957-729-898-0(平裝)

Modern Business English
by Mary Huang
Copyright 2012 by Mary Huang
Published by BestWise Co., Ltd.
智勝網址: www.bestwise.com.tw